Methods in Cell Biology

The Zebrafish: Cellular and Developmental Biology, Part A Cellular Biology

Volume 133

Series Editors

Leslie Wilson
Department of Molecular, Cellular and Developmental Biology
University of California
Santa Barbara, California

Phong Tran
University of Pennsylvania
Philadelphia, USA &
Institut Curie, Paris, France

Methods in Cell Biology

The Zebrafish: Cellular and Developmental Biology, Part A Cellular Biology

Volume 133

Edited by

H. William Detrich, III
Northeastern University Marine Science Center,
Nahant, MA, United States

Monte Westerfield
University of Oregon, Eugene, OR, United States

Leonard I. Zon
Harvard University, Boston, MA, United States

AMSTERDAM • BOSTON • HEIDELBERG • LONDON
NEW YORK • OXFORD • PARIS • SAN DIEGO
SAN FRANCISCO • SINGAPORE • SYDNEY • TOKYO

Academic Press is an imprint of Elsevier

Academic Press is an imprint of Elsevier
50 Hampshire Street, 5th Floor, Cambridge, MA 02139, USA
525 B Street, Suite 1800, San Diego, CA 92101-4495, USA
125 London Wall, London EC2Y 5AS, UK
The Boulevard, Langford Lane, Kidlington, Oxford OX5 1GB, UK

Fourth edition 2016

Copyright © 2016, 2010, 2004, 1998 Elsevier Inc. All rights reserved.

No part of this publication may be reproduced or transmitted in any form or by any means, electronic or mechanical, including photocopying, recording, or any information storage and retrieval system, without permission in writing from the publisher. Details on how to seek permission, further information about the Publisher's permissions policies and our arrangements with organizations such as the Copyright Clearance Center and the Copyright Licensing Agency, can be found at our website: www.elsevier.com/permissions.

This book and the individual contributions contained in it are protected under copyright by the Publisher (other than as may be noted herein).

Notices

Knowledge and best practice in this field are constantly changing. As new research and experience broaden our understanding, changes in research methods, professional practices, or medical treatment may become necessary.

Practitioners and researchers must always rely on their own experience and knowledge in evaluating and using any information, methods, compounds, or experiments described herein. In using such information or methods they should be mindful of their own safety and the safety of others, including parties for whom they have a professional responsibility.

To the fullest extent of the law, neither the Publisher nor the authors, contributors, or editors, assume any liability for any injury and/or damage to persons or property as a matter of products liability, negligence or otherwise, or from any use or operation of any methods, products, instructions, or ideas contained in the material herein.

ISBN: 978-0-12-803475-0
ISSN: 0091-679X

For information on all Academic Press publications
visit our website at https://www.elsevier.com

Publisher: Zoe Kruze
Acquisition Editor: Zoe Kruze
Editorial Project Manager: Sarah Lay
Production Project Manager: Malathi Samayan
Designer: Victoria Pearson

Typeset by TNQ Books and Journals

Len, Monte, and I dedicate the 4th Edition of Methods in Cell Biology: The Zebrafish *to the postdoctoral fellows and graduate students who conducted the genetic screens that established the zebrafish as a preeminent vertebrate model system for analysis of development.*

Contents

Contributors xi
Preface xv

CHAPTER 1 Embryonic Cell Culture in Zebrafish 1
C.A. Ciarlo, L.I. Zon
Introduction 2
1. Methods 3
Conclusion 10
References 10

CHAPTER 2 Cellular Dissection of Zebrafish Hematopoiesis 11
D.L. Stachura, D. Traver
Introduction 12
1. Zebrafish Hematopoiesis 12
2. Hematopoietic Cell Transplantation 24
3. Enrichment of HSCs 32
4. In Vitro Culture and Differentiation of Hematopoietic Progenitors 35
Conclusions 46
References 46

CHAPTER 3 Second Harmonic Generation Microscopy in Zebrafish 55
D.C. LeBert, J.M. Squirrell, A. Huttenlocher, K.W. Eliceiri
Introduction 56
1. Materials 59
2. Methods 62
3. Notes 66
Acknowledgments 66
Supplementary Data 67
References 67

CHAPTER 4 Imaging Blood Vessels and Lymphatic Vessels in the Zebrafish 69
H.M. Jung, S. Isogai, M. Kamei, D. Castranova, A.V. Gore, B.M. Weinstein
Introduction 70
1. Imaging Vascular Gene Expression 71

    2. Nonvital Blood Vessel and Lymphatic Vessel Imaging............ 74
    3. Vital Imaging of Blood and Lymphatic Vessels 83
       Conclusion 98
       References... 99

CHAPTER 5   An Eye on Light-sheet Microscopy 105
D. Kromm, T. Thumberger, J. Wittbrodt
       Introduction 106
       History 106
    1. Principle Behind Selective Plane Illumination Microscopy 107
    2. The Microscope for Your Sample or the Sample for Your Microscope?.. 114
    3. Data Acquisition and Handling... 117
    4. Challenges and Perspectives... 119
       Acknowledgment... 120
       References... 121

CHAPTER 6   Single Neuron Morphology In Vivo with Confined Primed Conversion... 125
M.A. Mohr, P. Pantazis
       Introduction 126
    1. Photoconvertible Fluorescent Proteins............................... 127
    2. Confined Primed Conversion... 128
    3. Unraveling Single Neuron Morphology With Confined Primed Conversion 129
       Conclusion and Outlook.. 135
       References... 136

CHAPTER 7   Visualizing Retinoic Acid Morphogen Gradients 139
T.F. Schilling, J. Sosnik, Q. Nie
       Introduction 140
    1. Challenges for Morphogen Gradient Studies 140
    2. Feedback Allows Retinoic Acid to Act as a Graded Morphogen 142
    3. Cyp26s as Key Regulators of Retinoic Acid Gradient Formation 143
    4. Visualizing the Retinoic Acid Gradient 146
    5. Crabps and Retinoic Acid Signal Robustness 150
    6. Sharpening Boundaries of Gene Expression in Response to Retinoic Acid Gradients 151
    7. Noise—Both Good and Bad.. 156

8. Other Boundaries and Other Morphogens........................... 157
   Conclusions and Perspectives.. 158
   Acknowledgments... 159
   References... 159

CHAPTER 8 Using Fluorescent Lipids in Live Zebrafish Larvae: from Imaging Whole Animal Physiology to Subcellular Lipid Trafficking................................ 165
J.L. Anderson, J.D. Carten, S.A. Farber

Introduction.. 166
The Need for Whole Animal Studies of Lipid Metabolism..... 166
1. Forward Genetic Screening With Fluorescent Lipids 168
2. Visualizing Lipid Metabolism Using BODIPY Fatty Acid Analogs.. 171
   Summary.. 174
   Acknowledgments... 175
   References... 175

CHAPTER 9 Analysis of Cilia Structure and Function in Zebrafish 179
E. Leventea, K. Hazime, C. Zhao, J. Malicki

Introduction.. 180
1. Cilia in Zebrafish Organs... 181
2. Analytical Tools for Cilia Morphology and Motility............. 190
3. Phenotypes of Cilia Mutants in Zebrafish 208
4. Future Directions 215
   Acknowledgments... 216
   References... 216

CHAPTER 10 Functional Calcium Imaging in Zebrafish Lateral-line Hair Cells... 229
Q.X. Zhang, X.J. He, H.C. Wong, K.S. Kindt

Introduction.. 230
1. Calcium Indicator Selection and Comparison..................... 231
2. Imaging Systems and Optimal Parameters 237
3. Image Processing 242
   Summary.. 248
   Discussion.. 248
   Acknowledgments... 249
   References... 249

CHAPTER 11 Physiological Recordings from the Zebrafish Lateral Line 253
  J. Olt, A.J. Ordoobadi, W. Marcotti, J.G. Trapani
  Introduction 254
  1. Common Methods for Lateral Line Electrophysiology 255
  2. Stimulation of Neuromast Hair Cells 258
  3. Recording Microphonic Potentials 261
  4. In Vivo Hair Cell Physiology 265
  5. Afferent Neuron Action Currents 270
  6. Summary 275
  Discussion 275
  Acknowledgments 276
  References 276

Volumes in Series 281
Index 295

Contributors

J.L. Anderson
Carnegie Institution for Science, Baltimore, MD, United States

J.D. Carten
Carnegie Institution for Science, Baltimore, MD, United States

D. Castranova
National Institute of Child Health and Human Development, Bethesda, MD, United States

C.A. Ciarlo
Harvard Medical School and Children's Hospital, Boston, MA, United States

K.W. Eliceiri
University of Wisconsin—Madison, Madison, WI, United States; Morgridge Institute for Research, Madison, WI, United States

S.A. Farber
Carnegie Institution for Science, Baltimore, MD, United States

A.V. Gore
National Institute of Child Health and Human Development, Bethesda, MD, United States

K. Hazime
The University of Sheffield, Sheffield, United Kingdom

X.J. He
National Institute on Deafness and Other Communication Disorders, NIH, Bethesda, MD, United States

A. Huttenlocher
University of Wisconsin—Madison, Madison, WI, United States

S. Isogai
Iwate Medical University, Morioka, Japan

H.M. Jung
National Institute of Child Health and Human Development, Bethesda, MD, United States

M. Kamei
South Australian Health and Medical Research Institute, Adelaide, SA, Australia

K.S. Kindt
National Institute on Deafness and Other Communication Disorders, NIH, Bethesda, MD, United States

D. Kromm
Centre for Organismal Studies, Heidelberg University, Heidelberg, Germany

D.C. LeBert
University of Wisconsin—Madison, Madison, WI, United States

E. Leventea
The University of Sheffield, Sheffield, United Kingdom

J. Malicki
The University of Sheffield, Sheffield, United Kingdom

W. Marcotti
University of Sheffield, Sheffield, United Kingdom

M.A. Mohr
Eidgenössische Technische Hochschule Zurich (ETH Zurich), Basel, Switzerland

Q. Nie
University of California, Irvine, CA, United States

J. Olt
University of Sheffield, Sheffield, United Kingdom

A.J. Ordoobadi
Amherst College, Amherst, MA, United States

P. Pantazis
Eidgenössische Technische Hochschule Zurich (ETH Zurich), Basel, Switzerland

T.F. Schilling
University of California, Irvine, CA, United States

J. Sosnik
University of California, Irvine, CA, United States

J.M. Squirrell
University of Wisconsin—Madison, Madison, WI, United States

D.L. Stachura
California State University, Chico, Chico, CA, United States

T. Thumberger
Centre for Organismal Studies, Heidelberg University, Heidelberg, Germany

J.G. Trapani
Amherst College, Amherst, MA, United States

D. Traver
University of California, San Diego, San Diego, CA, United States

B.M. Weinstein
National Institute of Child Health and Human Development, Bethesda, MD, United States

J. Wittbrodt
Centre for Organismal Studies, Heidelberg University, Heidelberg, Germany

H.C. Wong
National Institute on Deafness and Other Communication Disorders, NIH, Bethesda, MD, United States

Q.X. Zhang
National Institute on Deafness and Other Communication Disorders, NIH, Bethesda, MD, United States

C. Zhao
The University of Sheffield, Sheffield, United Kingdom; Ocean University of China, Qingdao, China

L.I. Zon
Children's Hospital and Dana Farber Cancer Institute, Boston, MA, United States; Harvard University, Cambridge, MA, United States

Preface

Len, Monte, and I are pleased to introduce the fourth edition of *Methods in Cell Biology: The Zebrafish*. The advantages of the zebrafish, *Danio rerio*, are numerous, including its short generation time and high fecundity, external fertilization, and the optical transparency of the embryo. The ease of conducting forward genetic screens in the zebrafish, based on the pioneering work of George Streisinger, culminated in screens from the laboratories of Wolfgang Driever, Mark C. Fishman, and Christiane Nüsslein-Volhard, published in a seminal volume of *Development* (Volume 123, December 1, 1996) that described a "candy store" of mutants whose phenotypes spanned the gamut of developmental processes and mechanisms. Life for geneticists who study vertebrate development became *really* fine.

Statistics derived from ZFIN (The Zebrafish Model Organism Database; http://zfin.org) illustrate the dramatic growth of research involving zebrafish. The zebrafish genome has been sequenced, and as of 2014, more than 25,000 genes have been placed on the assembly. Greater than 15,500 of these genes have been established as orthologs of human genes. The zebrafish community has grown from ∼1400 researchers in 190 laboratories as of 1998 to ∼7000 in 930 laboratories in 2014. The annual number of publications based on the zebrafish has risen from 1,913 to 21,995 in the same timeframe. Clearly, the zebrafish has arrived as a vertebrate biomedical model system *par excellence*.

When we published the first edition (Volumes 59 and 60) in 1998, our goal was to encourage biologists to adopt the zebrafish as a genetically tractable model organism for studying biological phenomena from the cellular through the organismal. Our goal today remains unchanged, but the range of subjects and the suite of methods have expanded rapidly and significantly in sophistication over the years. With the second and third editions of *MCB: The Zebrafish* (Volumes 76 and 77 in 2004; Volumes 100, 101, 104, and 105 in 2010–11), we documented this extraordinary growth, again relying on the excellent chapters contributed by our generous colleagues in the zebrafish research community.

When Len, Monte, and I began planning the fourth edition, we found that the zebrafish community had once more developed and refined novel experimental systems and technologies to tackle challenging biological problems across the spectrum of the biosciences. We present these methods following the organizational structure of the third edition, with volumes devoted to *Cellular and Developmental Biology*, to *Genetics, Genomics, and Transcriptomics*, and to *Disease Models and Chemical Screens*. Here we introduce the first two volumes, *Cellular and Developmental Biology, Parts A and B*.

CDB Part A is devoted to cellular techniques. Here the reader will find methods for culturing zebrafish cells and advanced light-microscopic strategies for imaging cells and multicellular structure in living or fixed embryos. Additional chapters cover the direct measurement of morphogen gradients, the analysis of lipid metabolism, methods to determine the structure and function of cilia, and techniques for functional calcium imaging and measuring sensory transduction in lateral line hair cells.

In *CDB Part B*, we transition to the study of developmental phenomena. A chapter on germ cell specification leads off, followed by several devoted to various aspects of central and peripheral nervous system development and function. Chapters on the inner ear, lateral line, and retina provide excellent examples of the relevance of the zebrafish to vertebrate developmental biology. Methods for studying organogenesis in the zebrafish cover the cardiovascular system, kidney, pancreas, liver, cartilage, and bone. The final chapters consider cancer from an organ perspective, associative learning and memory, and methods for working with larval and juvenile zebrafish.

We anticipate that you, our readership, will apply these methods successfully in your own zebrafish research programs and will develop your own technical advances that may be considered for a future edition of *Methods in Cell Biology: The Zebrafish*. The zebrafish is a remarkable experimental system—the preeminent vertebrate model for mechanistic studies of cellular and developmental processes in vivo.

We thank the series editors, Leslie Wilson and Phong Tran, and the staff of Elsevier/Academic Press, especially Zoe Kruze and Sarah Lay, for their enthusiastic support of our fourth edition. Their help, patience, and encouragement are profoundly appreciated.

H. William Detrich, III
Monte Westerfield
Leonard I. Zon

CHAPTER 1

Embryonic cell culture in zebrafish

C.A. Ciarlo*,[1], L.I. Zon[§,¶]

*Harvard Medical School and Children's Hospital, Boston, MA, United States
[§]Children's Hospital and Dana Farber Cancer Institute, Boston, MA, United States
[¶]Harvard University, Cambridge, MA, United States
[1]Corresponding author: E-mail: caciarlo@enders.tch.harvard.edu

CHAPTER OUTLINE

Introduction 2
1. Methods 3
   1.1 Blastomere Cell Culture 3
       1.1.1 Overview 3
       1.1.2 Materials and reagents 4
       1.1.3 Plating zebrafish blastomeres 4
       1.1.4 Representative results 5
   1.2 Neural Crest Cell Culture 6
       1.2.1 Overview 6
       1.2.2 Materials and reagents 7
       1.2.3 Plating zebrafish neural crest cells 7
       1.2.4 Representative results 9
Conclusion 10
References 10

Abstract

Zebrafish embryonic cell cultures have many useful properties that make them complementary to intact embryos for a wide range of studies. Embryonic cell cultures allow for maintenance of transient cell populations, control of chemical and mechanical cues received by cells, and facile chemical screening. Zebrafish cells can be cultured in either heterogeneous or homogeneous cultures from a wide range of developmental time points. Here we describe two methods with particular applicability to chemical screening: a method for the culture of blastomeres for directed differentiation toward the myogenic lineage and a method for the culture of neural crest cells in heterogeneous cultures from early somitogenesis embryos.

INTRODUCTION

While the zebrafish has proved a versatile model organism for studying in vivo biology, cultured zebrafish cells have also provided a platform for controlled genetic and chemical manipulation. More than 20 years ago, cells isolated from blastula stage embryos were reported to retain embryonic stem cell (ESC) characteristics in culture (Sun, Bradford, Ghosh, Collodi, & Barnes, 1995). Since then zebrafish embryonic cell culture has retained its relevance as a complementary system to in vivo studies, with diverse applications including chemical biology, disease modeling, and studies of cellular differentiation. While avoiding drawbacks of cell lines such as adaptation to culture, zebrafish primary cell culture allows for precise control of chemical cues received by cells, generation of homogenous populations of cells, maintenance of transient cell populations, and facile chemical screening.

Zebrafish cell cultures facilitate studies of cellular differentiation with a defined set of chemical cues (Huang, Lindgren, Wu, Liu, & Lin, 2012; Norris, Neyt, Ingham, & Currie, 2000; Xu et al., 2013). Like human ESCs, cultured zebrafish blastomeres, the cells that make up blastula stage embryos, can retain long-term pluripotency and germline competency (Fan et al., 2004; Ho et al., 2014). However, these cells can be induced to differentiate much more quickly than human ESCs and can be maintained without feeder layers (Ho et al., 2014; Huang et al., 2012; Myhre & Pilgrim, 2010; Xu et al., 2013). While many developmental studies can be conducted in vivo in zebrafish, blastomere cultures or cells derived from blastomere cultures provide a distinct advantage in certain situations. Studies of subcellular structure or localization, particularly in a specific tissue, can be difficult to conduct in vivo. Furthermore for studies in which a large amount of material is needed, a homogenous population of cells, for example of a specific cell type, can be generated from blastomeres. Genetic homogeneity can also be achieved by culture of blastomeres from single embryos and is essential for any study of embryonic lethal mutants in which heterozygous carriers are used (Myhre & Pilgrim, 2010).

Another advantage of zebrafish cell cultures is the potential to maintain transient cell populations. For example, neural crest progenitor cells can be isolated from early somitogenesis embryos and maintained in culture for at least 3 days (Kinikoglu, Kong, & Liao, 2013). These cells are capable of differentiation over the course of 2–3 weeks into neural crest-derived cell types including neurons, smooth muscle, pigment cells, and chondrocytes, using similar chemical cues as are used for differentiation of human ESCs or induced pluripotent stem cells (Kinikoglu et al., 2013). This culture system can be used to study the direct effects of chemicals, growth factors, or genetic mutations on neural crest development, proliferation, and migration.

Finally zebrafish cell culture allows for facile chemical screening. Unlike intact zebrafish embryos, zebrafish cells can be screened quickly in a completely automated manner with standard robotics and image analysis software. Screening can be conducted in either heterogeneous or homogeneous cell cultures. While heterogeneous cell cultures provide cell-type-specific information and do not require sorting, homogeneous cell cultures provide optimal signal strength.

Chemical screening of zebrafish blastomeres has been used to identify chemicals that promote endothelial and myogenic cell fate. Huang et al. used cells from gastrula-stage embryos to identify chemicals that promote vascular endothelial cell differentiation (Huang et al., 2012). Transgenic *flk1:GFP* embryos were plated with chemicals at 80% epiboly and cultured for 5 days before analysis of GFP fluorescence. A subset of hit chemicals from this screen were validated in promoting endothelial differentiation of murine ESCs, proliferation of human endothelial cells in culture, and angiogenesis in adult fish. The authors also demonstrated expression of transgenes marking various other cell types in cultures from late-blastula or early gastrula embryos, including *gata1* and *scl* for hematopoetic cells, *cmlc2* for cardiomyocytes, and *insulin* for pancreatic beta cells.

In 2013, Xu et al. identified signaling pathways that promote the myogenic differentiation of pluripotent cells, which were then validated in human iPS cells. Blastomeres from *myf5:GFP;mylz2:mCherry* fish were plated with chemicals, and fluorescence was determined after only one day, identifying basic fibroblast growth factor (bFGF), inhibitors of GSK3β, and forskolin as promoting muscle differentiation. Previous attempts to differentiate iPS cells into skeletal muscle had met with limited success, but using a combination of the factors identified in the zebrafish screen, engraftable myogenic progenitors were generated. The success of using zebrafish cell culture in these cases can be attributed to the ability to quickly screen many compounds, including combinations of growth factors and chemicals. Using such culture systems, signals required for differentiation can be addressed in an unbiased manner.

This chapter describes methods for the culture of zebrafish blastomeres, in particular for directed differentiation to muscle, and culture of cells from early somitogenesis embryos, in particular for maintenance of neural crest progenitor cells in heterogeneous cultures. Both of these methods are ideal for application to chemical screening.

1. METHODS
1.1 BLASTOMERE CELL CULTURE
1.1.1 Overview
Blastomeres can be used for short- or long-term studies of pluripotent cells or for differentiation to a desired cell type. Here we describe a simple general method for blastomere culture in which the whole embryo is plated at the oblong stage (3.7 hpf). Addition of 1 ng/mL bFGF to the culture medium at the time of plating results in differentiation into muscle, as presented in representative results.

1.1.2 Materials and reagents

E3 embryo medium (Nüsslein-Volhard & Dahm, 2002)
0.05% bleach solution in E3
2.5 mg/mL pronase solution in E3
agarose-coated petri dishes
100 μm cell strainer
0.1% gelatin solution, suitable for cell culture
tissue culture plates of desired size
Zebrafish embryonic stem cell (zESC) medium (Table 1)

1.1.3 Plating zebrafish blastomeres

1. Fish were allowed to mate and embryos were collected, cleaned, and incubated at 28.5°C for 3–4 h.
2. zESC medium was prepared as indicated in Table 1 by combining additives and bringing the solution to the final volume with base medium. Media was sterile filtered and used within 1 week of preparation.
3. Gelatin-coated cell culture plates were prepared. Gelatin solution was warmed to 37°C and used to cover the bottom of tissue culture plates. Plates were

Table 1 Zebrafish Embryonic Stem Cell Medium

Base Medium	Stock Concentration	Volume (Per 100 mL)	Final Concentration
Leibowitz's L-15 medium, with sodium pyruvate, no phenol red	1x	35 mL	35%
DMEM high glucose, HEPES, no phenol red	1x	21 mL	21%
DMEM/F12 1:1 high glucose, HEPES, no phenol red	1x	14 mL	14%
RTS34st conditioned medium[a]	1x	30 mL	30%
Additives			
Sodium bicarbonate	75 g/L	24 μL	0.18 g/L
HEPES	1 M	500 μL	5 mM
L-glutamine	200 mM	1 mL	2 mM
Sodium selenite	1 μM	1 μL	10 nM
Primocin (InVivoGen)	100%	200 μL	0.2%
N2 supplement	100%	1 mL	1%
B27 supplement	100%	2 mL	2%

Unless otherwise indicated, media components were obtained from Invitrogen.
[a] Prepare by incubating Leibowitz's L-15 plus 15% FBS on a confluent culture of RTS34st cells for 3 days.

incubated in a cell culture incubator at 37°C for 30 min, at which point the gelatin solution was aspirated. Plates were used immediately or stored in a sterile environment.
4. Embryos were bleached for decontamination. Embryos were transferred to a clean petri dish and incubated with 0.05% bleach solution in E3 for 2 min. Embryos were then washed three times with E3.
5. Embryos were dechorionated using pronase. When embryos had reached the high stage (~3.3 hpf), unfertilized eggs were removed and the desired number of embryos were transferred to agarose-coated petri dishes to avoid adherence of embryos to the surface of the dish. Embryos were incubated in a 2.5 mg/mL pronase solution in E3 at 28.5°C for 5 min and then washed three times with E3.
6. Embryos were counted, and proper staging was verified. For myogenic cultures, embryos should be homogenized at the oblong stage and plated at a density of 1 embryo per 5.6 mm^2 (1 well of a standard 384 well plate). For example, for plating a full 384 well plate, 400 embryos would be homogenized in 12 mL of zESC medium, and 30 μL of the resulting cell suspension would be plated in each well.
7. Embryos were transferred to an appropriate tube, ie, a 1.5-mL microcentrifuge tube for small volumes, or 15/50 mL Falcon tube for larger volumes. E3 was removed, along with any remaining chorions. Transferring chorions should be avoided, as they will clog the filter. For large numbers of embryos, they can be homogenized in a smaller volume and brought up to the final plating volume after filtering. However, no more than 500 embryos should be homogenized in every milliliter of medium.
8. zESC medium was added to the embryos, and they were dissociated by rapidly inverting the tube approximately 5 times or until a homogenous cell suspension was observed, ie, no clumps of cells were visible. For directed differentiation toward muscle, zESC medium containing 1 ng/ml recombinant human bFGF should be used.
9. The cell suspension was passed through a 100 μm cell strainer and brought up to the desired final volume.
10. The cell suspension was pipetted into gelatin-coated tissue culture plates and incubated at 28.5°C, 5% CO_2 in a humidified chamber until analysis.

1.1.4 Representative results

As shown in Fig. 1, blastomeres cultured without bFGF form clusters after 1 day of culture and do not express muscle-specific reporters such as *myf5:GFP*, marking myogenic progenitors and *mylz2:mCherry*, marking differentiated muscle cells. Blastomeres cultured with 1 ng/ml bFGF have a more flattened morphology and express *myf5:GFP* and *mylz2:mCherry*. Yolk droplets are also visible as uniform spheres in both cultures. As Fig. 2 demonstrates, myogenic cells differentiate in culture over a 2-day period. As determined by qPCR, *myf5* expression peaks during day 1 of culture, while *mylz2* expression peaks during day 2.

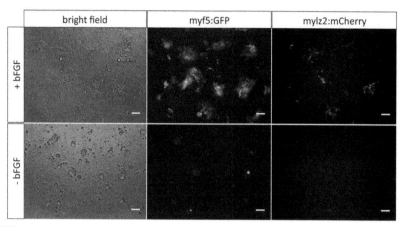

FIGURE 1

Addition of basic fibroblast growth factor (bFGF) to zebrafish embryonic stem cell medium alters the morphology and reporter gene expression of cultured blastomeres. Cells were plated using the described method and cultured for 26 h. While bFGF-treated cultures express both *myf5:GFP* and *mylz2:mCherry* and appear flattened and elongated, cultures without bFGF do not express these transgenes and form clusters of rounded cells. Yolk droplets are visible as uniform spheres in the cultures. Scale bar represents 100 μm.

1.2 NEURAL CREST CELL CULTURE

1.2.1 Overview

Zebrafish neural crest cells can be grown in heterogeneous cultures to facilitate studies of the effects that chemicals or growth factors have on neural crest migration,

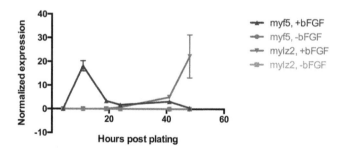

FIGURE 2

Muscle gene expression changes over the course of a 2-day culture. Blastomeres were plated as described with or without basic fibroblast growth factor for the indicated period of time. Expression of *myf5* or *mylz2* was determined in whole cultures by qPCR and normalized to β-actin. Data points represent the average and standard error of the mean of three technical replicates.

differentiation, proliferation, and survival. This technique avoids costly and time-consuming cell sorting and retains information about cell-type specificity. Furthermore relatively few embryos can be used: 250 embryos provide enough material for a full 384 well plate. Here we describe a method for the short-term culture of neural crest cells in heterogeneous cultures from early somitogenesis embryos. *Crestin:GFP* reporter fish can be used to identify neural crest cells in the culture.

1.2.2 Materials and reagents

E3 embryo medium (Nüsslein-Volhard & Dahm, 2002)
0.05% bleach solution in E3
2.5mg/mL pronase in E3
40 μm cell strainer
rotor-stator homogenizer (OMNI International TH-115) and sterilized probes
tissue culture plates of desired size
500 μg/mL rat tail collagen type I or 10 μg/mL fibronectin (optional)
sterile phosphate-buffered saline (PBS)
neural crest medium (Table 2)

1.2.3 Plating zebrafish neural crest cells

1. Fish expressing *crestin:GFP* or another transgenic marker were allowed to mate and embryos were collected, cleaned, and incubated at 28.5°C for 10–11 h. For convenience, embryos may also be incubated at room temperature for 16–18 h.

Table 2 Neural Crest Medium

Base Medium	Stock Concentration	Volume (per 100 mL)	Final Concentration
DMEM/F12 1:1 with L-glutamine and 2.438 g/L sodium bicarbonate	1x	100 mL	100%
Additives			
FBS	100%	12 mL	12%
N2 supplement	100%	1 mL	1%
Recombinant human insulin (Gemini Bioproducts)	2 mg/mL	1 mL	20 μg/mL
Recombinant human EGF (R&D systems)	500 μg/mL	4 μL	20 ng/mL
Recombinant human fibroblast growth factor basic (R&D systems)	100 μg/mL	20 μL	20 ng/mL
Primocin (InVivoGen)	100%	200 μL	0.2%

Unless otherwise indicated, media components were obtained from Invitrogen.

2. Neural crest medium was prepared as indicated in Table 2 by combining additives and bringing the solution to the final volume with base medium. Media was sterile filtered and used within 1 week of preparation.
3. Coated cell culture plates were prepared (optional). Neural crest cells can be cultured in heterogeneous cultures with or without collagen or fibronectin coating. Coating of plates with collagen or fibronectin results in a more adherent culture. Without coating, neural crest cells, as marked by *crestin: GFP*, are attached and highly migratory, but can be easily washed away. Collagen or fibronectin solution was warmed to 37°C, used to cover the bottom of tissue culture wells, and incubated in a cell culture incubator at 37°C for 30 min. The solution was aspirated, and wells were washed twice with sterile PBS. Plates were used immediately or stored in a sterile environment.
4. Embryos were bleached for decontamination. When embryos had reached the two somite stage, they were transferred to a clean petri dish and incubated with 0.05% bleach solution in E3 for 2 min. Embryos were then washed three times with E3.
5. For dechorionation, embryos were incubated in a 2.5 mg/mL pronase solution in E3 at 28.5°C for 5 min and then washed three times with E3.
6. Embryos were counted, and proper staging was verified. Neural crest cells can be plated from 5 to 15 somite-stage embryos at a density of 0.6–1 embryo per 5.6 mm^2 (1 well of a standard 384 well plate). For example, for plating a full 384 well plate, 250 embryos would be homogenized in 12.3 mL of neural crest medium, and 30 μL of the resulting cell suspension would be plated in each well. Embryo number can also be estimated by volume. For example, 500 chorionated embryos take up approximately 0.9 mL.
7. Embryos were transferred to appropriate tube for mechanical homogenization, eg., a round-bottom FACS tube. E3 was removed along with any remaining chorions. Transferring chorions should be avoided, as they will clog the filter. For large numbers of embryos, embryos can be homogenized in a smaller volume and brought up to the final plating volume after filtering. However, no more than 500 embryos should be homogenized in every milliliter of medium.
8. Embryos were homogenized in neural crest medium using a rotor–stator homogenizer at 5000 rpm for 10–15 s. The resulting cell suspension was pipetted up and down twice to break up clumps. During this step, the cell suspension was observed to confirm that the embryos had been sufficiently homogenized. If clumps of cells were observed, the suspension was homogenized for another 10–15 s. If embryos have not been sufficiently homogenized, cells will be removed during filtering and clog the filter.
9. The cell suspension was passed through a 40-μm cell strainer. The homogenization tube and cell strainer were then rinsed with an additional 1 mL of neural crest medium. The cell suspension can also be diluted to facilitate passage through the cell strainer.

10. The cell suspension was brought to the desired final volume, pipetted into tissue culture plates, and incubated at 28.5°C, 5% CO_2 in a humidified chamber until analysis. Neural crest cells cultured in this manner should remain viable for at least 3 days.

1.2.4 Representative results

As shown in Fig. 3A, the percentage of *crestin:GFP* + cells increases over a period of 1 day in the culture conditions described. While at the time of plating, GFP + cells represent 3% of the total, after 1 day of culture, GFP + cells represent 18% of the total. *Crestin:GFP* + cells are adherent, migratory, and extend projections in heterogeneous whole embryo cultures (Fig. 3B).

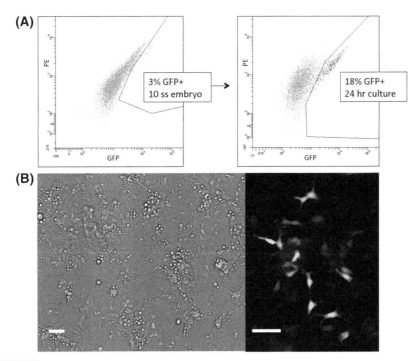

FIGURE 3

(A) FACS analysis of *crestin:GFP* embryos before and after 1 day of culture demonstrate promotion of neural crest cell fate in neural crest medium. Cells were plated as described at the 10 somite stage at a density of 0.6 embryos per 5.6 mm^2. FACS plots show GFP (x-axis) versus PE (y-axis) to control for autofluorescence. (B) *Crestin:GFP* + cells are adherent and extend projections in heterogeneous cultures. Left panel: GFP and bright field overlay, 24 h culture. Right panel: magnification of GFP only, 13 h culture. Scale bar represents 100 µm. (See color plate)

CONCLUSION

Zebrafish embryonic cell culture has many uses, including generating hypotheses for in vivo validation, performing mechanistic experiments in a controlled environment, and facilitating chemical screening. Zebrafish embryonic cells can be cultured from blastula-, gastrula-, or somitogenesis-stage embryos in either a homogeneous or heterogeneous context. As described in this chapter, zebrafish blastomeres can be cultured for differentiation into myogenic progenitors, and zebrafish neural crest cells can be maintained in a heterogeneous context without the need for sorting. These and similar cell culture techniques promise to inform stem and progenitor cell biology in the future.

REFERENCES

Fan, L., Crodian, J., Liu, X., Aleström, A., Aleström, P., & Collodi, P. (2004). Zebrafish embryo cells remain pluripotent and germ-line competent for multiple passages in culture. *Zebrafish, 1*(1), 21–26. http://dx.doi.org/10.1089/154585404774101644.

Ho, S. Y., Goh, C. W. P., Gan, J. Y., Lee, Y. S., Lam, M. K. K., Hong, N. ... Shu-Chien, A. C. (2014). Derivation and long-term culture of an embryonic stem cell-like line from zebrafish blastomeres under feeder-free condition. *Zebrafish, 11*(5), 407–420. http://dx.doi.org/10.1089/zeb.2013.0879.

Huang, H., Lindgren, A., Wu, X., Liu, N.-A., & Lin, S. (2012). High-throughput screening for bioactive molecules using primary cell culture of transgenic zebrafish embryos. *Cell Reports, 2*(3), 695–704. http://dx.doi.org/10.1016/j.celrep.2012.08.015.

Kinikoglu, B., Kong, Y., & Liao, E. C. (2013). Characterization of cultured multipotent zebrafish neural crest cells. *Experimental Biology and Medicine (Maywood, N.J.)*. http://dx.doi.org/10.1177/1535370213513997.

Myhre, J. L., & Pilgrim, D. B. (2010). Cellular differentiation in primary cell cultures from single zebrafish embryos as a model for the study of myogenesis. *Zebrafish, 7*(3), 255–266. http://dx.doi.org/10.1089/zeb.2010.0665.

Norris, W., Neyt, C., Ingham, P. W., & Currie, P. D. (2000). Slow muscle induction by Hedgehog signalling in vitro. *Journal of Cell Science, 113*(Pt 15), 2695–2703.

Nüsslein-Volhard, C., & Dahm, R. (2002). *Zebrafish*. USA: Oxford University Press.

Sun, L., Bradford, C. S., Ghosh, C., Collodi, P., & Barnes, D. W. (1995). ES-like cell cultures derived from early zebrafish embryos. *Molecular Marine Biology and Biotechnology, 4*(3), 193–199.

Xu, C., Tabebordbar, M., Iovino, S., Ciarlo, C., Liu, J., Castiglioni, A. ... Zon, L. I. (2013). A zebrafish embryo culture system defines factors that promote vertebrate myogenesis across species. *Cell, 155*(4), 909–921. http://dx.doi.org/10.1016/j.cell.2013.10.023.

CHAPTER

Cellular dissection of zebrafish hematopoiesis

2

D.L. Stachura*,[1], D. Traver[§]

California State University, Chico, Chico, CA, United States
[§]*University of California, San Diego, San Diego, CA, United States*
[1]*Corresponding author: E-mail: dstachura@csuchico.edu*

CHAPTER OUTLINE

Introduction	12
1. Zebrafish Hematopoiesis	12
1.1 Primitive Hematopoiesis	13
1.2 Definitive Hematopoiesis	13
1.3 Adult Hematopoiesis	18
2. Hematopoietic Cell Transplantation	24
2.1 Embryonic Donor Cells	24
2.1.1 Protocol for isolating hematopoietic cells from embryos	26
2.1.2 Transplanting purified cells into embryonic recipients	28
2.1.3 Transplanting cells into blastula recipients	28
2.1.4 Transplanting cells into 48 hpf embryos	29
2.2 Adult Donor Cells	29
2.2.1 Protocols for isolating hematopoietic cells from adult zebrafish	30
2.2.2 Transplanting WKM	31
2.2.3 Transplanting cells into irradiated adult recipients	31
2.2.4 Irradiation	31
2.2.5 Transplantation	32
3. Enrichment of HSCs	32
4. In Vitro Culture and Differentiation of Hematopoietic Progenitors	35
4.1 Stromal Cell Culture Assays	36
4.1.1 Generation of ZKS cells	37
4.1.2 Maintenance and culture of ZKS cells	37
4.1.3 Generation of ZEST cells	38
4.1.4 Maintenance and culture of ZEST cells	38
4.1.5 Protocols for in vitro proliferation and differentiation assays on ZKS and ZEST cells	40
4.2 Clonal Methylcellulose-Based Assays	42

        4.2.1 Methylcellulose.. 44
        4.2.2 Methylcellulose stock preparation 44
        4.2.3 Methylcellulose clonal assays... 44
        4.2.4 Enumeration of colony forming units 45
        4.2.5 Picking and analyzing colonies from methylcellulose 45
Conclusions.. 46
References 46

Abstract

Zebrafish as a model system have been instrumental in understanding early vertebrate development, especially of the hematopoietic system. The external development of zebrafish and their genetic amenability have allowed in-depth studies of multiple blood cell types and their respective genetic regulation. This chapter highlights some new data in zebrafish hematopoiesis regarding primitive and definitive hematopoiesis in the embryonic and adult fish, allowing the isolation of prospective progenitor subsets. It also highlights assays developed to examine the function of these progenitors in vivo and in vitro, allowing an evolutionary understanding of the hematopoietic system and how zebrafish can be better utilized as a model system for a multitude of hematopoietic disorders.

INTRODUCTION

Over the past 25 years, the development of forward genetic approaches in the zebrafish has provided unprecedented power in understanding the molecular basis of vertebrate blood development. Establishment of cellular and hematological approaches to better understand the biology of resulting blood mutants, however, has lagged behind these efforts. In this chapter, recent multiple advances in zebrafish hematology will be reviewed, with an emphasis on prospective strategies for isolation of both embryonic and adult hematopoietic stem cells and the development of assays with which to rigorously test their function.

1. ZEBRAFISH HEMATOPOIESIS

Developmental hematopoiesis in mammals and teleosts occurs in four sequential waves (Fig. 1). The first two waves are termed "primitive," and each generates transient precursors that give rise to embryonic macrophages and erythrocytes, respectively (Keller, Lacaud, & Robertson, 1999; Palis, Robertson, Kennedy, Wall, & Keller, 1999). The next two waves consist of "definitive" hematopoietic precursors, defined as multipotent progenitors of adult cell types. The first to arise are erythromyeloid progenitors (EMPs) that give rise to erythroid and myeloid lineages (Bertrand et al., 2005a; 2007; Palis et al., 1999, 2001), followed by multipotent hematopoietic stem cells (HSCs), which have the potential to both self-renew and generate all adult hematopoietic cell types (reviewed in Cumano & Godin, 2007; Eaves, 2015; Orkin & Zon, 2008).

1.1 PRIMITIVE HEMATOPOIESIS

Primitive hematopoiesis has been extensively studied in the mouse, where primitive macrophages and erythroid cells are generated in the extraembryonic yolk sac (YS) (Bertrand et al., 2005a; Palis et al., 1999). In the zebrafish, primitive macrophages develop in an anatomically distinct area known as the rostral blood island (RBI) (Fig. 1A). Transcripts for *tal1* (also known as *scl*), *lmo2*, *gata2a*, and *fli1a* are found in the RBI between the three- and five-somite stages (Brown et al., 2000; Liao et al., 1998; Thompson et al., 1998). This is quickly followed by expression of *spi1* (also known as *pu.1*) in a subset of these precursors (Bennett et al., 2001; Lieschke et al., 2002). Between the 11- and 15-somite stages, $spi1^+$ macrophages are detectable, and they migrate toward the head midline (Herbomel, Thisse, & Thisse, 1999; Lieschke et al., 2002; Ward et al., 2003), and across the yolk ball (Fig. 1C). Some of these precursors enter circulation, while others migrate into the head (Herbomel et al., 1999). By 28−32 hours post fertilization (hpf) macrophages are found in circulation and dispersed throughout the embryo.

Primitive erythroid cell generation begins in the murine YS blood islands at day 7.5 post coitum (E7.5) (Fig. 1B) (reviewed in Palis, Malik, McGrath, & Kingsley, 2010). These blood islands consist of nucleated erythroid cells that express embryonic *globin* genes surrounded by endothelial cells. Although it was previously believed that mammalian primitive erythroid cells uniquely remained nucleated (similar to the nucleated erythrocytes of birds, fish, and amphibians), it is now accepted that mammalian primitive red blood cells do, in fact, enucleate into reticulocytes and prenocytes (Fraser, Isern, & Baron, 2007; Kingsley, Malik, Fantauzzo, & Palis, 2004; McGrath et al., 2008). The zebrafish has an equivalent anatomical site to mammalian blood islands, known as the intermediate cell mass (ICM), where two stripes of mesodermal cells expressing *tal1*, *lmo2*, and *gata1a* converge to the midline of the zebrafish embryo and are surrounded by endothelial cells that become the cardinal vein (Al-Adhami & Kunz, 1977; Detrich et al., 1995) (Fig. 1A and B). Although the ICM is intraembryonic in zebrafish, it has a cellular architecture similar to the mammalian YS blood islands (Al-Adhami & Kunz, 1977; Willett, Cortes, Zuasti, & Zapata, 1999).

The development of transgenic zebrafish expressing fluorescent markers under the control of early mesodermal, prehematopoietic promoters (Table 1) now allow testing of primitive fate potentials by prospective isolation strategies and the functional assays outlined in this chapter.

1.2 DEFINITIVE HEMATOPOIESIS

Similar to primitive hematopoiesis, definitive hematopoiesis initiates through two distinct precursor subsets. In the mouse, multilineage hematopoiesis is first evident in the YS (Bertrand et al., 2005a; Palis et al., 1999, 2001; Yoder et al., 1997a; Yoder, Hiatt, & Mukherjee, 1997b) and placenta (Gekas, Dieterlen-Lievre, Orkin, &

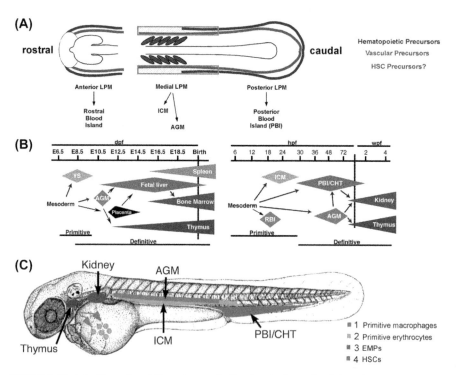

FIGURE 1 Model of hematopoietic ontogeny in the developing zebrafish embryo.

(A) Different regions of lateral plate mesoderm (LPM) give rise to anatomically distinct regions of blood cell precursors. Anatomical regions of embryo responsible for generation of hematopoietic precursors (red), vasculature (blue), and pre-hematopoietic stem cells (HSCs) (green) are highlighted. Cartoon is a five-somite stage embryo, dorsal view. (B) Timing of mouse and zebrafish hematopoietic development. In mouse (left), primitive hematopoiesis initiates in the yolk sac (YS; yellow), producing primitive erythroid cells and macrophages. Later, definitive erythromyeloid progenitors (EMPs) emerge in the YS. HSCs are specified in the aorta, gonad, and mesonephros (AGM, teal) region. These HSCs eventually seed the fetal liver (orange), the main site of embryonic hematopoiesis. Adult hematopoiesis occurs in the thymus (blue), spleen (green), and bone marrow (red). Zebrafish hematopoiesis is similar: temporal analogy to mouse hematopoiesis shown in (B, right), spatial locations shown in (C). Numbers in (C) correspond to timing of distinct precursor waves. (C) Embryonic hematopoiesis occurs through four independent waves of precursor production. First, primitive macrophages arise in cephalic mesoderm, migrate onto the yolk ball, and spread throughout the embryo (purple, 1). Then, primitive erythrocytes develop in the intermediate cell mass (ICM; yellow, 2). The first definitive progenitors are EMPs, which develop in the posterior blood island (PBI; orange, 3). Later, HSCs arise in the AGM region (teal, 4), migrate to the CHT (later name for the PBI, orange), and eventually seed the thymus and kidney (blue; red). Similar hematopoietic

Mikkola, 2005; Ottersbach & Dzierzak, 2005) by E9.5. Multilineage precursors in both tissues can be isolated and distinguished by the expression of CD41, an integrin molecule that labels early hematopoietic progenitors. CD41$^+$ cells differentiate into both myeloid and erythroid lineages, but conspicuously lack lymphoid potential (Bertrand et al., 2005b; Yokota et al., 2006). These studies suggest that the definitive hematopoietic program in the developing mouse begins with committed EMPs. Studies in the zebrafish demonstrated evolutionary conservation of EMPs as the first definitive precursor formed, and expanded upon findings in the mouse YS. EMPs can be isolated from the zebrafish posterior blood island between 26 and 36 hpf (Fig. 1C) by their coexpression of fluorescent transgenes driven by the *lmo2* and *gata1a* promoters (Fig. 2A) (Bertrand et al., 2007). In vitro differentiation experiments (Fig. 2B and C) indicate that EMPs are capable of only erythroid and myeloid differentiation (Bertrand et al., 2007), even when grown in culture conditions that support lymphoid differentiation (Stachura et al., 2009). Accordingly, in vivo transplantation of EMPs did not generate a lymphoid readout (Bertrand et al., 2007). Studies performed in *mindbomb* mutant zebrafish lacking Notch signaling showed that EMP specification and differentiation are not affected by loss of the Notch pathway (Bertrand, Cisson, Stachura, & Traver, 2010a), whereas HSCs are completely absent (Burns, Traver, Mayhall, Shepard, & Zon, 2005), further distinguishing these two definitive progenitors.

The zebrafish allows the discrimination of EMPs from HSCs, both of which have similar cell-surface markers and differ only in their differentiation and self-renewal potentials (Bertrand et al., 2005a, 2005b). Unlike the murine system, these two progenitors arise in separate anatomical locations (summarized in Fig. 1C) (Bertrand et al., 2007), and are easily distinguishable. Fate-mapping studies in the zebrafish demonstrated that EMPs arise from posterior mesodermal derivatives that express the *lmo2* gene. This finding, in combination with lineage tracing studies demonstrating EMPs to completely lack T lymphoid potential, indicates that EMPs and HSCs are unique populations independently derived during development.

The final wave of hematopoiesis culminates with the formation of HSCs, which self-renew and give rise to all definitive blood cell lineages, including lymphocytes. HSCs arise in an area of the midgestation mouse bounded by the aorta, gonads, and mesonephros (AGM) at E10-10.5 (Fig. 1B) (Cumano & Godin, 2007; Dzierzak, 2005). Many studies also suggested that transplantable HSCs are present in the YS on E9 (Lux et al., 2008; Weissman, Papaioannou, & Gardner, 1978; Yoder et al., 1997a,b) and later in the placenta by E11 (Gekas et al., 2005; Ottersbach &

---

events in mouse and fish are color matched between right and left panels of (B). Hematopoietic sites (B) and locations (C) are also color matched. *dpf*, days post fertilization; *E*, embryonic day; *hpf*, hours post fertilization; *RBI*, rostral blood island; *wpf*, weeks post fertilization. (See color plate)

Table 1 List of Relevant Transgenic Zebrafish Lines Currently Available for Hematopoietic Studies, Indicating the Promoter:Gene Expressed and the Cell Population(s) Identified

Transgene	Tissue	Transgene	Tissue
lmo2:GFP (Zhu et al., 2005)	Prehematopoietic, vasculature	kdrl:EGFP (Cross, Cook, Lin, Chen, & Rubinstein, 2003)	Prehematopoietic, vasculature
lmo2:mCherry[a]	Prehematopoietic, vasculature	kdrl:DsRed (Jin et al., 2007)	Prehematopoietic, vasculature
lmo2:DsRed (Lin et al., 2005)	Prehematopoietic, vasculature	fli1a:EGFP (Lawson & Weinstein, 2002)	Prehematopoietic, vasculature
itga2b:GFP (Lin et al., 2005)	EMPs, HSCs, thrombocytes	fli1a:DsRed (Jin et al., 2007)	Prehematopoietic, vasculature
itga2b:mCherry[a]	EMPs, HSCs, thrombocytes	gata3:AmCyan (Bertrand et al., 2008)	Kidney
itga2b:CFP[a]	EMPs, HSCs, thrombocytes	rag2:EGFP (Langenau et al., 2003)	Immature B and T cells
ptprc:DsRed (Bertrand et al., 2008)	Panleukocyte	lck:EGFP (Langenau et al., 2004)	Mature T cells
ptprc:CFP[a]	Panleukocyte	il7r:mCherry[a]	Lymphoid precursors, T cells
ptprc:AmCyan[a]	Panleukocyte	mhcII:GFP (Wittamer, Bertrand, Gutschow, & Traver, 2011)	B cells, macrophages, dendritic cells
gata1a:GFP (Long et al., 1997)	Red blood cells	mhcII:AmCyan[a]	B cells, macrophages, dendritic cells
gata1a:DsRed (Traver et al., 2003a)	Red blood cells	lyz:EGFP (Hall, Flores, Storm, Crosier, & Crosier, 2007)	Neutrophils
cmyb:GFP (Bertrand et al., 2008)	HSCs, neural	lyz:DsRed (Hall et al., 2007)	Neutrophils
mpx:EGFP (Renshaw et al., 2006)	Neutrophils	runx1P1:GFP (Lam et al., 2008)	EMPs
gata2a:EGFP (Traver et al., 2003a)	Eosinophils	runx1P2:GFP (Lam et al., 2008)	HSCs
Ighm:EGFP (Page et al., 2013)	B cells	ccr9a:cfp[a]	Lymphoid precursors
Ighz:EGFP[a]	B cells	mpeg1:GAL4 (Ellett, Pase, Hayman, Andrianopoulos, & Lieschke, 2010)	Embryonic macrophages
		rag2:DsRed (Langenau et al., 2007)	Immature B and T cells

EMP, erythromyeloid progenitor; HSC, hematopoietic stem cell. This list is not comprehensive; other transgenic animals are being constantly generated, but these are a few of the essential tools currently being used in zebrafish hematopoiesis laboratories.
[a] Unpublished transgenic animals generated in the Traver laboratory.

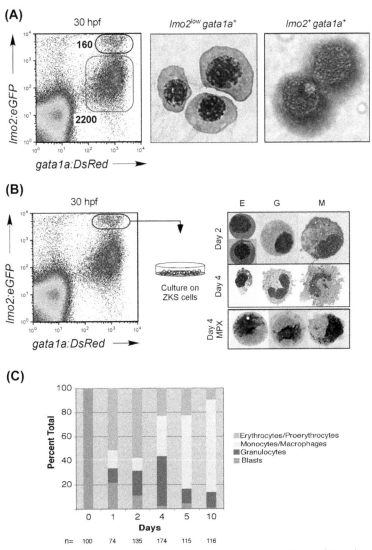

FIGURE 2 Functional in vitro differentiation studies demonstrate that $gata1^+lmo2^+$ cells are committed erythromyeloid progenitors.

(A) Purified erythromyeloid progenitors at 30 hpf ($lmo2^+gata1a^+$, black gate) have the immature morphology of early hematopoietic progenitors. As a comparison, purified primitive erythroblasts are shown ($lmo2^{low}gata1a^+$, red gate). Magnification, 1000×. (B) Short-term in vitro culture of $lmo2^+gata1^+$ cells atop zebrafish kidney stromal (ZKS) cells demonstrate erythroid (E), granulocytic (G), and monocytic/macrophage (M) differentiation potentials. Cultured cells were stained with May-Grünwald/Giemsa and for myeloperoxidase (MPX) activity. $lmo2^{low}gata1a^+$ cells only differentiated into erythroid cells (not shown). Magnification, 1000×. (C) Lineage differentials of cell types produced from cultured erythromyeloid progenitors. n, number of cells counted from each time point. (See color plate)

Dzierzak, 2005). Whereas these results suggest that HSCs may arise in distinctly different locations in the developing mouse embryo, it is clear that HSCs originate from arterial endothelium. Recent studies in the E10.5 mouse (Boisset et al., 2010) and 36–52 hpf zebrafish embryo (Bertrand et al., 2010b; Kissa & Herbomel, 2010) demonstrated directly the birth of HSCs from aortic endothelium with confocal imaging. A commonality among all vertebrate embryos thus seems to be the generation of HSCs from hemogenic endothelium lining the aortic floor (de Bruijn et al., 2002; Ciau-Uitz, Walmsley, & Patient, 2000; Jaffredo, Gautier, Eichmann, & Dieterlen-Lievre, 1998; North et al., 2002; Oberlin, Tavian, Blazsek, & Peault, 2002; and reviewed in Ciau-Uitz, Monteiro, Kirmizitas, & Patient, 2014). Similar studies need to be performed to determine if additional embryonic sites can also autonomously generate HSCs, including the YS and placenta. In all locations, however, it is clear that HSCs are present only transiently; by E11 the fetal liver (FL) is populated by circulating HSCs (Houssaint, 1981; Johnson & Moore, 1975) and becomes the predominant site of blood production during midgestation, producing the first full complement of definitive, adult-type effector cells. Shortly afterward, hematopoiesis is evident in the fetal spleen, and occurs in bone marrow throughout adulthood (Keller et al., 1999).

The zebrafish possesses an anatomical site that closely resembles the mammalian AGM (Fig. 1B and C). Between the dorsal aorta and cardinal vein between 28 and 48 hpf, $cmyb^+$ and $runx1^+$ blood cells appear in intimate contact with the dorsal aorta (Burns et al., 2002; Kalev-Zylinska et al., 2002; Thompson et al., 1998). Lineage tracing of $CD41^+$ HSCs derived from this ventral aortic region show their ability to colonize the thymus (Bertrand et al., 2007; Kissa et al., 2008) and pronephros (Bertrand, Kim, Teng, & Traver, 2008; Murayama et al., 2006), which are the sites of adult hematopoiesis in the fish (Jin, Xu, & Wen, 2007; Murayama et al., 2006). After 48 hpf, blood production appears to shift to the caudal hematopoietic tissue (Fig. 1C), and later the pronephros, which serves as the definitive hematopoietic organ for the remainder of life.

The development of transgenic zebrafish expressing fluorescent markers under the control of definitive hematopoietic promoters such as *itga2b* (also known as *cd41*), *cmyb*, and *runx1* (see Table 1) now allow testing of fate potentials by prospective isolation strategies and functional assays outlined later in this chapter.

1.3 ADULT HEMATOPOIESIS

Previous genetic screens in zebrafish were successful in identifying mutants that affected primitive erythropoiesis. These screens scored visual defects in circulating blood cells during early embryogenesis, so mutants defective in definitive hematopoiesis but displaying normal primitive blood cell development were likely missed. Current screens aimed at identifying mutants with defects in the generation of definitive HSCs in the AGM should reveal new genetic pathways required for multilineage hematopoiesis. Recent studies in zebrafish show that nearly all adult hematopoietic cells derive from HSCs born from the aortic endothelium (Bertrand et al., 2010b), consistent with findings in the murine system (Chen, Yokomizo, Zeigler, Dzierzak, & Speck, 2009; Zovein et al., 2008). Therefore, mutational screens designed to identify defects in hemogenic endothelium may yield information about the full repertoire of hematopoietic regulation, specification, maintenance, and differentiation over the organism's life span. Understanding the biology of mutants isolated using these approaches, however, first requires the characterization of normal, definitive hematopoiesis and the development of assays to study the biology of zebrafish blood cells more precisely. To this end, we have established several tools to characterize the definitive blood-forming system of adult zebrafish.

Blood production in adult zebrafish, like other teleosts, occurs in the kidney, which supports both renal functions and multilineage hematopoiesis (Zapata, 1979). Similar to mammals, T lymphocytes develop in the thymus (Trede & Zon, 1998; Willett et al., 1999) (Fig. 3A), which exists in two bilateral sites in zebrafish (Hansen & Zapata, 1998; Willett, Zapata, Hopkins, & Steiner, 1997). The teleostean kidney is a sheath of tissue that runs along the spine (Fig. 3B and E); the anterior portion, or head kidney, shows a higher ratio of blood cells to renal tubules than does the posterior portion (Zapata, 1979), termed the trunk kidney (Fig. 3B and C). All mature blood cell types are found in the kidney and morphologically resemble their mammalian counterparts (Fig. 3G, Fig. 4), with the exceptions that erythrocytes remain nucleated and thrombocytes perform the clotting functions of platelets (Jagadeeswaran, Sheehan, Craig, & Troyer, 1999). Histologically, the zebrafish spleen (Fig. 3D) has a simpler structure than its mammalian counterpart in that germinal centers have not been observed (Zapata & Amemiya, 2000). The absence of immature precursors in the spleen, or any other adult tissue, suggests that the kidney is the predominant hematopoietic site in adult zebrafish. The cellular composition of whole kidney marrow (WKM), spleen, and blood are shown in Fig. 3F–H. Morphological examples of all hematopoietic cell types are presented in Fig. 4.

Analysis of WKM by fluorescence-activated cell sorting (FACS) showed that several distinct populations could be resolved by light scatter characteristics (Fig. 5A). Forward scatter (FSC) is directly proportional to cell size, and side scatter (SSC) proportional to cellular granularity (Shapiro, 2002). Using combined scatter profiles, the major blood lineages can be isolated to purity from WKM following two rounds of cell sorting (Traver, 2003a). Mature erythroid cells were found exclusively within two FSC^{low} fractions (Populations R1 and R2, Fig. 5A and D), lymphoid cells within a FSC^{int} SSC^{low} subset (Population R3, Fig. 5A and E), immature precursors within a FSC^{high} SSC^{int} subset (Population R4, Fig. 5A and F), and myelomonocytic cells within only a FSC^{high}, SSC^{high} population (Population R5, Fig. 5A and G). Interestingly, two distinct populations of mature erythroid cells exist (Fig. 5A, R1, R2 gates). Attempts at sorting either of these subsets reproducibly resulted in approximately equal recovery of both (Fig. 5D). This is likely due to the elliptical nature of zebrafish red blood cells, because sorting of all other populations yielded cells that fell within the original sorting gates upon reanalysis. Examination of splenic (Fig. 5B) and peripheral blood (Fig. 5C) suspensions showed each to have distinct profiles from WKM,

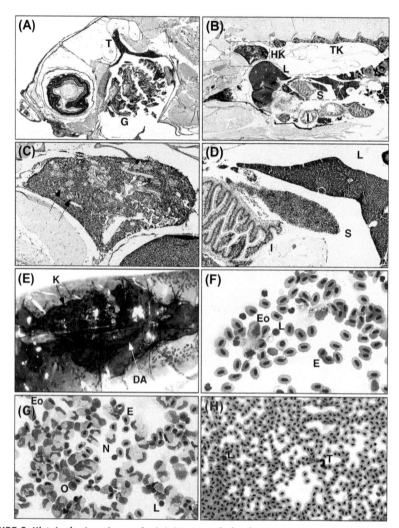

FIGURE 3 Histological analyses of adult hematopoietic sites.

(A) Sagittal section showing location of the thymus (T), which is dorsal to the gills (G). (B) Midline sagittal section showing location of the kidney, which is divided into the head kidney (HK), and trunk kidney (TK), and spleen (S). The head kidney shows a higher ratio of blood cells to renal tubules (*black arrows*), as shown in a close up view of the HK in (C). (D) Close up view of the spleen, which is positioned between the liver (L) and the intestine (I). (E) Light microscopic view of the kidney (K), over which passes the dorsal aorta (DA, *white arrow*). (F) Cytospin preparation of splenic cells, showing erythrocytes (E), lymphocytes (L), and an eosinophil (Eo). (G) Cytospin preparation of kidney cells showing cell types as noted earlier plus neutrophils (N) and erythroid precursors (O, orthochromic erythroblast). (H) Peripheral blood smear showing occasional lymphocytes and thrombocytes (T) clusters among mature erythrocytes. (A–D) Hematoxylin and eosin stains; (F–H) May-Grünwald/Giemsa stains. (See color plate)

1. Zebrafish hematopoiesis 21

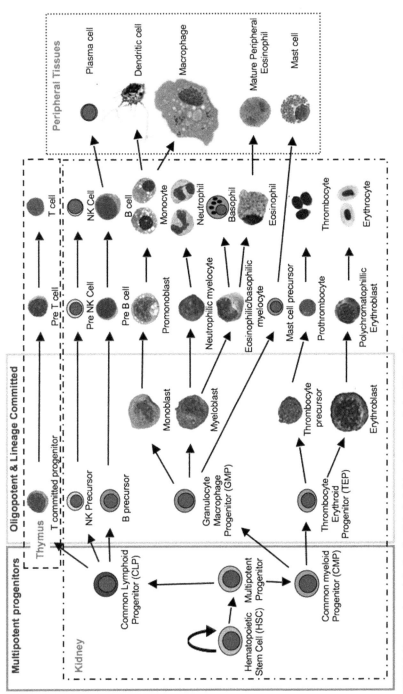

FIGURE 4 Proposed model of zebrafish definitive hematopoietic differentiation.

Isolated, cytospun, and stained blood cells from the zebrafish kidney, thymus, and peripheral tissues and their proposed upstream progenitors. Proposed lineage relationships are based on those demonstrated in clonogenic murine studies. Multipotent and lineage restricted progenitors likely reside in the kidney marrow, but their existence has never been experimentally proved due to a paucity of in vitro assays. (See color plate)

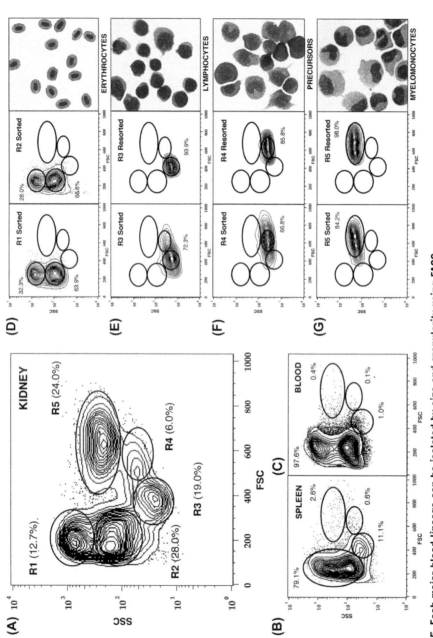

FIGURE 5 Each major blood lineage can be isolated by size and granularity using FACS.

(A) Scatter profile for whole kidney marrow (WKM). Mature erythrocytes are found within R1 and R2 gates, lymphocytes within the R3 gate, immature precursors within the R4 gate, and myeloid cells within the R5 gate. Mean percentages of each population within WKM are shown. Scatter profiling can also be utilized for analyzing spleen (B) and peripheral blood (C). Purification of each WKM fraction by FACS (D–G). (D) Sorting of populations R1 or R2 yields both upon reanalysis. This appears to be due to the elliptical shape of erythrocytes (right panel). (E) Isolation of lymphoid cells. (F) Isolation of precursor fraction. (G) Isolation of myeloid cells. FACS profiles following one round of sorting are shown in left panels, after two rounds in middle panels, and morphology of double-sorted cells shown in right panels (E–G).

each being predominantly erythroid. It should be noted that, due to differences in the fluidics and beam size, erythroid cells run on BD FACScan, FACS Caliber, or FACS Aria I and II flow cytometers are not discretely detectable by FSC and SSC alone; fluorescent transgenes must be utilized to separate them from other cells (Table 1). However, FACS Vantage and LSR-II flow cytometers have different fluidics systems and are well suited for separating erythroid cells by size and granularity. Sorting of each scatter population from spleen and blood showed each to contain only erythrocytes, lymphocytes, or myelomonocytes in a manner identical to those in the kidney. Immature precursors were not observed in either tissue. Percentages of cells within each scatter population closely matched those obtained by morphological cell counts, demonstrating that flow cytometric assays are accurate in measuring the relative percentages of each of the major blood lineages.

Many transgenic zebrafish lines have been created using proximal promoter elements from genes that demonstrate lineage-affiliated expression patterns in the mouse. These include *gata1a:GFP* (Long et al., 1997), *gata2a:EGFP* (Jessen et al., 1998; Traver et al., 2003a), *rag2:EGFP* (Langenau et al., 2003), *lck:EGFP* (Langenau et al., 2004), *spi1:EGFP* (Hsu et al., 2004; Ward et al., 2003), and *itga2b*:*EGFP* (Lin et al., 2005; Traver et al., 2003a) stable transgenic lines. In the adult kidney, we have demonstrated that each of these animals expresses GFP in the expected kidney scatter fractions (Traver et al., 2003a). For example, all mature erythrocytes express GFP in *gata1a:GFP* transgenic animals, as do erythroid progenitors within the precursor population. High expression levels of Gata2 are seen only within eosinophils, Rag2 and Lck only within cells in the lymphoid fraction, and Spi1 in both myeloid cells and rare lymphoid cells. The development of *itga2b:EGFP* transgenic animals has demonstrated that rare thrombocytic cells are found within the kidney, with thrombocyte precursors appearing in the precursor scatter fraction and mature thrombocytes in the lymphoid fraction. Without fluorescent reporter genes, rare populations such as thrombocytes cannot be resolved by light scatter characteristics alone. By combining the simple technique of scatter separation with fluorescent transgenesis, specific hematopoietic cell subpopulations can now be isolated to a relatively high degree of purity for further analyses.

FACS profiling can also serve as a diagnostic tool in the examination of zebrafish blood mutants. The majority of blood mutants identified to date are those displaying defects in embryonic erythrocyte production (Traver et al., 2003b). Most of these mutants are recessive and many are embryonic lethal when homozygous. Most have not been examined for subtle defects as heterozygotes. Several heterozygous mutants, such as *retsina*, *riesling*, and *merlot* showed haploinsufficiency as evidenced by aberrant kidney erythropoiesis (Traver et al., 2003a). All mutants displayed anemia with concomitant increases in erythroid precursors. These findings suggest that many of the gene functions required to make embryonic erythrocytes are similarly required in their adult counterparts at full gene dosage for normal function.

2. HEMATOPOIETIC CELL TRANSPLANTATION

In mammals, cellular transplantation has been used extensively to functionally test putative hematopoietic stem and progenitor cell (HSPC) populations, precursor/progeny relationships, and cell autonomy of mutant gene function. To address similar issues in zebrafish, several different varieties of hematopoietic cell transplantation (HCT) have been developed (Fig. 6).

2.1 EMBRYONIC DONOR CELLS

Although flow cytometry has proved very useful in analyzing and isolating specific blood lineages from the adult kidney, it cannot be used to enrich blood cells from dissociated developing embryos. To study the biology of the earliest blood-forming cells in the embryo, we have made use of transgenic zebrafish expressing fluorescent proteins. As discussed earlier, hematopoietic precursors appear to be specified from mesodermal derivatives that express *lmo2*, *kdrl* (also known as *flk1*), and *gata2a*. The proximal promoter elements from each of these genes have been shown to be sufficient to recapitulate their endogenous expression patterns. Using germline transgenic animals expressing GFP under the control of each of these promoters, blood cell precursors can be isolated by flow cytometry from embryonic and larval animals for transplantation into wild-type recipients. For example, GFP$^+$ cells in *lmo2:EGFP* embryos can be visualized by FACS by 8–10 somites (Traver, 2004). These cells can be sorted to purity and tested for functional potential in a variety

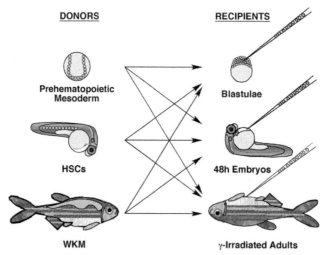

FIGURE 6 Methods of hematopoietic cell transplantation in the zebrafish.

See text for experimental details.

of transplantation (Fig. 6) or in vitro culture assays (see Section 4, In vitro Culture and Differentiation of Zebrafish Hematopoietic Progenitors).

We have used two types of heterochronic transplantation strategies to address two fundamental questions in developmental hematopoiesis. The first is if cells that express Lmo2 at 8–12 somites have hemangioblastic potential, ie, can generate both blood and vascular cells. We reasoned that purified cells should be placed into a relatively naive environment to provide the most permissive conditions to assess their full fate potentials. Therefore, we attempted transplantation into 1000-cell stage blastulae recipients. Transplanted cells appear to survive this procedure well and GFP^+ cells could be found over several days later in developing embryos and larvae. By isolating GFP^+ cells from *lmo2:EGFP* animals also carrying a *gata1a:DsRed* transgene, both donor-derived endothelial and erythroid cells can be independently visualized in green and red, respectively. Using this approach we have shown that $Lmo2^+$ cells from 8- to 12-somite stage (ss) embryos can generate robust regions of donor endothelium and intermediate levels of circulating erythrocytes (D. Traver, C. E. Burns, H. Zhu, and L. I. Zon, unpublished results). We are currently generating additional transgenic lines that express DsRED or mCherry under ubiquitous promoters to test the full fate potentials of $Lmo2^+$ cells upon transplantation. In addition, although these studies demonstrate that $Lmo2^+$ cells can generate at least blood and endothelial cells at the population level, single-cell fate-mapping studies need to be performed to assess whether clonogenic hemangioblasts can be identified in vivo.

The second question addressed through transplantation is whether the earliest identifiable primitive blood precursors can generate definitive hematopoietic cells that arise later in embryogenesis. It has been previously reported that the embryonic lethal *vlad tepes* mutant dies from erythropoietic failure due to a defect in the *gata1a* gene (Lyons et al., 2002). This lethality can be rescued by transplantation of WKM from wild-type adults into mutant recipients at 48 hpf (Traver et al., 2003a). We therefore tested if cells isolated from 8–12 ss *lmo2:EGFP* embryos could give rise to definitive cell types and rescue embryonic lethality in *vlad tepes* recipients. Following transplantation of GFP^+ cells at 48 hpf, approximately half of the cells in circulation were GFP^+ and the other half were $DsRed^+$ one day posttransplantation. Three days later, analyses of the same animals showed that the vast majority of cells in circulation were $DsReD^+$, apparently due to the differentiation of $Lmo2^+$ precursors to the erythroid fates. Compared to untransplanted control animals which all died by 12 days post fertilization (dpf), some mutant recipients survived for 1–2 months following transplantation. We observed no proliferation of donor cells at any time point following transplantation, however, and survivors analyzed over 1 month post transplantation showed no remaining cells in circulation (D. Traver, C. E. Burns, H. Zhu, and L. I. Zon, unpublished results). Therefore, these data indicate that mutant survivors were only transiently rescued by short-lived, donor-derived erythrocytes. Thus, within the context of this transplantation setting, it does not appear that $Lmo2^+$ hematopoietic precursors can seed definitive hematopoietic organs to give rise to enduring repopulation of the host blood forming system.

2.1.1 Protocol for isolating hematopoietic cells from embryos

This simple physical dissociation procedure is effective in producing single-cell suspensions from early embryos (8–12 ss) as well as from embryos as late as 24 hpf.

1. Stage and collect embryos. Approximately 200 cells can be isolated per 10–12 ss *lmo2:EGFP* embryo. It is recommended that as many embryos are collected as possible since subsequent transplantation efficiency depends largely upon cell concentration. At least 500–1000 embryos are recommended.
2. Transfer embryos to 1.5-mL Eppendorf centrifuge tubes. Add embryos until they sediment to the 0.5 mL mark. Remove E3 embryo medium since it is not optimal for cellular viability.
3. Wash 2X with 0.9X Dulbecco's phosphate-buffered saline (DPBS) (Gibco; 500 mL 1X(DPBS) + 55 mL ddH$_2$O).
4. Remove 0.9X DPBS and add 750 μL ice-cold staining medium (SM; 0.9X DPBS + 5% fetal calf serum (FCS)). Keep cells on ice from this point onward.
5. Homogenize with blue plastic pestle and pipette a few times with a p1000 tip.
6. Strain resulting cellular slurry through a 40 μm nylon cell strainer (Falcon 2340) atop a 50-mL conical tube. Rinse with additional SM to flush cells through the filter.
7. Gently mash remaining debris atop strainer with a plunger removed from a 3-mL syringe.
8. Rinse with more SM until the conical tube is filled to the 25 mL mark (this helps yolk removal).
9. Centrifuge for 5 min at 300g at 4°C. Remove supernatant until 1–2 mL remains.
10. Add 2–3 mL SM; resuspend by pipetting.
11. Strain again through 40-μm nylon mesh into a 5-mL Falcon 2054 tube. It is important to filter the cell suspension at least twice before running the sample by FACS; embryonic cells are sticky and will clog the nozzle if clumps are not properly removed beforehand.
12. Centrifuge again for 5 min at 300g at 4°C. Repeat steps 10–12 if necessary.
13. Remove supernatant and resuspend with 1–2 mL SM depending upon the number of embryos used.
14. Propidium iodide (PI) may be added at this point to 1 μg/mL to exclude dead cells and debris on the flow cytometer. When using, analyze samples with PI only and GFP only to set instrument compensations properly. Otherwise, the signal from PI may bleed into the GFP channel resulting in false-positive results. Alternatively, add 1:1000 Sytox Red (excited by a 633 nm laser), or Sytox Blue (excited by a 405 nm laser) for dead cell discrimination as they have no spectral overlap with GFP.

While simple physical dissociation is effective for dissociating early-stage embryos, embryos older than 24 hpf require enzymatic dissociation to generate single-cell suspensions. It is also advised to treat embryos with 1-phenyl 2-thiourea (see Westerfield, 2000) to prevent pigmentation, which negatively affects fluorescence discrimination by FACS.

1. Stage and collect embryos as described earlier. As embryos are larger and more cellular at these stages, it is possible to use fewer embryos for isolation. This depends on the transgene used and the downstream application; the following protocol is optimized for approximately 200 embryos—it can be scaled up or down accordingly. We have successfully employed this procedure to examine and isolate cells from as few as five embryos (Stachura et al., 2013).
2. Transfer embryos to 1.5-mL Eppendorf centrifuge tubes. Add embryos until they sediment to the 0.5 mL mark. Lay tube horizontally, and let embryos sit in E3 medium with 10 mM dithiothreitol (DTT) for 30 min; the DTT helps to remove the outer mucosal layer surrounding the embryo.
3. Wash 2X with 0.9X DPBS with Ca^{2+} and Mg^{2+} (Gibco; 500 mL 1X DPBS + 55 mL ddH_2O). DPBS must contain Ca^{2+} and Mg^{2+} for the enzymatic dissociation to work properly.
4. Add 1000 μL ice-cold DPBS with Ca^{2+} and Mg^{2+}. Add Liberase TM (Roche; final concentration in tube should be 50 μg/mL).
5. Lay tubes horizontally on a vortexer, and tape them down. Shake very gently (so that the platform is barely rotating) at 37°C for 1 h, or until embryos are dissolved. Check on their progress every 10–15 min; do not allow embryos to overdigest, as the cells will begin to die.
6. Dilute solution with ice-cold SM (with no Ca^{2+} or Mg^{2+}). Strain resulting cellular slurry through a 40-μm nylon cell strainer (Falcon 2340) atop a 50-mL conical tube. Rinse with additional SM to flush cells through the filter.
7. Rinse with more SM until the conical tube is filled to the 25 mL mark (this helps remove embryo debris).
8. Centrifuge for 5 min at 300g at 4°C. Remove supernatant until 1–2 mL remains.
9. Add 2–3 mL SM; resuspend by pipetting.
10. Strain again through a 40-μm nylon mesh into a 5-mL Falcon 2054 tube. Centrifuge again for 5 min at 300g at 4°C. Repeat steps 10–12 if the solution is cloudy or cells stick to each other.
11. Remove supernatant and resuspend with 1–2 mL SM depending upon the number of embryos used.
12. PI may be added at this point to 1 μg/mL to exclude dead cells and debris on the flow cytometer. When using, analyze samples with PI only and GFP only to set instrument compensations properly. Otherwise, the signal from PI may bleed into the GFP channel resulting in false-positive results. Alternatively, add 1:1000 Sytox Red, or Sytox Blue for dead cell discrimination as they have no spectral overlap with GFP.

Embryonic cells are now ready for analysis or sorting by FACS. It is often difficult to visualize GFP$^+$ cells when the expression is low or the target population is rare, so one should always prepare age-matched GFP-negative embryos in parallel with transgenic embryos to indicate where the sorting gates should be drawn to sort bona fide GFP$^+$ cells. If highly purified cells are desired, one should check the purity of the desired cell populations after the original sort. Depending on the

FACS machine and settings used, many times the purity is not acceptable for sensitive assays and two successive rounds of cell sorting are required. Cells should be kept ice-cold during the sorting procedure.

2.1.2 Transplanting purified cells into embryonic recipients

After sorting, centrifuge the cells for 5 min at 300g and 4°C. Carefully remove all the supernatant. Resuspend the cell pellet in 5–10 μL of ice-cold SM containing 3 units (U) heparin and 1 U DnaseI to prevent coagulation and lessen aggregation. Preventing the cells from aggregating or adhering to the glass capillary needle used for transplantation is critical. Mix the cells by gently pipetting with a 10-μL pipette tip. Keep on ice. For transplantation, we use the same needle-pulling parameters used to make needles for nucleic acid injections, the only difference being the use of filament-free capillary tubes to maintain cell viability. We use standard air-powered injection stations used for nucleic acid injections.

2.1.3 Transplanting cells into blastula recipients

1. Stage embryonic recipients to reach the 500- to 1000-cell stage at the time of transplantation.
2. Prepare plates for transplantation by pouring a thin layer of 2% agarose made in E3 medium into a 6-cm Petri dish. Drop transplantation mold [similar to the embryo injection mold described in Chapter 5 of *The Zebrafish Book* (Westerfield, 2000) but having individual depressions rather than troughs; (Adaptive Science Tools mold PT-1) atop molten agarose and let solidify.
3. Dechorionate blastulae in 1–2% agarose-coated Petri dishes by light pronase treatment or manually with watchmaker's forceps.
4. Transfer individual blastulae into individual wells of the transplantation plate that has been immersed in 1X Hank's Balanced Salt Solution (HBSS) (Gibco). Position the animal pole upward.
5. Using glass, filament-free, fine-pulled capillary needles (1.0 mm outer diameter) backload 3–6 μL of cell suspension after breaking the needle on a bevel to an opening of ~20 μm. Load into a needle holder and force cells to the injection end by positive pressure using a pressurized air injection station.
6. Gently insert the needle into the center of the embryo and expel cells using either very gentle pressure bursts or slight positive pressure. Transplanting cells near the marginal zone of the blastula leads to higher blood cell yields since embryonic fate maps show blood cells to derive from this region in later-gastrula-stage embryos.
7. Carefully transfer the embryos to agarose-coated Petri dishes using glass transfer pipettes.
8. Place them into E3 embryo medium and incubate at 28°C. Many embryos will not survive the transplantation procedure, so clean periodically to prevent microbial outgrowth.
9. Monitor by fluorescence microscopy for donor cell types.

2.1.4 Transplanting cells into 48 hpf embryos

1. All procedures are performed as described earlier except that dechorionated 48 hpf embryos are staged and used as transplant recipients.
2. Fill the transplantation plate with 1X HBSS containing 1X penicillin/streptomycin and 1X buffered tricaine, pH 7.0. Do not use E3, as it is suboptimal for cellular viability. Anesthetize recipients in tricaine and then array individual embryos into individual wells of the transplantation plate. Position the head at bottom of well, yolk side up.
3. Load cells as described earlier. Insert the injection needle into the sinus venosus/duct of Cuvier and gently expel the cells by positive pressure or gentle pressure bursts. Take care not to rupture the yolk ball membrane. A very limited volume can be injected into each recipient. It is thus important to use very concentrated cell suspensions in order to reconstitute the host blood-forming system. If using WKM as donor cells, concentrations of 5×10^5 cells/μL can be achieved if care is taken to filter and prevent coagulation of the sample.
4. Allow animals to recover at 28°C in E3. Keep clean and visualize daily by microscopy for the presence of donor-derived cells.

2.2 ADULT DONOR CELLS

Whereas the first HSCs transdifferentiate from the embryonic aortic endothelium, multilineage hematopoiesis is not fully apparent until the kidney becomes the site of blood cell production. The kidney appears to be the main site of adult hematopoiesis, and we have previously demonstrated that it contains HSCs capable of the long-term repopulation of embryonic (Traver et al., 2003a) and adult (Langenau et al., 2004; Traver et al., 2003a) recipients. For HSC enrichment strategies, both high-dose transplants and limiting dilution assays are required to gauge the purity of input cell populations. In embryonic recipients, we estimate that the maximum number of cells that can be transplanted is approximately 5×10^3, but the quantitation of transplanted cell numbers is difficult. To circumvent both issues, we have developed HCT with adult recipients.

For transplantation into adult recipients, myeloablation is necessary for successful engraftment of donor cells. We have found γ-irradiation to be the most consistent way to deplete zebrafish hematopoietic cells. The minimum lethal dose of 40 Gy specifically ablates cells of the blood-forming system and can be rescued by transplantation of one kidney equivalent (10^6 WKM cells). Thirty-day survival of transplanted recipients is approximately 75% (Traver et al., 2004). An irradiation dose of 20–25 Gy is sublethal, and the vast majority of animals survive this treatment despite having nearly total depletion of all leukocyte subsets 1 week following irradiation (Traver et al., 2004). We have shown that this dose is necessary and sufficient for transfer of a lethal T-cell leukemia (Traver et al., 2004), and for long-term (>6 month) engraftment of thymus repopulating cells (Langenau et al., 2004). That this dose is sufficient for robust engraftment, for long-term repopulation, and yields extremely high survival suggests that 20–25 Gy may be the optimal dose for myeloablative conditioning prior to transplantation. Short-term engraftment and long-term survival of transplant recipients is also greatly improved in clonal strains of zebrafish (Smith et al., 2010; Mizgirev & Revskoy, 2010a, 2010b) and fish with matching major histocompatibility complex loci (de Jong et al., 2011; de Jong & Zon, 2012).

2.2.1 Protocols for isolating hematopoietic cells from adult zebrafish

1. Anesthetize adult animals in 0.02% tricaine in fish water.
2. For peripheral blood collection, dry the animal briefly on tissue then place on a flat surface with head to the left, dorsal side up. Coat a 10-μL pipette tip with heparin (3 U/μL) then insert tip just behind the pectoral fin and puncture the skin. Direct the tip into the heart cavity, puncture the heart, and aspirate up to 10 μL blood by gentle suction. Immediately perform blood smears or place into 0.9X PBS containing 5% FCS and 1 U/μL heparin. Mix immediately to prevent clotting. Blood from several animals may be pooled in this manner for later use by flow cytometry. Red blood cells may be removed using hypotonic lysis solution (Sigma; 8.3 g/L ammonium chloride in 0.01 M Tris–HCl, pH 7.5) on ice for 5 min. Add 10 volumes of ice-cold SM then centrifuge at 300g for 5 min at 4°C. Resuspended blood leukocytes can then be analyzed by flow cytometry or cytocentrifuge preparations.
3. For collection of other hematopoietic tissues, place the fish on ice for several minutes following tricaine. Make a ventral, midline incision starting at the urogenital opening that splits and runs up along the two sides of the body cavity toward the head using fine scissors under a dissection microscope.
    - For spleen collection, locate the spleen just dorsal to the major intestinal loops and tease out with watchmaker's forceps. Place into ice-cold SM. Dissect any nonsplenic tissue away and place on a 40-μm nylon cell strainer (Falcon 2340) atop a 50-mL conical tube. Gently mash the spleen using a plunger removed from a 3-mL syringe and rinse with SM to flush cells through the filter. Up to 10 spleens can be processed through each filter. Centrifuge at 300g for 5 min at 4°C. Filter again through a 40-μm nylon mesh if performing FACS. For kidney collection, remove all internal organs using forceps and a dissection microscope. Take care during dissection; ruptured intestines or gonads will contaminate the kidney preparation. Using watchmaker's forceps, tease the entire kidney away from the body wall starting at the head of the kidney and working toward the rear. Place into ice-cold SM. Aspirate vigorously with a 1-mL pipetteman to separate hematopoietic cells (WKM) from renal cells. Filter through a 40-μm nylon mesh, wash, centrifuge, and repeat. Perform the last filtration step into a Falcon 2054 tube if using for FACS. It is important to filter the WKM cell suspension at least twice before running the sample. PI may be added at this point to 1 μg/mL to exclude dead cells and debris on the flow cytometer. When using, analyze samples with PI only and GFP only to set

instrument compensations properly. Otherwise, the signal from PI may bleed into the GFP channel resulting in false-positive results. Alternatively, add 1:1000 Sytox Red, or Sytox Blue for dead cell discrimination as they have no spectral overlap with GFP.

2.2.2 Transplanting WKM

After filtering and washing the WKM suspension three times, centrifuge cells for 5 min at 300g and 4°C. Carefully remove all the supernatant. Resuspend cell pellet in 5–10 μL of ice-cold SM containing 3 U heparin and 1 U DnaseI to prevent coagulation and lessen aggregation. Preventing the cells from aggregating or adhering to the glass capillary needle used for transplantation is critical. Mix the cells by gently pipetting with a 10-μL pipette tip. Keep on ice. For blastulae and embryo transplantation, perform following previous protocols. Between 5×10^2 and 5×10^3 cells can be injected into each 48 hpf embryo if the final cell concentration is approximately 5×10^5 cells/μL.

2.2.3 Transplanting cells into irradiated adult recipients

For irradiation of adult zebrafish, we have used a ^{137}Cesium source irradiator typically used for the irradiation of cultured cells (Gammacell 1000). We lightly anesthetize five animals at a time and then irradiate in sealed Petri dishes filled with fish water (without tricaine). We performed careful calibration of the irradiator using calibration microchips to obtain the dose rate at the height within the irradiation chamber nearest to the ^{137}Cesium point source. We found the dose rate to be uniform among calibration chips placed within euthanized animals in Petri dishes under water, under water alone, or in air alone, verifying that the tissue dosage was accurate.

Injecting cells directly into the heart or into the retroorbital sinus (Pugach, Li, White, & Zon, 2009) is the most efficient way to transplant cells into circulation. We perform intracardiac transplantation using pulled filament-free capillary needles as described earlier, but break the needles at a larger bore size of approximately 50 μm. The needle assembly can be handheld and used with a standard air-powered microinjection station. We have also had limited success transplanting cells intraperitoneally using a 10-μL Hamilton syringe. Engraftment efficiency for WKM is only marginal using this method, but transplantation of T-cell leukemia or solid tumor suspensions is highly efficient following irradiation at 20 Gy (Traver et al., 2004).

2.2.4 Irradiation

1. Briefly anesthetize adult zebrafish in 0.02% tricaine in fish water.
2. Place five fish at a time into 60 mm × 15 mm Petri dishes (Falcon) containing fish water. Wrap dish with Parafilm and irradiate for length of time necessary to achieve desired dose.
3. Return irradiated animals to clean tanks containing fish water. We have successfully transplanted irradiated animals from 12 to 72 h following irradiation. Using a 20-Gy dose, the nadir of host hematopoietic cells occurs at approximately 72 h postirradiation.

2.2.5 Transplantation

1. Prepare the cells to be transplanted as described earlier, taking care to remove particulates/contaminants by multiple filtration and washes. When using WKM as donor cells, we typically make final cell suspensions at 2×10^5 cells/µL. Keep the cells on ice.
2. Anesthetize an irradiated animal in 0.02% tricaine in fish water.
3. Transfer ventral side up into a well cut into a sponge wetted with fish water. Under a dissection microscope, remove the scales covering the pericardial region with fine forceps.
4. Fill an injection needle with ~20 µL of cell suspension. Force cells to end of the needle with positive pressure and adjust pressure balance to be neutral. Hold the needle assembly in one hand while placing gentle pressure on the abdomen of the recipient with the index finger of the other hand. This will position the heart adjacent to the skin and allow visualization of the heartbeat. Insert the needle through the skin and into the heart. If the needle is positioned within the heart, and the pressure balance is neutral, blood from the heart will enter the needle and the meniscus will rise and fall with the heartbeat. Inject approximately 5–10 µL by gentle pressure bursts.
5. Return the recipient to fresh fish water. Repeat for each additional recipient. Do not feed until the next day to lessen the chance of infection.

3. ENRICHMENT OF HSCs

The development of many different transplantation techniques now permits the testing of cell autonomy of mutant gene function, oncogenic transformation, and stem cell enrichment strategies in the zebrafish. For HSC enrichment strategies, fractionation techniques can be used to divide WKM into distinct subsets for functional testing via transplantation. The most successful means of HSC enrichment in the mouse has resulted from the subfractionation of whole bone marrow cells with monoclonal antibodies (mAbs) and flow cytometry (Spangrude, Heimfeld, & Weissman, 1988). We have attempted to generate mAbs against zebrafish leukocytes by repeated mouse immunizations using both live WKM and purified membrane fractions followed by standard fusion techniques. Many resulting hybridoma supernatants showed affinity to zebrafish WKM cells in FACS analyses (Fig. 7A). All antibodies showed one of two patterns, however. The first showed binding to all WKM cells at similar levels. The second showed binding to all kidney leukocyte subsets but not to kidney erythrocytes, similar to the pattern shown in the left panel of Fig. 7A. We found no mAbs that specifically bound only to myeloid cells, lymphoid cells, etc. when analyzing positive cells by their scatter profiles. We reasoned that these nonspecific binding affinities might be due to different oligosaccharide groups present on zebrafish blood cells. If the glycosylation of zebrafish membrane proteins were different from the mouse, then the murine immune system would likely mount an immune response against

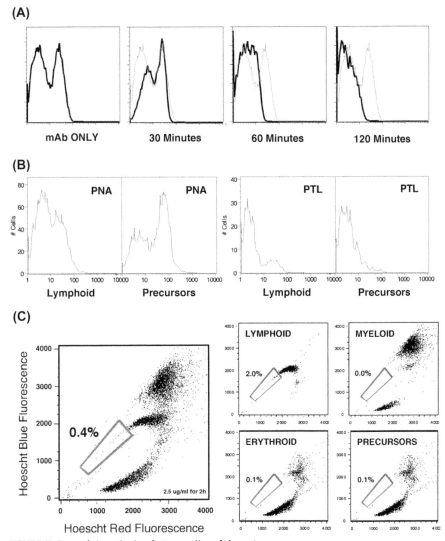

FIGURE 7 Potential methods of stem cell enrichment.
(A) Mouse monoclonal antibodies (mAbs) generated against zebrafish whole kidney marrow (WKM) cells react against oligosaccharide epitopes. Deglycosylation enzymes result in time-dependent loss of antibody binding (bold histograms) compared to no enzyme control (left panel and gray histograms). (B) Differential binding of lectins to WKM scatter fractions. Peanut agglutinin (PNA) splits both the lymphoid and precursor fraction into positive and negative populations (left panels). Potato lectin (PTL) shows a minor positive fraction only within the lymphoid fraction (right panels). (C) Hoechst 33342 dye reveals a side population within WKM. About 0.4% of WKM cells appear within the verapamil-sensitive side population gate (left panel). Only the lymphoid fraction, where kidney HSCs reside, contains appreciable numbers of side population cells (right panels).

these epitopes. To test this hypothesis, we removed both O-linked and N-linked sugars from WKM using a deglycosylation kit (Prozyme), and then incubated these cells with previously positive mAbs. All mAbs tested in this way showed a time-dependent decrease in binding, with nearly all binding disappearing following 2 h of deglycosylation (Fig. 7A). It thus appears that standard immunization approaches using zebrafish WKM cells elicit a strong immune response against oligosaccharide epitopes. This response is likely to be extremely robust, because we did not recover any mAbs that reacted with specific blood cell lineages. Similar approaches by other investigators using blood cells from frogs or other teleost species have yielded similar results (L. du Pasquier, M. Flajnik, personal communications). In an attempt to circumvent the glycoprotein issue, new series of immunizations using deglycosylated kidney cell membrane preparations may be effective.

Previous studies have shown that specific lectins can be used to enrich HSPC subsets in the mouse (Huang & Auerbach, 1993; Lu, Wang, & Auerbach, 1996; Visser, Bauman, Mulder, Eliason, & de Leeuw, 1984). We found that FITC-labeled lectins such as peanut agglutinin (PNA) and potato lectin (PTL) differentially bind to zebrafish kidney subsets; PNA enriches dendritic cells in the myelomonocytic scatter gate (Lugo-Villarino et al., 2010). As shown in Fig. 7B, PNA binds to a subset of cells both within the lymphoid and precursor kidney scatter fractions. Staining with PTL also shows that a minor fraction of lymphoid cells binds PTL, whereas the precursor (and other) scatter fractions are largely negative (Fig. 7B). We are currently testing both positive and negative fractions in transplantation assays to determine whether these different binding affinities can be used to enrich HSCs.

We have previously demonstrated that long-term HSCs reside in the adult kidney (Traver et al., 2003a). We therefore isolated each of the kidney scatter fractions from *gata1:EGFP* transgenic animals and transplanted cells from each into 48 hpf recipients to determine which subset contains HSC activity. The only population that could generate GFP$^+$ cells for over 3 weeks in wild-type recipients was the lymphoid fraction. This finding is in accord with mouse and human studies that have shown purified HSCs to have the size and morphological characteristics of inactive lymphocytes (Morrison, Uchida, & Weissman, 1995).

Another method that has been extremely useful in isolating stem cells from whole bone marrow is differential dye efflux. Dyes such as rhodamine 123 (Mulder & Visser, 1987; Visser & de Vries, 1988) or Hoechst 33342(Goodell, Brose, Paradis, Conner, & Mulligan, 1996) allow the visualization and purification of a "side population" (SP) that is highly enriched for HSCs. This technique appears to take advantage of the relatively high activity of multidrug resistance transporter proteins in HSCs that actively pump each dye out of the cell in a verapamil-sensitive manner (Goodell et al., 1996). Other cell types lack this activity and become positively stained, allowing isolation of the negative SP fraction by FACS. Our preliminary studies of SP cells in the zebrafish kidney demonstrated a typical SP profile when stained with 2.5 μg/mL of Hoechst 33342 for 2 h at 28°C (Fig. 7C). This population disappears when verapamil is added to the incubation. Interestingly, the vast majority of SP cells appear within the lymphoid scatter fraction (Fig. 7C). Further study of

SP cells indicated that they expressed ABCG2/Abcg2, an ATP-binding cassette transporter that may be a useful marker for HSCs (Kobayashi et al., 2008). Further examination of whether SP cells are enriched for HSC activity in transplantation assays is warranted.

Finally, there are many other methods to enrich HSPCs from WKM including sublethal irradiation, cytoreductive drug treatment, and use of transgenic lines expressing fluorescent reporter genes (see Table 1). We have shown following 20-Gy doses of γ-irradiation that nearly all hematopoietic lineages are depleted within 1 week (Traver et al., 2004). Examination of kidney cytocentrifuge preparations at this time shows that the vast majority of cells are immature precursors. That this dose does not lead to death of the animals demonstrates that HSCs are spared and are likely highly enriched 5–8 days following exposure. We have also shown that cytoreductive drugs such as Cytoxan and 5-fluorouracil have similar effects on kidney cell depletion, although the effects were more variable than those achieved with sublethal irradiation (A. Winzeler, D. Traver, and L. I. Zon unpublished). Because HSCs are contained within the kidney lymphoid fraction, they can be further enriched by HSC-specific or lymphocyte-specific transgenic markers. Studies indicate that transplanted CD41:GFP^{low} cells lead to long-term engraftment, indicating that this transgenic marker enriches HSCs (Ma, Zhang, Lin, Italiano, & Handin, 2011). Possible examples of other transgenic promoters that may mark HSCs are *lmo2*, *gata2a*, and *cmyb*, while *ccr9a*, *il7r*, *rag2*, *lck*, and B-cell receptor genes could be used to exclude lymphocytes from this subset (see Table 1).

4. IN VITRO CULTURE AND DIFFERENTIATION OF HEMATOPOIETIC PROGENITORS

Hematopoiesis is one of the best-studied models of developmental differentiation because of the multitude of experimental methods developed over the past 60 years to assess the proliferation, differentiation, and maintenance of its cellular constituents. Stem and progenitor cell transplantation into lethally irradiated animal recipients (Ford, Hamerton, Barnes, & Loutit, 1956; McCulloch & Till, 1960) were the first in vivo assays to be developed, followed shortly thereafter by the clonal growth of bone marrow progenitors in vitro (Bradley & Metcalf, 1966). Although these techniques have been substantially refined over the past decades, they still remain the foundation for analyzing the hierarchical organization of vertebrate HSPCs.

In vitro cultures to assess HSPC biology generally fall into two categories: growth of progenitor cells on a supportive stromal cell layer, and clonal growth of cells in a semisolid medium with the addition of supplemental cytokines or growth factors. Most stromal culture assays largely derive from the modification and refinement of Dexter cultures (Dexter, Allen, & Lajtha, 1977; Dexter, Moore, & Sheridan, 1977), whereby stromal cells from hematopoietic organs support the differentiation of HSCs and their downstream progenitors. These early studies were instrumental for the development of cobblestone-area-forming cell (Ploemacher, van der Sluijs, van Beurden, Baert, & Chan, 1991) and long-term culture initiating cell assays, which have been used to examine murine (Lemieux, Rebel, Lansdorp, & Eaves, 1995) and human (Sutherland, Eaves, Lansdorp, Thacker, & Hogge, 1991) multilineage hematopoietic differentiation. The development of stromal cells from the calvaria of macrophage colony stimulating factor—deficient mice (Nakano, Kodama, & Honjo, 1994, 1996) were instrumental for examining hematopoietic differentiation of embryonic stem cells down the hematopoietic pathway, and for differentiation of hematopoietic precursors into multiple mature blood cell types. With refinement, these OP9 cells have proved to be an efficient tool to study T-cell lineage commitment and development (Schmitt et al., 2004; Schmitt & Zuniga-Pflucker, 2002), once an extremely difficult process to study.

To assess the progenitor capacity of normal and mutant zebrafish hematopoietic cells functionally, we have created primary zebrafish kidney stromal (ZKS) cells derived from the main site of hematopoiesis in the adult fish. Culture of hematopoietic progenitor cells on these stromal cells resulted in their continued maintenance (Stachura et al., 2009) and differentiation (Bertrand et al., 2007, 2010a; Stachura et al., 2009). It also allowed investigation and rescue of a genetic block in erythroid maturation, confirming the utility of these assays (Stachura et al., 2009). Finally, the ZKS culture system has been used to investigate the molecular events underlying the progression of T-lymphoblastic lymphoma to acute T-lymphoblastic leukemia (Feng et al., 2010).

Recently, we created primary zebrafish embryonic stromal trunk (ZEST) cells derived from tissue surrounding the dorsal aorta, the site of de novo HSC production in vertebrates. Interestingly, ZEST cells also support HSPC proliferation and differentiation, as well as express similar transcripts when compared to ZKS cells (Campbell et al., 2015). These findings indicate that even though these ZKS and ZEST cells are from distinct ontological locations and temporal windows of hematopoiesis, they share functionality and transcript expression. These studies should allow further investigation of hematopoietic-supportive signaling from the zebrafish niche, as well as comparison to mammalian systems.

4.1 STROMAL CELL CULTURE ASSAYS

To create a suitable in vitro environment for the culture of zebrafish hematopoietic cells, we isolated the stromal fraction of the zebrafish kidney, the main site of hematopoiesis in adult fish (Zapata, 1979). We also generated another cell line from tissue surrounding the embryonic dorsal aorta, the site of HSC emergence in embryonic fish (Bertrand et al., 2010b; Kissa & Herbomel, 2010). The benefit of utilizing hematopoietic stromal layers is twofold. First, performing culture assays in the zebrafish has been hampered by a paucity of defined and purified hematopoietic cytokines. Most zebrafish cytokines have poor sequence homology to their mammalian counterparts, and as a consequence, have not been well described, characterized, or rigorously tested. Secondly, some hematopoietic cell types, especially T cells, require physical cell—cell interaction for their differentiation.

4. In vitro culture and differentiation of hematopoietic progenitors

4.1.1 Generation of ZKS cells

To create ZKS cells, kidney was isolated from AB* wild-type fish as described earlier (also see Stachura et al., 2009). The kidney tissue was sterilized by washing for 5 min in 0.000525% sodium hypochlorite (Fisher Scientific), then rinsed in sterile 0.9X DPBS. Tissue was then mechanically dissociated by trituration and filtered through a 40-µm filter (BD Biosciences). Flow-through cells (WKM) were discarded, and the remaining kidney tissue was cultured in vacuum plasma treated vented flasks (Corning Incorporated Life Sciences) at 32°C and 5% CO_2.

4.1.2 Maintenance and culture of ZKS cells

ZKS cells are maintained in the following tissue culture medium:

500 mL	L-15
350 mL	Dulbecco's Modified Eagle Medium (DMEM) (high glucose)
150 mL	Ham's F-12
150 mg	Sodium bicarbonate
15 mL	4-(2-hydroxyethyl)-1-piperazineethanesulfonic acid (HEPES) (1 M stock)
20 mL	Penicillin/streptomycin (5000 U/mL penicillin, 5000 µg/mL streptomycin stock)
10 mL	L-glutamine (200 mM stock)
100 mL	Fetal bovine serum (FBS)
2 mL	Gentamicin sulfate (50 mg/mL stock)

Medium is made by first adding sodium bicarbonate to the mixture of L-15, DMEM, and Ham's F-12. Warm the medium to 37°C, and allow the sodium bicarbonate to dissolve. Then, add other medium components and filter-sterilize with 0.22 µm vacuum apparatus.

All medium components are available from Mediatech. We use FBS from the American Type Culture Collection, but one can use FBS from other sources. It is important to note, however, that different manufacturing lots of FBS investigated in the laboratory differ wildly in their support of hematopoietic progenitor differentiation and proliferation. Once a manufacturing lot is tested and shown to be supportive, we recommend buying a large quantity to minimize experimental variation.

ZKS cells are maintained at 32°C and 5% CO_2 in a humidified incubator. Cells are grown in vacuum plasma treated 75 cm^2 vented flasks (T-75; Corning Incorporated Life Sciences) in 10 mL of medium until 60–80% confluent before passaging. Medium is then removed, and 2 mL trypsin-EDTA (0.25%; Invitrogen) is added to cover the stromal cells. Allow cells to incubate for 5 min at 32°C. Add 8 mL medium to cells to stop trypsinization, pipetting up and down to achieve a single cell suspension. Spin cells for 5 min at 300g, aspirate supernatant and resuspend pellet gently in 10 mL of medium. Take 1 mL of the cell solution, add 9 mL of medium, and move to a new flask. Cells should not be split more than 1:10 to passage, as they are somewhat density-dependent.

ZKS cells may be frozen and thawed at a later time. Even though we have never experienced senescence or a decrease in hematopoietic differentiation capacity of

ZKS cells in culture, it is useful to perform critical experiments with similar passages of cells to avoid experimental variation. To freeze ZKS cells, first trypsinize a T-75 flask. Spin cells for 5 min at 300g, aspirate supernatant and resuspend pellet gently in 2 mL of medium. Prepare freezing medium (500 μL of tissue culture certified DMSO and 1.5 mL of FBS) and aliquot 500 μL into four cryopreservation tubes keeping everything on ice. Add 500 μL of cells to each tube, invert to gently mix, and place tubes into isopropanol-jacketed freezing chamber. Place freezing chamber at −80°C for 24 h. Remove tubes from freezing chamber and place into liquid nitrogen for long-term storage.

To thaw ZKS cells at a later date, remove tube from nitrogen, and quickly warm in 37°C water bath. Wear eye protection; if nitrogen seeped into the freezing tubes the rapid warming may cause the tube to violently rupture. Remove liquid from tube, add slowly to 10 mL of medium in a 15 mL conical tube, and spin at 300g for 5 min. Carefully aspirate medium to remove all traces of DMSO. Resuspend cells in 10 mL of fresh medium, and place into T-75 flask. Early the next morning change medium and determine whether the cells are ready to be passaged or require another day to recover from thawing.

As with all tissue culture, strict attention to sterility and cleanliness should be adhered to at all times. All procedures should be performed in a tissue culture laminar flow hood, and it is recommended that vented, filtered flasks be used to prevent airborne contamination during culture in incubators.

4.1.3 Generation of ZEST cells

To create ZEST cells, the dorsal aorta and surrounding tissue were surgically removed from 48 hpf zebrafish embryos. Approximately 200 AB* wild-type embryos were rinsed several times in sterile E3 medium in 10 cm^2 petri dishes. Tissue anterior to the yolk tube extension was removed with a scalpel, as was tissue anterior to the yolk tube extension (including the embryo's head and its large yolk ball). The remaining trunk of the embryo was finely minced and grown in ZKS medium in vacuum plasma treated vented flasks (Corning Incorporated Life Sciences) at 32°C and 5% CO_2 (also see Campbell et al., 2015).

4.1.4 Maintenance and culture of ZEST cells

ZEST cells are maintained in the same tissue culture medium as ZKS cells:

500 mL	L-15
350 mL	DMEM (high glucose)
150 mL	Ham's F-12
150 mg	Sodium bicarbonate
15 mL	HEPES (1 M stock)
20 mL	Penicillin/streptomycin (5000 U/mL Penicillin, 5000 μg/mL streptomycin stock)
10 mL	L-glutamine (200 mM stock)
100 mL	Fetal bovine serum (FBS)
2 mL	Gentamicin sulfate (50 mg/mL stock)

Medium is made by first adding sodium bicarbonate to the mixture of L-15, DMEM, and Ham's F-12. Warm the medium to 37°C, and allow the sodium bicarbonate to dissolve. Then, add other medium components and filter-sterilize with 0.22-μm vacuum apparatus.

All medium components are available from Mediatech. We use FBS from the American Type Culture Collection, but one can use FBS from other sources. Different manufacturing lots of FBS investigated in our laboratory differ in their support of hematopoietic progenitor differentiation and proliferation; once a manufacturing lot is tested and shown to be supportive, we recommend buying a large quantity to minimize experimental variation.

ZEST cells are maintained at 32°C and 5% CO_2 in a humidified incubator. Cells are grown in vacuum plasma treated 75 cm^2 vented flasks (T-75; Corning Incorporated Life Sciences) in 10 mL of medium until 60–80% confluent before passaging. Medium is then removed, and 2 mL trypsin-EDTA (0.25%; Invitrogen) is added to cover the stromal cells. Allow cells to incubate for 5 min at 32°C. Add 8 mL medium to cells to stop trypsinization, pipetting up and down to achieve a single cell suspension. Spin cells for 5 min at 300g, aspirate supernatant and resuspend pellet gently in 10 mL of the medium. Take 1 mL of the cell solution, add 9 mL of medium, and move to a new flask. Cells should not be split more than 1:5 to passage, as they are somewhat density dependent.

ZEST cells may be frozen and thawed at a later time. Even though we have never experienced senescence or a decrease in hematopoietic differentiation capacity of ZEST cells in culture, it is useful to perform critical experiments with similar passages of cells to avoid experimental variation. To freeze ZEST cells, first trypsinize a T-75 flask. Spin cells for 5 min at 300g, aspirate supernatant and resuspend pellet gently in 2 mL of medium. Prepare freezing medium (500 μL of tissue culture certified DMSO and 1.5 mL of FBS) and aliquot 500 μL into four cryopreservation tubes keeping everything on ice. Add 500 μL of cells to each tube, invert to gently mix, and place tubes into isopropanol-jacketed freezing chamber. Place freezing chamber at −80°C for 24 h. Remove tubes from freezing chamber and place into liquid nitrogen for long-term storage.

To thaw ZEST cells at a later date, remove tube from nitrogen, and quickly warm in 37°C water bath. Wear eye protection; if nitrogen seeped into the freezing tubes the rapid warming may cause the tube to violently rupture. Remove liquid from tube, add slowly to 10 mL of medium in a 15-mL conical tube, and spin at 300g for 5 min. Carefully aspirate the medium to remove all traces of DMSO. Resuspend cells in 10 mL of fresh medium, and place into T-75 flask. Early the next morning change the medium and determine whether the cells are ready to be passaged or require another day to recover from thawing.

As with all tissue culture, strict attention to sterility and cleanliness should be adhered to at all times. All procedures should be performed in a tissue culture laminar flow hood, and it is recommended that vented, filtered flasks be used to prevent airborne contamination during culture in incubators.

4.1.5 Protocols for in vitro proliferation and differentiation assays on ZKS and ZEST cells

Purify prospective progenitors by FACS as described earlier. Plate cells on confluent ZKS/ZEST at a density of 1×10^4 cells/well in a 12-well tissue culture plate, using 2 mL complete medium per well. Lower density of progenitors is not recommended; if using fewer cells, reduce the size of the tissue culture well and volume of medium. If testing or investigating the effects of growth factors or small molecules, add them to the medium, being sure to have a vehicle-only control as well as different concentrations of your experimental factor. Twenty-four–well tissue culture plates are extremely useful in this regard, as one can easily plate out a multitude of experimental conditions on one plate.

1. Morphological assessment of hematopoietic cells after in vitro culture
   a. Gently aspirate hematopoietic cells from the ZKS/ZEST cultures, taking care not to disturb the stromal underlayer. ZEST cells do not adhere as tightly to flasks, so be more careful while washing them.
   b. Cytocentrifuge up to 200 μL of the hematopoietic cells at $250g$ for 5 min onto glass slides using a Shandon Cytospin 4 (Thermo Fischer Scientific). It is possible to concentrate the cells before cytocentrifugation at $300g$ for 10 min. Cytocentrifugation of over 200 μL of cell suspensions is not advised if using the standard cytofunnels.
   c. Perform May-Grünwald/Giemsa staining by allowing slides to air dry briefly. Then, submerge the slide in May-Grünwald staining solution (Sigma Aldrich) for 10 min. Transfer the slide to 1:5 dilution of Giemsa stain (Sigma Aldrich; make fresh immediately before use) in dH_2O for an additional 20 min. Rinse slide in dH_2O, and allow to air dry. Coverslip slide with cytoseal XYL mounting medium (Richard-Allan Scientific) and Corning no.1 18 mm square cover glass (Corning). Allow slides to completely dry before visualization on upright microscope, especially if using an oil immersion lens.
2. Proliferation assessment of hematopoietic cells after in vitro culture
   a. Gently aspirate hematopoietic cells as described earlier.
   b. Count cells with the use of a bright line hemacytometer (Hausser Scientific) using Trypan Blue dye (Invitrogen) exclusion to assess viability.
3. Reverse transcription polymerase chain reaction (RT-PCR) analysis of hematopoietic cells after in vitro culture
   a. Gently aspirate hematopoietic cells as described earlier.
   b. Isolate RNA from hematopoietic cells using either Trizol (Invitrogen) or RNAeasy kit (Qiagen).
   c. Generate cDNA with oligo dT primers and the Superscript RT-PCR kit (Invitrogen).
   d. Perform PCR with desired zebrafish DNA primers.
4. Quantitative RT-PCR (qRT-PCR) analysis of hematopoietic cells after in vitro culture

a. Gently aspirate cells as described earlier.
   b. Isolate RNA from hematopoietic cells using either Trizol (Invitrogen) or RNAeasy kit (Qiagen).
   c. Generate cDNA with oligo dT primers and the Superscript RT-PCR kit (Invitrogen). If using low amounts of cells, a random hexamer primer kit may be used.
   d. Perform qRT-PCR with SYBR Green Supermix (Biorad) and desired primers. Note: Make sure primers are located on different exons of your gene of interest, or else you may detect genomic copies of the gene and not just cDNA. Alternatively, use DNAse to remove any contaminating genomic DNA (see step b).
5. Cell labeling and cell division determination of hematopoietic cells after in vitro culture
   a. Prior to plating cells on ZKS/ZEST monolayer, wash cells twice with 0.1% bovine serum albumin (BSA) to remove FBS from the medium.
   b. Resuspend cells in 0.1% BSA and 2 μL/mL of 5 mM carboxyfluorescein succinimidyl ester (CFSE; Invitrogen) at room temperature for 10 min, in the dark.
   c. Wash cells with complete medium supplanted with an additional 10% FBS twice.
   d. Save 1/10 of the culture and perform FACS (Day 0 time point). Culture remaining cells in complete medium as described earlier.
   e. For analysis, remove hematopoietic cells from culture at desired time points as described earlier and perform FACS. CFSE is read in the FL-1 channel (on FACS Caliber) or with most GFP filters (FACS Aria I and II, LSR-II), and will decrease in fluorescence intensity as cells divide. Compare divisions to Day 0 time point with FloJo software (TreeStar, Ashland, OR). We recommend using the BD LSR-II flow cytometer, as the different scatter profile of mature cells is easily distinguished and directly comparable to profiles shown in Fig. 5.
   f. Optional: Alternatively, if measuring the proliferation of transgenic cells marked with GFP fluorescence, one can use PKH-26, a red cell membrane dye (Sigma).
      i. Centrifuge 1×10^6 cells at 300g for 5 min at room temperature.
      ii. Aspirate supernatant, and resuspend cells in 500 μL of Diluent C (provided with kit) to create a 2x Cell Solution.
      iii. Prepare 2x Dye Solution by adding 2 μL of PKH-26 to 500 μL of Diluent C in a separate polypropylene tube.
      iv. Rapidly mix the Cell Solution and the Dye Solution by pipetting. The staining is instantaneous, so it is essential that this is done quickly and thoroughly.
      v. Incubate for 1–5 min, and stop staining by adding 1 mL of FBS. Allow solution to sit for 1 min.

vi. Save 1/10 of the culture and perform FACS to determine the Day 0 fluorescence expression time point. Culture the remaining 9/10 of the cells in medium for examination at a later time point.

4.2 CLONAL METHYLCELLULOSE-BASED ASSAYS

Although stromal in vitro culture methods have been instrumental for the investigation of hematopoiesis, culturing bulk populations of progenitor cells on stroma cannot distinguish between homogeneous multipotent progenitor populations or heterogeneous lineage-restricted populations without performing limiting dilution assays. The development of clonal in vitro cultures by Metcalf and colleagues allowed not only the growth of murine bone marrow progenitors (Bradley & Metcalf, 1966), but also the study and quantitation of progenitor numbers during hematological disease (Bradley, Robinson, & Metcalf, 1967) and exposure to irradiation (Robinson, Bradley, & Metcalf, 1967). These assays were used to investigate the ontogeny of the developing murine hematopoietic system (Moore & Metcalf, 1970), and refined to study human hematopoietic progenitors dysregulated during leukemogenesis (Moore, Williams, & Metcalf, 1973a, 1973b). The utilization of clonal assays was instrumental for the identification and validation of colony stimulating factors, secreted proteins that stimulate the differentiation of specific hematopoietic lineages. The ability to isolate, recombinantly produce, and test these factors was a huge advance in hematological research, allowing the sensitive analysis of progenitor differentiation, proliferation, and restriction in the murine and human blood system.

This capability to grow progenitors in vitro to test their differentiation capacity in an unbiased manner has greatly advanced the current understanding of hematopoietic lineage restriction. The isolation of putative lineage-restricted daughter cells by FACS coupled with in vitro clonal analysis was pivotal in identifying multipotent (Akashi, Traver, Miyamoto, & Weissman, 2000; Kondo, Weissman, & Akashi, 1997), oligopotent (Akashi et al., 2000), and monopotent progenitor (Mori et al., 2008; Nakorn, Miyamoto, & Weissman, 2003) intermediates downstream of HSCs in the murine system.

While zebrafish likely possess multipotent, oligopotent, and monopotent progenitor cells, their existence had never been proved, and remains speculative (see Fig. 4). To investigate if these cells existed in zebrafish, we developed assays to investigate progenitors in the zebrafish in a clonal manner by modifying existing methylcellulose culture techniques, exogenously adding zebrafish recombinant cytokines erythropoietin (Epo) (Paffett-Lugassy et al., 2007) and granulocyte colony stimulating factor (Gcsf) (Liongue, Hall, O'Connell, Crosier, & Ward, 2009) to quantitate the number of myeloid and erythroid progenitors in adult kidney marrow scatter fractions (Stachura et al., 2011). We used these assays to investigate differences in the function of paralogous cytokines in zebrafish (Stachura et al., 2013). Recently, we discovered and recombinantly generated zebrafish thrombopoietin (Tpo), allowing analysis of thrombocytes (Svoboda et al., 2014). We now have the ability to examine and quantitate erythroid, myeloid, and thrombocytic cell subsets with clonal assays (Fig. 8). This level of precision allows the further testing of prospective hematopoietic progenitors in normal and mutant zebrafish, allowing more careful investigation of lineage determination and its conservation among vertebrate animals. In addition, it allows comparison of hematopoietic progenitor cells and their response to cytokines, furthering our understanding of cytokine signaling. Furthermore, the ability to examine blocks in hematopoietic differentiation, aberrant gene expression, and proliferative regulation is now possible in mutant fish already (and currently being) generated. Finally, it also allows the rapid

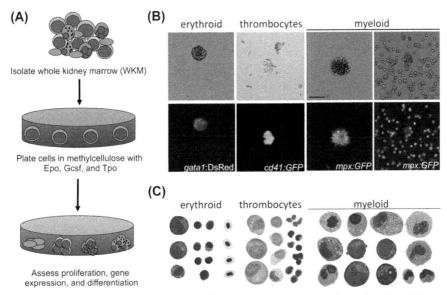

FIGURE 8 Recombinantly generated and purified granulocyte colony stimulating factor (Gcsf), thrombopoietin (Tpo), and erythropoietin (Epo) encourage myeloid, thrombocytic, and erythroid differentiation, respectively, from zebrafish hematopoietic progenitors in clonal methylcellulose assays.

(A) Experimental schematic for isolation and culture of cells from adult whole kidney marrow. To look at specific subsets of hematopoietic stem and progenitor cells, see techniques described in text. (B) Bright-field images (top row) and fluorescence images (second row) of colonies grown in Epo, Tpo, and Gcsf. All images in taken at 100×. (C) Colonies isolated from methylcellulose cultures cytospun and stained with May-Grünwald/Giemsa.
Colonies were isolated from cultures with carp serum and Epo, Tpo, and Gcsf. Erythroid colonies contain erythrocytes and progenitors (left column), thrombocytic colonies contain thrombocyte precursors and mature thrombocytes (middle column), while ruffled and spread colonies contain myeloid cells (right column). All images were taken at 1000×. (See color plate)

screening of small molecules, blocking antibodies, and other drug compounds that may affect lineage differentiation, maturation, and proliferation.

4.2.1 Methylcellulose

To develop a clonal assay to further enumerate and characterize progenitor cells in the zebrafish, we used methylcellulose, a semisolid, viscous cell culture medium used in murine and human progenitor studies. The nature of methylcellulose culture allows individual progenitor cells to develop isolated colonies within the medium, where they can be enumerated after several days in culture. In addition, the use of methylcellulose allows examination of colony morphology and subsequent isolation for further characterization by morphological examination and gene expression. For isolating carp serum, an essential component of these assays, as well as detailed, complementary explanation of these clonal assays, see (Svoboda et al., 2015).

4.2.2 Methylcellulose stock preparation

Prepare 2.0% methylcellulose by adding 20 g of methylcellulose powder (Sigma Aldrich) to 450 mL of autoclaved H_2O and boiling for 3 min. Allow the mixture to cool to room temperature before adding 2x L-15 medium powder (Mediatech). Then, adjust the weight of the methylcellulose mixture to 1000 g with sterile water. Methylcellulose should be allowed to thicken at 4°C overnight before being aliquoted and stored at −20°C.

4.2.3 Methylcellulose clonal assays

Complete methylcellulose medium:

10 mL	2.0% Methylcellulose stock
4.9 mL	DMEM (high glucose)
2.1 mL	Ham's F-12
2 mL	FBS
300 μL	HEPES (1 M stock)
200 μL	Penicillin/streptomycin (5000 U/mL and 5000 μg/mL stock, respectively)
200 μL	L-Glutamine (200 mM stock)
40 μL	Gentamicin sulfate (50 mg/mL)

To perform experiments in triplicate, add 3.5 mL of complete methylcellulose to sterile round-bottom 14-mL tubes (Becton Dickinson) with 5-mL syringes and 16-gauge needles for each condition. Cells of interest (prospective progenitors) should be isolated, counted, and resuspended in 100 μL of ZKS/ZEST medium and added to complete methylcellulose, along with cytokines, small molecules, or other agents to be investigated.

For myeloid differentiation, add 1% Carp serum and 0.3 μg/mL recombinant zebrafish Gcsf to methylcellulose medium. For erythroid differentiation, add 1% Carp serum and 0.1 μg/mL recombinant zebrafish Epo to methylcellulose medium.

For thrombocytic differentiation, add 1% Carp serum and 30 ng/mL Tpo. To examine multilineage progenitors, add desired combinations of these factors. Cytokines and additives should not total more than 10% of the total volume, as the medium will not be viscous enough to discern individual colonies.

To observe separable, individual colonies, cells should be resuspended at $1 \times 10^4 - 5 \times 10^4$ cells/mL. Tightly cap tubes, and gently vortex the solution to mix. In triplicate, aliquot 1 mL of solution into 35-mm Petri dishes (Becton Dickinson). Swirl Petri dishes to distribute the methylcellulose culture evenly, and place plates in a humidified 15-cm dish (made by placing a plate of sterile dH_2O inside the 15-cm dish) at 32°C and 5% CO_2. Plates should be removed 7–10 days after plating for microscopic examination, colony isolation, and gene expression analyses.

As with all tissue culture, strict attention to sterility and cleanliness should be adhered to at all times. All procedures should be performed in a tissue culture laminar flow hood.

4.2.4 Enumeration of colony forming units

Colony forming units (CFUs) are a measurement of how many progenitors are present in a given population of cells; if an individual cell has the capability to proliferate and divide into mature blood cells under certain growth conditions, it will make an individual colony. For example, if 100 putative myeloid progenitor cells are plated under conditions suitable for myeloid differentiation and one myeloid colony arises, 1:100 of the cells plated was a myeloid CFU.

To perform enumeration of CFUs, observe and count colonies on an inverted microscope after 7 days in culture. Counting with a 5× or 10× objective (50×–100× magnification) is recommended, with the aperture closed down slightly to grant high contrast, which aids in the visualization of colonies. Be careful not to disturb the dish; even though methylcellulose is viscous, excessive movement of the plates will cause colonies to move, complicating further analysis. Depending on the cytokines, growth factors, and other culture additives, colony shape, size, and color will be different; be sure to note and record this information. Inclusion of a transgenic marker will aid in identification of colony type and morphology as shown in Fig. 8B, whereby erythroid colonies express DsRed driven by the erythroid-specific *gata1a* promoter, thrombocytic colonies express GFP driven by the *cd41* promoter, and myeloid colonies express GFP driven by the myeloid-specific *mpx* promoter.

If cultures are to be returned to the incubator do not remove Petri plate lids, and take care to wipe down surfaces with 70% ethanol before starting experiment.

4.2.5 Picking and analyzing colonies from methylcellulose

Hematopoietic colonies can be carefully plucked from methylcellulose cultures with a pipetteman, preferably a p20 with a fine tip. Pay attention to pick only individual colonies, placing them into 1.5 mL Eppendorf tubes with 200 μL of PBS. Pipette up and down gently in the PBS to remove traces of methylcellulose from your tip, and to break up the colony. It is possible to pool colonies of similar morphology for analyses, especially when large cell numbers are required. Colonies may be cytospun and stained with May-Grünwald Giemsa (Sigma Aldrich) as described earlier (Fig. 8C). In addition, colonies may be subjected to RT-PCR or qRT-PCR analysis for mature lineage gene transcripts as described earlier.

CONCLUSIONS

Over the past decade, the zebrafish has rapidly become a powerful model system in which to elucidate the molecular mechanisms of vertebrate blood development through forward genetic screens. In this review, we have described the cellular characterization of the zebrafish blood forming system and provided detailed protocols for the isolation, transplantation, and culture of hematopoietic cells. Through the development of lineal subfractionation techniques, transplantation technology, and in vitro hematopoietic assays, a hematological framework now exists for the continued study of the genetics of hematopoiesis. By adapting these experimental approaches that have proved to be powerful in the mouse, the zebrafish is uniquely positioned to address fundamental questions regarding the biology of HSPCs.

REFERENCES

Akashi, K., Traver, D., Miyamoto, T., & Weissman, I. L. (2000). A clonogenic common myeloid progenitor that gives rise to all myeloid lineages. *Nature, 404*, 193–197.

Al-Adhami, M. A., & Kunz, Y. W. (1977). Ontogenesis of haematopoietic sites in *Brachydanio rerio*. *Development Growth and Differentiation, 19*, 171–179.

Bennett, C. M., et al. (2001). Myelopoiesis in the zebrafish, *Danio rerio*. *Blood, 98*, 643–651.

Bertrand, J. Y., et al. (2005a). Three pathways to mature macrophages in the early mouse yolk sac. *Blood, 106*, 3004–3011.

Bertrand, J. Y., et al. (2005b). Characterization of purified intraembryonic hematopoietic stem cells as a tool to define their site of origin. *Proceedings of the National Academy of Sciences of the United States of America, 102*, 134–139.

Bertrand, J. Y., et al. (2007). Definitive hematopoiesis initiates through a committed erythromyeloid progenitor in the zebrafish embryo. *Development (Cambridge, England), 134*, 4147–4156.

Bertrand, J. Y., Cisson, J. L., Stachura, D. L., & Traver, D. (2010a). Notch signaling distinguishes 2 waves of definitive hematopoiesis in the zebrafish embryo. *Blood, 115*, 2777–2783.

Bertrand, J. Y., et al. (2010b). Haematopoietic stem cells derive directly from aortic endothelium during development. *Nature, 464*, 108–111.

Bertrand, J. Y., Kim, A. D., Teng, S., & Traver, D. (2008). CD41$^+$ cmyb$^+$ precursors colonize the zebrafish pronephros by a novel migration route to initiate adult hematopoiesis. *Development (Cambridge, England), 135*, 1853–1862.

Boisset, J. C., et al. (2010). In vivo imaging of haematopoietic cells emerging from the mouse aortic endothelium. *Nature, 464*, 116–120.

References

Bradley, T. R., & Metcalf, D. (1966). The growth of mouse bone marrow cells in vitro. *Australian Journal of Experimental Biology and Medical Science, 44*, 287−299.

Bradley, T. R., Robinson, W., & Metcalf, D. (1967). Colony production in vitro by normal polycythaemic and anaemic bone marrow. *Nature, 214*, 511.

Brown, L. A., et al. (2000). Insights into early vasculogenesis revealed by expression of the ETS-domain transcription factor Fli-1 in wild-type and mutant zebrafish embryos. *Mechanisms of Development, 90*, 237−252.

de Bruijn, M. F., et al. (2002). Hematopoietic stem cells localize to the endothelial cell layer in the midgestation mouse aorta. *Immunity, 16*, 673−683.

Burns, C. E., et al. (2002). Isolation and characterization of runxa and runxb, zebrafish members of the runt family of transcriptional regulators. *Experimental Hematology, 30*, 1381−1389.

Burns, C. E., Traver, D., Mayhall, E., Shepard, J. L., & Zon, L. I. (2005). Hematopoietic stem cell fate is established by the Notch-Runx pathway. *Genes and Development, 19*, 2331−2342.

Campbell, C., et al. (2015). Zebrafish embryonic stromal trunk (ZEST) cells support hematopoietic stem and progenitor cell (HSPC) proliferation, survival, and differentiation. *Experimental Hematology*. http://dx.doi.org/10.1016/j.exphem.2015.09.001.

Chen, M. J., Yokomizo, T., Zeigler, B. M., Dzierzak, E., & Speck, N. A. (2009). Runx1 is required for the endothelial to haematopoietic cell transition but not thereafter. *Nature, 457*, 887−891.

Ciau-Uitz, A., Monteiro, R., Kirmizitas, A., & Patient, R. (2014). Developmental hematopoiesis: ontogeny, genetic programming and conservation. *Experimental Hematology, 42*, 669−683. http://dx.doi.org/10.1016/j.exphem.2014.06.001.

Ciau-Uitz, A., Walmsley, M., & Patient, R. (2000). Distinct origins of adult and embryonic blood in Xenopus. *Cell, 102*, 787−796.

Cross, L. M., Cook, M. A., Lin, S., Chen, J. N., & Rubinstein, A. L. (2003). Rapid analysis of angiogenesis drugs in a live fluorescent zebrafish assay. *Arteriosclerosis, Thrombosis, and Vascular Biology, 23*, 911−912.

Cumano, A., & Godin, I. (2007). Ontogeny of the hematopoietic system. *Annual Review of Immunology, 25*, 745−785.

Detrich, H. W., 3rd, et al. (1995). Intraembryonic hematopoietic cell migration during vertebrate development. *Proceedings of the National Academy of Sciences of the United States of America, 92*, 10713−10717.

Dexter, T. M., Allen, T. D., & Lajtha, L. G. (1977). Conditions controlling the proliferation of haemopoietic stem cells in vitro. *Journal of Cellular Physiology, 91*, 335−344.

Dexter, T. M., Moore, M. A., & Sheridan, A. P. (1977). Maintenance of hemopoietic stem cells and production of differentiated progeny in allogeneic and semiallogeneic bone marrow chimeras in vitro. *Journal of Experimental Medicine, 145*, 1612−1616.

Dzierzak, E. (2005). The emergence of definitive hematopoietic stem cells in the mammal. *Current Opinion in Hematology, 12*, 197−202.

Eaves, C. J. (2015). Hematopoietic stem cells: concepts, definitions, and the new reality. *Blood, 125*, 2605−2613. http://dx.doi.org/10.1182/blood-2014-12-570200.

Ellett, F., Pase, L., Hayman, J. W., Andrianopoulos, A., & Lieschke, G. J. (2010). mpeg1 promoter transgenes direct macrophage-lineage expression in zebrafish. *Blood, 117*.

Feng, H., et al. (2010). T-lymphoblastic lymphoma cells express high levels of BCL2, S1P1, and ICAM1, leading to a blockade of tumor cell intravasation. *Cancer Cell, 18*, 353−366.

Ford, C. E., Hamerton, J. L., Barnes, D. W., & Loutit, J. F. (1956). Cytological identification of radiation-chimaeras. *Nature, 177*, 452−454.

Fraser, S. T., Isern, J., & Baron, M. H. (2007). Maturation and enucleation of primitive erythroblasts during mouse embryogenesis is accompanied by changes in cell-surface antigen expression. *Blood, 109*, 343–352.

Gekas, C., Dieterlen-Lievre, F., Orkin, S. H., & Mikkola, H. K. (2005). The placenta is a niche for hematopoietic stem cells. *Developmental Cell, 8*, 365–375.

Goodell, M. A., Brose, K., Paradis, G., Conner, A. S., & Mulligan, R. C. (1996). Isolation and functional properties of murine hematopoietic stem cells that are replicating in vivo. *Journal of Experimental Medicine, 183*, 1797–1806.

Hall, C., Flores, M. V., Storm, T., Crosier, K., & Crosier, P. (2007). The zebrafish lysozyme C promoter drives myeloid-specific expression in transgenic fish. *BMC Developmental Biology, 7*, 42.

Hansen, J. D., & Zapata, A. G. (1998). Lymphocyte development in fish and amphibians. *Immunological Reviews, 166*, 199–220.

Herbomel, P., Thisse, B., & Thisse, C. (1999). Ontogeny and behaviour of early macrophages in the zebrafish embryo. *Development (Cambridge, England), 126*, 3735–3745.

Houssaint, E. (1981). Differentiation of the mouse hepatic primordium. II. Extrinsic origin of the haemopoietic cell line. *Cell Differentiation, 10*, 243–252.

Hsu, K., et al. (2004). The pu.1 promoter drives myeloid gene expression in zebrafish. *Blood, 104*.

Huang, H., & Auerbach, R. (1993). Identification and characterization of hematopoietic stem cells from the yolk sac of the early mouse embryo. *Proceedings of the National Academy of Sciences of the United States of America, 90*, 10110–10114.

Jaffredo, T., Gautier, R., Eichmann, A., & Dieterlen-Lievre, F. (1998). Intraaortic hemopoietic cells are derived from endothelial cells during ontogeny. *Development (Cambridge, England), 125*, 4575–4583.

Jagadeeswaran, P., Sheehan, J. P., Craig, F. E., & Troyer, D. (1999). Identification and characterization of zebrafish thrombocytes. *British Journal of Haematology, 107*, 731–738.

Jessen, J. R., et al. (1998). Modification of bacterial artificial chromosomes through chi-stimulated homologous recombination and its application in zebrafish transgenesis. *Proceedings of the National Academy of Sciences of the United States of America, 95*, 5121–5126.

Jin, S. W., et al. (2007). A transgene-assisted genetic screen identifies essential regulators of vascular development in vertebrate embryos. *Developmental Biology, 307*, 29–42.

Jin, H., Xu, J., & Wen, Z. (2007). Migratory path of definitive hematopoietic stem/progenitor cells during zebrafish development. *Blood, 109*, 5208–5214.

Johnson, G. R., & Moore, M. A. (1975). Role of stem cell migration in initiation of mouse foetal liver haemopoiesis. *Nature, 258*, 726–728.

de Jong, J. L., et al. (2011). Characterization of immune-matched hematopoietic transplantation in zebrafish. *Blood, 117*, 4234–4242. http://dx.doi.org/10.1182/blood-2010-09-307488.

de Jong, J. L., & Zon, L. I. (2012). Histocompatibility and hematopoietic transplantation in the zebrafish. *Advances in Hematology, 2012*. http://dx.doi.org/10.1155/2012/282318, 282318.

Kalev-Zylinska, M. L., et al. (2002). Runx1 is required for zebrafish blood and vessel development and expression of a human RUNX1-CBF2T1 transgene advances a model for studies of leukemogenesis. *Development (Cambridge, England), 129*, 2015–2030.

Keller, G., Lacaud, G., & Robertson, S. (1999). Development of the hematopoietic system in the mouse. *Experimental Hematology, 27*, 777–787.

Kingsley, P. D., Malik, J., Fantauzzo, K. A., & Palis, J. (2004). Yolk sac-derived primitive erythroblasts enucleate during mammalian embryogenesis. *Blood, 104*, 19–25.

Kissa, K., et al. (2008). Live imaging of emerging hematopoietic stem cells and early thymus colonization. *Blood, 111*, 1147–1156.

Kissa, K., & Herbomel, P. (2010). Blood stem cells emerge from aortic endothelium by a novel type of cell transition. *Nature, 464*, 112–115.

Kobayashi, I., et al. (2008). Characterization and localization of side population (SP) cells in zebrafish kidney hematopoietic tissue. *Blood, 111*, 1131–1137. http://dx.doi.org/10.1182/blood-2007-08-104299.

Kondo, M., Weissman, I. L., & Akashi, K. (1997). Identification of clonogenic common lymphoid progenitors in mouse bone marrow. *Cell, 91*, 661–672.

Lam, E. Y., et al. (2008). Zebrafish runx1 promoter-EGFP transgenics mark discrete sites of definitive blood progenitors. *Blood, 133*.

Langenau, D. M., et al. (2003). Myc-induced T-cell leukemia in transgenic zebrafish. *Science, 299*, 887–890.

Langenau, D. M., et al. (2004). In vivo tracking of T cell development, ablation, and engraftment in transgenic zebrafish. *Proceedings of the National Academy of Sciences of the United States of America, 101*, 7369–7374.

Langenau, D. M., et al. (2007). Effects of RAS on the genesis of embryonal rhabdomyosarcoma. *Genes and Development, 21*, 1382–1395. http://dx.doi.org/10.1101/gad.1545007.

Lawson, N. D., & Weinstein, B. M. (2002). In vivo imaging of embryonic vascular development using transgenic zebrafish. *Developmental Biology, 248*, 307–318.

Lemieux, M. E., Rebel, V. I., Lansdorp, P. M., & Eaves, C. J. (1995). Characterization and purification of a primitive hematopoietic cell type in adult mouse marrow capable of lymphomyeloid differentiation in long-term marrow "switch" cultures. *Blood, 86*, 1339–1347.

Liao, E. C., et al. (1998). SCL/Tal-1 transcription factor acts downstream of cloche to specify hematopoietic and vascular progenitors in zebrafish. *Genes and Development, 12*, 621–626.

Lieschke, G. J., et al. (2002). Zebrafish SPI-1 (PU.1) marks a site of myeloid development independent of primitive erythropoiesis: implications for axial patterning. *Developmental Biology, 246*, 274–295.

Lin, H. F., et al. (2005). Analysis of thrombocyte development in CD41-GFP transgenic zebrafish. *Blood, 106*, 3803–3810.

Liongue, C., Hall, C. J., O'Connell, B. A., Crosier, P., & Ward, A. C. (2009). Zebrafish granulocyte colony-stimulating factor receptor signaling promotes myelopoiesis and myeloid cell migration. *Blood, 113*, 2535–2546.

Long, Q., et al. (1997). GATA-1 expression pattern can be recapitulated in living transgenic zebrafish using GFP reporter gene. *Development (Cambridge, England), 124*, 4105–4111.

Lu, L. S., Wang, S. J., & Auerbach, R. (1996). In vitro and in vivo differentiation into B cells, T cells, and myeloid cells of primitive yolk sac hematopoietic precursor cells expanded >100-fold by coculture with a clonal yolk sac endothelial cell line. *Proceedings of the National Academy of Sciences of the United States of America, 93*, 14782–14787.

Lugo-Villarino, G., et al. (2010). Identification of dendritic antigen-presenting cells in the zebrafish. *Proceedings of the National Academy of Sciences of the United States of America, 107*, 15850–15855. http://dx.doi.org/10.1073/pnas.1000494107.

Lux, C. T., et al. (2008). All primitive and definitive hematopoietic progenitor cells emerging before E10 in the mouse embryo are products of the yolk sac. *Blood, 111*, 3435–3438.

Lyons, S. E., et al. (2002). A nonsense mutation in zebrafish gata1 causes the bloodless phenotype in vlad tepes. *Proceedings of the National Academy of Sciences of the United States of America, 99*, 5454−5459.

Ma, D., Zhang, J., Lin, H. F., Italiano, J., & Handin, R. I. (2011). The identification and characterization of zebrafish hematopoietic stem cells. *Blood, 118*, 289−297. http://dx.doi.org/10.1182/blood-2010-12-327403.

McCulloch, E. A., & Till, J. E. (1960). The radiation sensitivity of normal mouse bone marrow cells, determined by quantitative marrow transplantation into irradiated mice. *Radiation Research, 13*, 115−125.

McGrath, K. E., et al. (2008). Enucleation of primitive erythroid cells generates a transient population of "pyrenocytes" in the mammalian fetus. *Blood, 111*, 2409−2417.

Mizgirev, I., & Revskoy, S. (2010a). Generation of clonal zebrafish lines and transplantable hepatic tumors. *Nature Protocols, 5*, 383−394. http://dx.doi.org/10.1038/nprot.2010.8.

Mizgirev, I. V., & Revskoy, S. (2010b). A new zebrafish model for experimental leukemia therapy. *Cancer Biology and Therapy, 9*, 895−902.

Moore, M. A., & Metcalf, D. (1970). Ontogeny of the haemopoietic system: yolk sac origin of in vivo and in vitro colony forming cells in the developing mouse embryo. *British Journal of Haematology, 18*, 279−296.

Moore, M. A., Williams, N., & Metcalf, D. (1973a). In vitro colony formation by normal and leukemic human hematopoietic cells: characterization of the colony-forming cells. *Journal of the National Cancer Institute, 50*, 603−623.

Moore, M. A., Williams, N., & Metcalf, D. (1973b). In vitro colony formation by normal and leukemic human hematopoietic cells: interaction between colony-forming and colony-stimulating cells. *Journal of the National Cancer Institute, 50*, 591−602.

Mori, Y., et al. (2008). Identification of the human eosinophil lineage-committed progenitor: revision of phenotypic definition of the human common myeloid progenitor. *Journal of Experimental Medicine, 206*.

Morrison, S. J., Uchida, N., & Weissman, I. L. (1995). The biology of hematopoietic stem cells. *Annual Reviews of Cell and Developmental Biology, 11*, 35−71.

Mulder, A. H., & Visser, J. W. M. (1987). Separation and functional analysis of bone marrow cells separated by Rhodamine-123 fluorescence. *Experimental Hematology, 15*, 99−104.

Murayama, E., et al. (2006). Tracing hematopoietic precursor migration to successive hematopoietic organs during zebrafish development. *Immunity, 25*, 963−975.

Nakano, T., Kodama, H., & Honjo, T. (1994). Generation of lymphohematopoietic cells from embryonic stem cells in culture. *Science, 265*, 1098−1101.

Nakano, T., Kodama, H., & Honjo, T. (1996). In vitro development of primitive and definitive erythrocytes from different precursors. *Science, 272*, 722−724.

Nakorn, T. N., Miyamoto, T., & Weissman, I. L. (2003). Characterization of mouse clonogenic megakaryocyte progenitors. *Proceedings of the National Academy of Sciences of the United States of America, 100*, 205−210.

North, T. E., et al. (2002). Runx1 expression marks long-term repopulating hematopoietic stem cells in the midgestation mouse embryo. *Immunity, 16*, 661−672.

Oberlin, E., Tavian, M., Blazsek, I., & Peault, B. (2002). Blood-forming potential of vascular endothelium in the human embryo. *Development (Cambridge, England), 129*, 4147−4157.

Orkin, S. H., & Zon, L. I. (2008). Hematopoiesis: an evolving paradigm for stem cell biology. *Cell, 132*, 631−644. http://dx.doi.org/10.1016/j.cell.2008.01.025.

Ottersbach, K., & Dzierzak, E. (2005). The murine placenta contains hematopoietic stem cells within the vascular labyrinth region. *Developmental Cell, 8*, 377–387.

Paffett-Lugassy, N., et al. (2007). Functional conservation of erythropoietin signaling in zebrafish. *Blood, 110*, 2718–2726.

Page, D. M., et al. (2013). An evolutionarily conserved program of B-cell development and activation in zebrafish. *Blood, 122*, e1–11. http://dx.doi.org/10.1182/blood-2012-12-471029.

Palis, J., et al. (2001). Spatial and temporal emergence of high proliferative potential hematopoietic precursors during murine embryogenesis. *Proceedings of the National Academy of Sciences of the United States of America, 98*, 4528–4533.

Palis, J., Malik, J., McGrath, K. E., & Kingsley, P. D. (2010). Primitive erythropoiesis in the mammalian embryo. *International Journal of Developmental Biology, 54*, 1011–1018.

Palis, J., Robertson, S., Kennedy, M., Wall, C., & Keller, G. (1999). Development of erythroid and myeloid progenitors in the yolk sac and embryo proper of the mouse. *Development (Cambridge, England), 126*, 5073–5084.

Ploemacher, R. E., van der Sluijs, J. P., van Beurden, C. A., Baert, M. R., & Chan, P. L. (1991). Use of limiting-dilution type long-term marrow cultures in frequency analysis of marrow-repopulating and spleen colony-forming hematopoietic stem cells in the mouse. *Blood, 78*, 2527–2533.

Pugach, E. K., Li, P., White, R., & Zon, L. (2009). Retro-orbital injection in adult zebrafish. *Journal of Visualized Experiments*. http://dx.doi.org/10.3791/1645.

Renshaw, S. A., et al. (2006). A transgenic zebrafish model of neutrophilic inflammation. *Blood, 108*, 3976–3978.

Robinson, W. A., Bradley, T. R., & Metcalf, D. (1967). Effect of whole body irradiation on colony production by bone marrow cells in vitro. *Proceedings of the Society for Experimental Biology and Medicine. Society for Experimental Biology and Medicine (New York, N.Y.), 125*, 388–391.

Schmitt, T. M., et al. (2004). Induction of T cell development and establishment of T cell competence from embryonic stem cells differentiated in vitro. *Nature Immunology, 5*, 410–417.

Schmitt, T. M., & Zuniga-Pflucker, J. C. (2002). Induction of T cell development from hematopoietic progenitor cells by delta-like-1 in vitro. *Immunity, 17*, 749–756.

Shapiro, H. M. (2002). *Practical flow cytometry* (4th ed.). Wiley-Liss.

Smith, A. C., et al. (2010). High-throughput cell transplantation establishes that tumor-initiating cells are abundant in zebrafish T-cell acute lymphoblastic leukemia. *Blood, 115*, 3296–3303. http://dx.doi.org/10.1182/blood-2009-10-246488.

Spangrude, G. J., Heimfeld, S., & Weissman, I. L. (1988). Purification and characterization of mouse hematopoietic stem cells. *Science, 241*, 58–62.

Stachura, D. L., et al. (2009). Zebrafish kidney stromal cell lines support multilineage hematopoiesis. *Blood, 114*, 279–289.

Stachura, D. L., Svoboda, O., Lau, R. P., Balla, K. M., Zon, L. I., Bartunek, P., & Traver, D. (2011 August 4). Clonal analysis of hematopoietic progenitor cells in the zebrafish. *Blood, 118*(5), 1274–1282.

Stachura, D. L., et al. (2013). The zebrafish granulocyte colony-stimulating factors (Gcsfs): 2 paralogous cytokines and their roles in hematopoietic development and maintenance. *Blood, 122*, 3918–3928. http://dx.doi.org/10.1182/blood-2012-12-475392.

Sutherland, H. J., Eaves, C. J., Lansdorp, P. M., Thacker, J. D., & Hogge, D. E. (1991). Differential regulation of primitive human hematopoietic cells in long-term cultures maintained on genetically engineered murine stromal cells. *Blood, 78*, 666–672.

Svoboda, O., et al. (2014). Dissection of vertebrate hematopoiesis using zebrafish thrombopoietin. *Blood, 124*, 220–228. http://dx.doi.org/10.1182/blood-2014-03-564682.

Svoboda, O., Stachura, D. L., Machonova, O., Zon, L. I., Traver, D., & Bartunek, P. (2015). Ex vivo tools for the clonal analysis of zebrafish hematopoiesis. *Nature Protocols* (in press).

Thompson, M. A., et al. (1998). The cloche and spadetail genes differentially affect hematopoiesis and vasculogenesis. *Developmental Biology, 197*, 248–269.

Traver, D., et al. (2003a). Transplantation and in vivo imaging of multilineage engraftment in zebrafish bloodless mutants. *Nature Immunology, 4*, 1238–1246.

Traver, D., et al. (2003b). *The zebrafish as a model organism to study development of the immune system* (Vol. 81). Academic Press.

Traver, D., et al. (2004 September 1). Biological effects of lethal irradiation and rescue by hematopoietic cell transplantation in zebrafish. *Blood, 104*(5), 1298–1305.

Traver, D. (2004). Cellular dissection of zebrafish hematopoiesis. *Methods in Cell Biology, 76*, 127–149.

Trede, N. S., & Zon, L. I. (1998). Development of T-cells during fish embryogenesis. *Developmental and Comparative Immunology, 22*, 253–263.

Visser, J. W., Bauman, J. G., Mulder, A. H., Eliason, J. F., & de Leeuw, A. M. (1984). Isolation of murine pluripotent hemopoietic stem cells. *Journal of Experimental Medicine, 159*, 1576–1590.

Visser, J. W., & de Vries, P. (1988). Isolation of spleen-colony forming cells (CFU-s) using wheat germ agglutinin and rhodamine 123 labeling. *Blood Cells, 14*, 369–384.

Ward, A. C., et al. (2003). The zebrafish spi1 promoter drives myeloid-specific expression in stable transgenic fish. *Blood, 102*, 3238–3240.

Weissman, I., Papaioannou, V., & Gardner, R. (1978). Fetal hematopoietic origins of the adult hematolymphoid system. In B. Clarkson, P. A. Marks, & J. E. Till (Eds.), *Differentiation of normal and neoplastic cells* (pp. 33–47). New York: Cold Spring Harbor Laboratory Press.

Westerfield, M. (2000). *The zebrafish book. A guide for the laboratory use of zebrafish (Danio rerio)* (4th ed.). Eugene: Univ. of Oregon Press.

Willett, C. E., Cortes, A., Zuasti, A., & Zapata, A. G. (1999). Early hematopoiesis and developing lymphoid organs in the zebrafish. *Developmental Dynamics, 214*, 323–336.

Willett, C. E., Zapata, A. G., Hopkins, N., & Steiner, L. A. (1997). Expression of zebrafish rag genes during early development identifies the thymus. *Developmental Biology, 182*, 331–341.

Wittamer, V., Bertrand, J. Y., Gutschow, P. W., & Traver, D. (2011). Characterization of the mononuclear phagocyte system in zebrafish. *Blood, 117*, 7126–7135. http://dx.doi.org/10.1182/blood-2010-11-321448.

Yoder, M. C., et al. (1997a). Characterization of definitive lymphohematopoietic stem cells in the day 9 murine yolk sac. *Immunity, 7*, 335–344.

Yoder, M., Hiatt, K., & Mukherjee, P. (1997b). In vivo repopulating hematopoietic stem cells are present in the murine yolk sac at day 9.0 postcoitus. *Proceedings of the National Academy of Sciences of the United States of America, 6780*(94), 6776.

Yokota, T., et al. (2006). Tracing the first waves of lymphopoiesis in mice. *Development (Cambridge, England), 133*, 2041–2051.

Zapata, A., & Amemiya, C. T. (2000). Phylogeny of lower vertebrates and their immunological structures. *Current Topics in Microbiology and Immunology, 248*, 67–107.

Zapata, A. (1979). Ultrastructural study of the teleost fish kidney. *Developmental and Comparative Immunology, 3*, 55–65.

Zhu, H., et al. (2005). Regulation of the lmo2 promoter during hematopoietic and vascular development in zebrafish. *Developmental Biology, 281*, 256–269.

Zovein, A. C., et al. (2008). Fate tracing reveals the endothelial origin of hematopoietic stem cells. *Cell Stem Cell, 3*, 625–636.

CHAPTER

Second harmonic generation microscopy in zebrafish

3

D.C. LeBert*,a, J.M. Squirrell*,a, A. Huttenlocher*,1, K.W. Eliceiri*,§,1

University of Wisconsin—Madison, Madison, WI, United States
§*Morgridge Institute for Research, Madison, WI, United States*
[1]*Corresponding authors: E-mail: huttenlocher@wisc.edu; eliceiri@wisc.edu*

CHAPTER OUTLINE

Introduction	56
1. Materials	59
1.1 Zebrafish Embryos	59
1.2 Microscope Supplies and Components	59
1.3 Buffers, Other Reagents, and Tools for Sample Preparation	61
1.4 Image Processing Components	61
2. Methods	62
2.1 Macrophage-Specific Protein Expression	62
2.2 Preparation of 35-mm Polystyrene Bottom Petri Dishes	62
2.3 Caudal Fin Amputation	62
2.4 Preparation of Fixed Samples	63
2.5 Image Acquisition	64
2.6 Image Processing and Analysis	65
3. Notes	66
Acknowledgments	66
Supplementary Data	67
References	67

---

[a]Authors contributed equally.

Abstract

Modern optical imaging has progressed rapidly with the ability to noninvasively image cellular and subcellular phenomena with high spatial and temporal resolution. In particular, emerging techniques such as second harmonic generation (SHG) microscopy can allow for the monitoring of intrinsic contrast, such as that from collagen, in live and fixed samples. When coupled with multiphoton fluorescence microscopy, SHG can be used to image interactions between cells and the surrounding extracellular environment. There is recent interest in using these approaches to study inflammation and wound healing in zebrafish, an important model for studying these processes. In this chapter we present the practical aspects of using second harmonic generation to image interactions between leukocytes and collagen during wound healing in zebrafish.

INTRODUCTION

Second harmonic generation (SHG) is a nonlinear optical event resulting from the interaction of light with noncentrosymmetric materials, including biological molecules such as collagen, which provides an endogenous source of contrast (Mohler, Millard, & Campagnola, 2003). This intrinsic property of molecular structures emits photons at exactly half the incident wavelength. The method allows a variety of incident wavelengths to be utilized. Therefore, SHG can be spectrally separated from fluorescent events, permitting the combination of SHG with a diversity of endogenous and exogenous fluorophores (Campagnola, Clark, Mohler, Lewis, & Loew, 2001; Zipfel et al., 2003). Additionally, the fact that SHG does not require fluorophore fusions ensures that observations are physiologically relevant and not a byproduct of the fusion construct. A primary use of SHG imaging of tissue has been the examination of collagen during cancer progression, where it is being studied as a potential diagnostic assessment tool since changes in collagen, and corresponding SHG patterns, have been shown to be involved in disease progression (Brisson et al., 2015; Conklin et al., 2011; Drifka et al., 2015; Keikhosravi, Bredfeldt, Sagar, & Eliceiri, 2014; Tilbury & Campagnola, 2015). We recently used SHG to show that collagen fibers in the zebrafish larval tail fin exhibit changes in organization in both chronic and acute inflammatory environments (LeBert et al., 2015). The SHG imaging modality we have used is backward SHG (Chen, Nadiarynkh, Plotnikov, & Campagnola, 2012) whereby the SHG signal is detected on the same side of the sample as the excitation source. A caveat of this method is that it does not distinguish between different types of fibrillar collagen (Chen et al., 2012) and therefore we refer to the fibers as simply "collagen fibers" rather than specifying collagen type.

The larval zebrafish is an increasingly popular vertebrate model for the study of both immunity and wound healing. The zebrafish innate immune system is highly conserved and represents a powerful model system (Deng & Huttenlocher, 2012; Meeker & Trede, 2008). Zebrafish are capable of rapidly regenerating many tissues, including the caudal fin, following tail transection. Additionally, due to their translucency during the first few days of development, they are highly amenable to noninvasive in vivo imaging of cellular behavior under physiologically relevant conditions. Although the innate immune system has been extensively studied during this stage of larval development, much less is known about collagen fiber organization in the developing zebrafish fin and its reorganization during wound healing (Duran, Mari-Beffa, Santamaria, Becerra, & Santos-Ruiz, 2011).

Our recent study (LeBert et al., 2015) identified a thickening of collagen fibers during fin regrowth following tail transection. Depletion of the matrix metalloprotease, MMP9, impairs wound healing and alters the collagen signature detected by SHG, suggesting that collagen reorganization likely plays an important role during fin regrowth. Studies in other systems also indicate that collagen changes during wound healing (Clore, Cohen, & Diegelmann, 1979; Madden & Peacock, 1971). However, little is known about potential interactions between the immune system and collagen structures during normal and aberrant wound healing. There is, however, evidence that macrophages may influence collagen synthesis and remodeling in damaged tissues (Hunt, Knighton, Thakral, Goodson, & Andrews, 1984; Madsen et al., 2013). The combination of SHG with the transparency, genetic malleability, and availability of fluorescently tagged immune cells in zebrafish models promises to provide significant new insight into the relationship between the immune system and extracellular matrix during wound healing and tissue regeneration in live animals.

Although our interests lie mainly within the realm of the immune response during wounding, inflammation, and infection, the SHG/zebrafish combination provides an accessible and relevant system for examining collagen changes in different tissues and organs under a variety of normal and disease states (Kieu, Mehravar, Gowda, Norwood, & Peyghambarian, 2013; Olivier et al., 2010; Sun et al., 2004). For example, SHG has been used in zebrafish for studying neutrophil transmigration through collagen during tumor cell invasion (He et al., 2012). SHG can also be utilized to assess other structured components in zebrafish, such as sarcomeres in the muscle during statin-induced myopathy (Huang et al., 2011) and subcellular components, such as mitotic spindles, during early embryonic development (Luengo-Oroz et al., 2012; Olivier et al., 2010; Sun et al., 2004).

The endogenous nature of SHG imaging lends itself readily to using live cells, tissues, and organisms. This is particularly powerful in zebrafish where it can be combined with the use of genetically encoded fluorescent tags. There is also the potential advantage of using multiphoton imaging (Denk, Strickler, & Webb, 1990) for long-term imaging of processes that involve changes in collagen organization. One obvious extension of the techniques presented here is the ability to perform in vivo imaging of macrophage—collagen interactions in real time, during wound healing. A challenge of long-term imaging during wound healing is the creation of application-specific devices that maintain larval viability and permit growth while accommodating the higher resolution lenses preferred to clearly delineate fine collagen fibers. Advances in SHG imaging technologies are permitting the simultaneous collection of both forward and backward SHG signal, which will allow more precise delineation of different types of fibrillar collagens (Tilbury, Lien, Chen, &

Campagnola, 2014), and as these imaging systems become widely available, could be applied to understanding collagen dynamics in vivo in zebrafish over time.

Herein we describe a newly established protocol for imaging (Fig. 1) and analyzing (Fig. 2) collagen fibers using multiphoton, SHG imaging, alone and in combination with transgenic, fluorescently labeled, macrophages (Fig. 3). By pairing the transparent larval zebrafish with this imaging technique, we are able to examine changes in collagen organization during normal development and after wounding. This technique allows for noninvasive assessment of both collagen fibers during development, wound healing, and macrophage response and, potentially, interactions between macrophages and collagen fibers. In the following sections, we provide protocols for performing SHG imaging on fixed transgenic zebrafish that expresses fluorescently-tagged proteins in macrophages. This includes protocols for preparation of transgenic fish, microscope setup, image acquisitions, and postacquisition analysis.

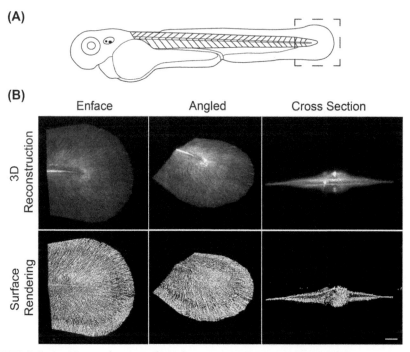

FIGURE 1 Second harmonic generation microscopy of the larval zebrafish tail.

(A) Diagram of zebrafish larvae showing region imaged. (B) Three views of 3D reconstruction of collagen fibers in an unwounded zebrafish larval tail (3 dpf) to illustrate organization of fibers. Image stitched in FIJI from z-stacks of 20 region of interests. Top row shows 3D reconstruction of SHG data while bottom row show surface rendering. (Supplemental video 1). Scale bar = 50 μm.

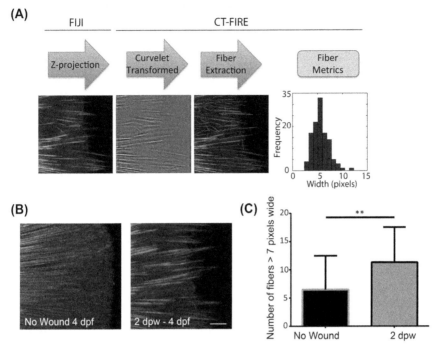

FIGURE 2 Second harmonic generation (SHG) imaging of fibers in the zebrafish tail reveal quantifiable changes in organization of collagen fibers during wound healing.

(A) Schematic illustrating steps involved in quantifying fiber metrics, specifically width, in zebrafish tails after wounding. (B) FIJI-generated projections of SHG z-stacks of the distal region of interest showing changes in fiber organization in wounded and unwounded tails from 4 dpf (2 dpw) larvae. Scale bar = 20 μm. (C) Graph showing quantitation of fiber width as determined using CT-FIRE fiber analysis software (Supplemental videos 2 and 3). Significance of $p < 0.01$ indicated by **.

1. MATERIALS
1.1 ZEBRAFISH EMBRYOS

1. Zebrafish husbandry was performed as described in The Zebrafish Book (Westerfield, 2000). Protocols can be accessed online at http://zfin.org/zf_info/zfbook/zfbk.html. Accessed August 17, 2015.

1.2 MICROSCOPE SUPPLIES AND COMPONENTS

The Laboratory for Optical and Computational Instrumentation (LOCI, http://loci.wisc.edu/) is an instrumentation laboratory advancing optical and computation techniques, thus much of the instrumentation, acquisition, and analysis software presented here was developed at LOCI. Therefore, we describe the setup used to

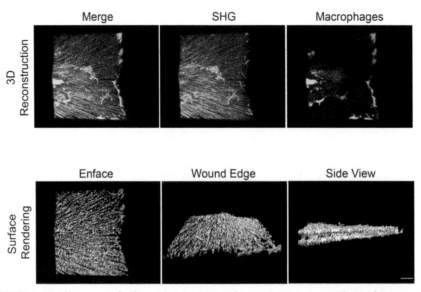

FIGURE 3 Multiphoton excitation of Dendra-expressing macrophages combined with second harmonic generation (SHG) imaging following wounding.

3D reconstructions, enface view of tail transection of multiphoton data of showing the merge of collagen fibers and fluorescently tagged macrophages as well as the two signals separated. Three views of 3D surface rendering more clearly demonstrate the spatial relationship between the macrophages and the fibers, with macrophages either accumulating at the wound edge or found within the wedge-shaped space between the layers of fibers. Scale bar = 20 μm (Supplemental video 4). (See color plate)

acquire the images presented in the figures. Commercial vendors also provide similar systems and all the components utilized are commercially available.

1. Multiphoton microscope (http://loci.wisc.edu/equipment/spectral-lifetime-multiphoton-microscope-slim):
   a. Microscope: Nikon Eclipse TE2000U inverted microscope.
   b. SHG and MP emission detector: Hamamatsu H7422P-40 GaAsP photomultiplier (PMT). Other PMTs can be used including the common multi-alkali PMTs available on all commercial systems.
   c. Transmitted light detector: An infrafred-transmitting Si photodiode (Hamamatsu K34213) was used in a BioRad MRC-1024 transmitted light detector box for detecting transmitted light. Any IR-transmitting photodiode or PMT could be used.
   d. Objectives: Nikon 10X S Fluor 0.5NA and Nikon CFI Apo Lambda S 40X long working distance water immersion 1.15NA.
   e. Laser: Coherent Chameleon Ultra II Ti:Sapphire laser—tuneable range 700–980 nm.

f. Scanning system: Cambridge Technologies 6210H galvanometer with Prairie Technologies (Bruker) galvanometer control box.
  g. Laser power control via Pockel Cell: Conoptics Model 350-80LA KD*P Series Electro-Optic modulator with Model 302RM amplifier. Other power control schemes may be used as well.
  h. Emission filters:
     i. For SHG: Semrock 445/20 nm BrightLine single-band bandpass filter.
     ii. For Dendra: Semrock 520/35 nm BrightLine single-band bandpass filter.
  i. Filter wheel: Sutter Instrument Lambda 10B Optical Filter Changer.
  j. Shutter: Uniblitz Model D122 shutter and driver system.
  k. Z and stage control: Applied Scientific Imaging MS-2000 xyz stage control.
  l. Vibration table: TMC CleanTopII vibration isolation system.
2. Multiphoton laser scanning acquisition system: WiscScan 7.2.2 (http://loci.wisc.edu/software/wiscscan). Commercial scanning packages that can control multiphoton microscopes are also sufficient.
3. Glass bottom petri dish: In vitro scientific 35-mm dish with 20-mm bottom well #1.5 (0.17 mm) cover glass thickness.

1.3 BUFFERS, OTHER REAGENTS, AND TOOLS FOR SAMPLE PREPARATION

1. 60x E3: To make 1 L, add 17.2 g NaCl, 2.9 g $CaCl_2$, and 4.9 g $MgSO_4$. Fill to 1 L with reverse osmosis (RO) water. Adjust pH to 7.2 with NaOH.
2. 1x E3: Dilute 60x E3 with RO water to 1x. Add 0.1% methylene blue.
3. E3/PTU: 0.003% 1-phenyl-2-thiourea in 1x E3.
4. 1X PBS.
5. #10 feather surgical blade for transection, #15 scalpel blade for sample preparation prior to imaging.
6. 20 gauge syringe needle.
7. FisherBrand 200-μL specialty pipette tips, large orifice.

1.4 IMAGE PROCESSING COMPONENTS

Image acquisition software usually contains some image viewing and/or imaging analysis functions. Additional commonly used software includes, but is not limited to, the following:

1. FIJI (free software; http://fiji.sc), accessed August 17, 2015 (Schindelin et al., 2012)
2. CTFIRE software: http://loci.wisc.edu/software/ctfire (Bredfeldt et al., 2014)
3. Bitplane Imaris (http://www.bitplane.com/go/products/imaris). Accessed August 17, 2015.

2. METHODS
2.1 MACROPHAGE-SPECIFIC PROTEIN EXPRESSION

For imaging purposes, transient mosaic expression is typically sufficient. However, in certain cases, high levels of expression of the transgene can be cytotoxic or have undesirable phenotypes. In this situation, generation of a transgenic line can be beneficial as lines with lower level expression can be identified. Preparation of transient or transgenic zebrafish that expresses the gene of interest in a macrophage-specific manner requires the use of a cell-specific promoter such as macrophage-expressed gene 1 (mpeg).

1. Plasmid construction: Use DNA expression vectors containing minimal Tol2 elements for efficient integration flanking the mpeg promoter, the gene of interest, and an SV40 polyadenylation sequence (Clonetech Laboratories Inc.).
2. Preparation of plasmid DNA: Transformation of the plasmid into regular cloning competent cells, eg, *Escherichia coli* DH5α is sufficient. DNA extraction and purification is achieved using a commercially available miniprep kit (Promega Wizard Plus SV Minipreps DNA purification system).
3. Preparation of Tol2 transposase mRNA: Transposase is prepared by in vitro transcription (Ambion mMESSAGE mMACHINE) followed by column purification (miRvana RNA isolation kit).
4. Microinjection of zebrafish embryos is a standard routine procedure. Different laboratories will have slightly different setups. For basic setup information and injection procedures, please refer to The Zebrafish Book.
5. Expression of the construct is obtained by injection of a 3 nL volume of 12.5 ng/μL DNA plasmid with 17.5 ng/μL Tol2 transposase mRNA into the cytoplasm of the 1-cell stage embryo.

2.2 PREPARATION OF 35-mm POLYSTYRENE BOTTOM PETRI DISHES

Protein coating of dishes is necessary because we find that larvae stick to polystyrene and glass surfaces without coating which can lead to wounds on the zebrafish larvae. The 35-mm dishes are submerged in 1% skim milk for 10 min and rinsed several times with RO water and air dried before use.

2.3 CAUDAL FIN AMPUTATION

The caudal fin region of the larval zebrafish is the preferred region for SHG imaging due to its relatively thin nature and lack of high numbers of pigment cells. Amputation of the caudal fin is also performed to assess wound healing as well as macrophage recruitment and resolution. In our hands, we find wounding of PTU-treated larvae at 2 to 3 dpf to be the most reproducible as larvae anesthetized with tricaine tend to lie flat at this stage of development. Zebrafish at 4 dpf lie at more of an angle, due to the orientation of the yolk sac, which makes amputating and imaging the caudal fin more difficult.

1. Place larvae in 3 mL of 0.2 mg/mL tricaine in E3/PTU in a protein coated 35-mm polystyrene plate. Let sit for at least 1 min.
2. Gently insert a feather no. 10 scalpel into the media being careful not to nick the fish. When the blade hits the surface of the media it creates currents that push the larvae along the bottom of the plate making wounding more difficult. It is recommended to keep the blade in the media once it has been inserted.
3. Rock the tip of the blade forward over the caudal fin, just distal to the notochord, before dragging the blade back over the fin. This dual-cutting motion ensures complete amputation and, in our hands, creates a smooth wound edge.
4. Once wounding is complete, rinse the larvae 3x in E3/PTU for 1 min.
5. Transfer the wounded larvae to a fresh milk-coated 35-mm polystyrene plate. The wounding plate will contain many small abrasions in its surface, which will cause unintended wounds to the larvae as they pass over them.

2.4 PREPARATION OF FIXED SAMPLES

Although fixation is not necessary for SHG imaging, fixation permits the imaging of larger numbers of samples from different conditions at various time points over multiple imaging days. This is of particular importance when first defining a phenomenon of interest.

1. For most applications, larvae can be fixed at set time points with the following 4% fixative solution: for 1 mL of fixative, add 100 μL of 10x PBS, 250 μL of fresh 16% paraformaldehyde (PFA), and 650 μL of ddH$_2$O. In situations in which the fluorophore dissipates rapidly upon fixation, we recommend using the following 1.5% fixative solution: for 1 mL of fixative, add 801.3 μL of ddH$_2$O, 100 μL PIPES (1 M), 93.7 μL PFA (16%), 4 μL EGTA (0.5 M), 1 μL MgSO$_4$ (1 M).
2. Samples are fixed directly in 35-mm dishes. Samples are transferred to 4°C O/N.
3. Fixed larvae are washed 3X in room temperature 1X PBS. The larvae can be stored in PBS at 4°C.
4. Using large orifice pipette tip on P200 pipette, two to four larvae are transferred in a drop (approximately 100 μL) of PBS to a 35-mm glass bottom dish. To ensure that the caudal fin lies evenly on the bottom of the glass bottom plate, the caudal fin is removed from the rest of the body using a scalpel blade, posterior to the yolk sac extension, such that a 1–2 mm piece of the caudal fin is retained.
5. The rest of the body is carefully moved out of the drop to the side of the dish using either a syringe needle or a fine tip pipette tip. It is important not to damage the yolk sack as yolk granules will contaminate the drop and interfere with imaging.
6. Because the fins are not attached in any way to the glass bottom, the dish must be transported carefully to the microscope to minimize jostling. Also for this

reason more than four tail fragments can cause crowding of the tail fins within the drop, resulting in overlap and potential difficulties in keeping track of which tails have or have not been imaged.

2.5 IMAGE ACQUISITION

Image acquisition requires the designation of the x, y, and z locations of the sample to be imaged as well as the establishment of imaging parameters (power levels, scan speed, detector gain, etc.) appropriate for the sample of interest.

1. In preparation for imaging, a drop of distilled water is placed on the 40X water immersion lens and then the lens is gently rotated out to permit the use of the 10X lens. The glass bottom dish is placed into the appropriate stage holder on the microscope and the 10X lens is used to initially locate the tails. To maintain consistency in fin orientation, the dish is rotated so that the notochord is parallel to the x axis of the image. The 40X water lens is rotated into place using proper microscopy technique and refocused on the fin.
2. Images are collected to include the tip of the notochord and the edge of the tail fin. For younger larvae or early postwound tails, this may be a single region of interest (ROI). In the case of older larvae or later postwound times this might include two or even three ROIs, though a greater number of ROIs could be defined if larger regions of the tail are of interest. These multiple regions of interest locations are marked for subsequent automated image collection. Then the z stack top and bottom are set to accommodate the ROI with the greatest z depth. Although this is usually the ROI about the tip of the notochord, fin tips that are curved, cupped, or distorted may require increased z-depth, so z-depth should be checked for all ROIs for a given fin. The z-interval was set at 1 or 2 µm.
3. Once regional information has been set, signal-specific parameters must be set. For a given experiment, imaging parameters should be identical or as similar as possible, with records kept of all settings to ensure consistency and comparability of data across treatments and imaging days. For SHG imaging we utilized 890-nm two-photon laser excitation and a 445-nm emission filter. Note that because of the nature of optical sectioning of angled surfaces such as the tail fin, any given individual z-image will only include a small region of collagen fibers. For Dendra image collection, an 890-nm excitation with 520-nm emission filter was used. Detector gain was kept consistent for both SHG and fluorophore image collection. Laser power was attenuated using a pockel cell and set to provide good contrast without signal saturation. Typically, SHG is a much weaker signal than exogenous fluorescence; thus our image acquisition software (WiscScan) permits differential Pockel cell settings for different emission filters. Scan speed was set, and maintained for all samples, to improve signal to noise while minimizing sample exposure. Automated image acquisition was then conducted for the z-stacks at multiple regions for both signals, with

sequential collection of the two signals. Note that our system includes simultaneous transmitted light collection so that brightfield features can be captured the same time as fluorescence and SHG.

2.6 IMAGE PROCESSING AND ANALYSIS (FIG. 2)

1. A number of software packages and plug-ins are available for a variety of image analyses. Our preference is to engage open sources software as much as possible, utilizing commercial software as needed when open source software is not readily available for the task required. Initial image processing for visualization was conducted using FIJI. If multiple ROIs were collected for a given tail, stitching of these ROIs was necessary for viewing the entire tail region. Additionally, because of the nature of the optical section and the angle of tail, SHG images of individual z-sections are often not very informative and benefit from z-projection as an initial processing for visual data assessment. ROI stitching was performed using the "Grid/Collection Stitching" plug-in in FIJI (Preibisch, Saalfeld, & Tomancak, 2009). This provides a stitched z-stack containing both the brightfield and emission channel. For data assessment, the brightfield and emission channel were separated and the emission channel was z-projected. This can be done manually or semi-automated with macros in FIJI (Schindelin et al., 2012). For data sets consisting of a single ROI, emission z-stacks were z-projected in FIJI. Similarly, simple macros can be written to batch process large number of images. If two labels are collected, either the z-stacks or the z-projections can be merged in FIJI to provide information about the relative location of the two signals.
2. The z-projections provide a simple, accessible method for visualizing these data sets, which can then be assessed for interesting changes in collagen fiber organization with different treatments in order to determine the appropriate quantitative measurements to be acquired. Although a variety of quantitative assessments could be accomplished using FIJI or other image analysis software packages, in our experiments we noticed a change in the fiber thickness as wound healing progressed (Fig. 2B). Therefore, we utilized the fiber analysis tool, CT-FIRE (Bredfeldt et al., 2014) (Fig. 2A) to assess fiber width in the z-projection of the SHG images. CT-FIRE extracts and analyzes a variety of metrics on images of fibers in tissues that can be statistically evaluated. The purpose of CT-FIRE is to allow users to automatically extract collagen fibers in an image and those quantify fibers with descriptive statistics, such as fiber angle, fiber length, fiber straightness, and fiber width.
3. Generating z-projections in FIJI and quantitative analysis in CT-FIRE are important tools for assessing changes in the SHG fiber organization in the larval tail during wound healing and regrowth under varying conditions. However, the tail and the process of wound healing are three dimensional (3D) entities and benefit from 3D visualization. To accomplish this, we opened z-stacks in

Bitplane Imaris (Figs. 1, 2 and 3). Z-stacks from all three channels (brightfield, SHG, and fluorescence) could be incorporated into a single 3D reconstruction. The different channels can be surface rendered to enhance the visualization. These 3D reconstructions can be rotated and resliced at different angles and locations in order to better understand the 3D nature and relationships within the healing tail.

3. NOTES

1. Glass bottom plates are commercially available, but an inexpensive and relatively easy alternative is to make them in house. Drill a 16–18 mm hole in the center of a 35-mm tissue culture dish bottom. Apply a thin layer of Norland Optical Adhesive 68 on the outside surface around the hole. Place a 22 mm round #1 cover slip on top of the adhesive and press down slightly. The adhesive is cured by UV light. Overnight exposure of UV light in a tissue culture hood or approximately 1 h on a UV transilluminator is sufficient to cure the adhesive.
2. To make a transgenic line, injected larvae displaying mosaic transgene expression are raised to sexual maturity (as early as 2-months). F0 founders are screened by outcrossing with wild-type zebrafish and checked for transgene expression in the F1 larvae at 2 to 3-dpf. The positive F1 progeny can then be used for imaging or propagation of the transgenic line.
3. Not all DNA extraction and purification procedures/kits yield DNA of high enough quality to be used for microinjection. Injecting poor quality DNA will result in embryo death. For small amounts of DNA preparation, the Promega Wizard Plus SV Minipreps DNA purification system yields good DNA quality in our experience.
4. The use of PTU treatment is crucial in ensuring high-quality image resolution. In our experience, the presence of even small amounts of pigmentation can result in tripping the detector and damage to the sample. The pigment cells absorb the infrared light generating heat. This heat is enough to damage the pigment cell and the surrounding cells. It is therefore important to fully pigment-inhibit as early as 15 hpf, though we find that beginning PTU treatment at 22–24 h is sufficient. The use of nonpigmented zebrafish lines maybe a viable alternative to PTU treatment.

ACKNOWLEDGMENTS

This work was supported by National Institutes of Health Grants GM074827 (A.H.) and GM102924 (A.H.). Further support came from the Laboratory for Optical and Computational Instrumentation (LOCI) and support from the UW-UWM Intercampus research program (K.W.E.).

SUPPLEMENTARY DATA

Supplementary data related to this article can be found online at http://dx.doi.org/10.1016/bs.mcb.2016.01.005

REFERENCES

Bredfeldt, J. S., Liu, Y., Pehlke, C. A., Conklin, M. W., Szulczewski, J. M., Inman, D. R. ... Eliceiri, K. W. (2014). Computational segmentation of collagen fibers from second-harmonic generation images of breast cancer. *Journal of Biomedical Optics, 19*, 16007.

Brisson, B. K., Mauldin, E. A., Lei, W., Vogel, L. K., Power, A. M., Lo, A. ... Volk, S. W. (2015). Type III collagen directs stromal organization and limits metastasis in a murine model of breast cancer. *American Journal of Pathology, 185*, 1471–1486.

Campagnola, P. J., Clark, H. A., Mohler, W. A., Lewis, A., & Loew, L. M. (2001). Second-harmonic imaging microscopy of living cells. *Journal of Biomedical Optics, 6*, 277–286.

Chen, X., Nadiarynkh, O., Plotnikov, S., & Campagnola, P. J. (2012). Second harmonic generation microscopy for quantitative analysis of collagen fibrillar structure. *Nature Protocols, 7*, 654–669.

Clore, J. N., Cohen, I. K., & Diegelmann, R. F. (1979). Quantitation of collagen types I and III during wound healing in rat skin. *Proceedings of the Society for Experimental Biology and Medicine, 161*, 337–340.

Conklin, M. W., Eickhoff, J. C., Riching, K. M., Pehlke, C. A., Eliceiri, K. W., Provenzano, P. P. ... Keely, P. J. (2011). Aligned collagen is a prognostic signature for survival in human breast carcinoma. *American Journal of Pathology, 178*, 1221–1232.

Deng, Q., & Huttenlocher, A. (2012). Leukocyte migration from a fish eye's view. *Journal of Cell Science, 125*, 3949–3956.

Denk, W., Strickler, J. H., & Webb, W. W. (1990). Two-photon laser scanning fluorescence microscopy. *Science, 248*, 73–76.

Drifka, C. R., Tod, J., Loeffler, A. G., Liu, Y., Thomas, G. J., Eliceiri, K. W., & Kao, W. J. (2015). Periductal stromal collagen topology of pancreatic ductal adenocarcinoma differs from that of normal and chronic pancreatitis. *Modern Pathology, 28*.

Duran, I., Mari-Beffa, M., Santamaria, J. A., Becerra, J., & Santos-Ruiz, L. (2011). Actinotrichia collagens and their role in fin formation. *Developmental Biology, 354*, 160–172.

He, S., Lamers, G. E., Beenakker, J. W., Cui, C., Ghotra, V. P., Danen, E. H. ... Snaar-Jagalska, B. E. (2012). Neutrophil-mediated experimental metastasis is enhanced by VEGFR inhibition in a zebrafish xenograft model. *Journal of Pathology, 227*, 431–445.

Huang, S. H., Hsiao, C. D., Lin, D. S., Chow, C. Y., Chang, C. J., & Liau, I. (2011). Imaging of zebrafish in vivo with second-harmonic generation reveals shortened sarcomeres associated with myopathy induced by statin. *PLoS One, 6*, e24764.

Hunt, T. K., Knighton, D. R., Thakral, K. K., Goodson, W. H., 3rd, & Andrews, W. S. (1984). Studies on inflammation and wound healing: angiogenesis and collagen synthesis stimulated in vivo by resident and activated wound macrophages. *Surgery, 96*, 48–54.

Keikhosravi, A., Bredfeldt, J. S., Sagar, A. K., & Eliceiri, K. W. (2014). Second-harmonic generation imaging of cancer. *Methods in Cell Biology, 123*, 531–546.

Kieu, K., Mehravar, S., Gowda, R., Norwood, R. A., & Peyghambarian, N. (2013). Label-free multi-photon imaging using a compact femtosecond fiber laser mode-locked by carbon nanotube saturable absorber. *Biomedical Optics Express, 4*, 2187−2195.

LeBert, D. C., Squirrell, J. M., Rindy, J., Broadbridge, E., Lui, Y., Zakrzewska, A. ... Huttenlocher, A. (2015). Matrix metalloproteinase 9 modulates collagen matrices and wound repair. *Development, 142*, 2136−2146.

Luengo-Oroz, M. A., Rubio-Guivernau, J. L., Faure, E., Savy, T., Duloquin, L., Olivier, N. ... Santos, A. (2012). Methodology for reconstructing early zebrafish development from in vivo multiphoton microscopy. *IEEE Transactions on Image Processing, 21*, 2335−2340.

Madden, J. W., & Peacock, E. E., Jr. (1971). Some thoughts on repair of peripheral nerves. *Southern Medical Journal, 64*, 17−21.

Madsen, D. H., Leonard, D., Masedunskas, A., Moyer, A., Jurgensen, H. J., Peters, D. E. ... Bugge, T. H. (2013). M2-like macrophages are responsible for collagen degradation through a mannose receptor-mediated pathway. *Journal of Cell Biology, 202*, 951−966.

Meeker, N. D., & Trede, N. S. (2008). Immunology and zebrafish: spawning new models of human disease. *Developmental and Comparative Immunology, 32*, 745−757.

Mohler, W., Millard, A. C., & Campagnola, P. J. (2003). Second harmonic generation imaging of endogenous structural proteins. *Methods, 29*, 97−109.

Olivier, N., Luengo-Oroz, M. A., Duloquin, L., Faure, E., Savy, T., Veilleux, I. ... Beaurepaire, E. (2010). Cell lineage reconstruction of early zebrafish embryos using label-free nonlinear microscopy. *Science, 329*, 967−971.

Preibisch, S., Saalfeld, S., & Tomancak, P. (2009). Globally optimal stitching of tiled 3D microscopic image acquisitions. *Bioinformatics, 25*, 1463−1465.

Schindelin, J., Arganda-Carreras, I., Frise, E., Kaynig, V., Longair, M., Pietzsch, T. ... Cardona, A. (2012). Fiji: an open-source platform for biological-image analysis. *Nature Methods, 9*, 676−682.

Sun, C. K., Chu, S. W., Chen, S. Y., Tsai, T. H., Liu, T. M., Lin, C. Y., & Tsai, H. J. (2004). Higher harmonic generation microscopy for developmental biology. *Journal of Structural Biology, 147*, 19−30.

Tilbury, K., & Campagnola, P. J. (2015). Applications of second-harmonic generation imaging microscopy in ovarian and breast cancer. *Perspectives in Medicinal Chemistry, 7*, 21−32.

Tilbury, K., Lien, C. H., Chen, S. J., & Campagnola, P. J. (2014). Differentiation of Col I and Col III isoforms in stromal models of ovarian cancer by analysis of second harmonic generation polarization and emission directionality. *Biophysical Journal, 106*, 354−365.

Westerfield, M. (2000). The zebrafish book. *A guide for the laboratory use of zebrafish (Danio rerio)* (4th ed.). Eugene: Univ. of Oregon Press.

Zipfel, W. R., Williams, R. M., Christie, R., Nikitin, A. Y., Hyman, B. T., & Webb, W. W. (2003). Live tissue intrinsic emission microscopy using multiphoton-excited native fluorescence and second harmonic generation. *Proceedings of the National Academy of Sciences of the United States of America, 100*, 7075−7080.

CHAPTER 4

Imaging blood vessels and lymphatic vessels in the zebrafish

H.M. Jung*, S. Isogai§, M. Kamei¶, D. Castranova*, A.V. Gore*, B.M. Weinstein*,[1]

*National Institute of Child Health and Human Development, Bethesda, MD, United States
§Iwate Medical University, Morioka, Japan
¶South Australian Health and Medical Research Institute, Adelaide, SA, Australia
[1]Corresponding author: E-mail: flyingfish@nih.gov

CHAPTER OUTLINE

Introduction	70
1. Imaging Vascular Gene Expression	71
2. Nonvital Blood Vessel and Lymphatic Vessel Imaging	74
2.1 Microdye and Microresin Injection	75
2.1.1 Resin injection method	75
2.1.2 Dye injection method	80
2.2 Alkaline Phosphatase Staining for 3 dpf Embryos	82
2.2.1 Materials	82
2.2.2 Protocol	83
2.2.3 Important notes	83
3. Vital Imaging of Blood and Lymphatic Vessels	83
3.1 Microangiography	86
3.1.1 Materials	87
3.1.2 Protocol	87
3.2 Imaging Blood and Lymphatic Vessels in Transgenic Zebrafish	90
3.2.1 Long-term mounting for time-lapse imaging	91
3.2.2 Short-term mounting for time-lapse imaging	96
3.2.3 Multiphoton time-lapse imaging	96
3.2.4 Imaging the zebrafish vasculature using light sheet microscopy	97
3.2.5 Imaging the zebrafish vasculature using superresolution microscopy	98
Conclusion	98
References	99

Abstract

Blood vessels supply tissues and organs with oxygen, nutrients, cellular, and humoral factors, while lymphatic vessels regulate tissue fluid homeostasis, immune trafficking, and dietary fat absorption. Understanding the mechanisms of vascular morphogenesis has become a subject of intense clinical interest because of the close association of both types of vessels with pathogenesis of a broad spectrum of human diseases. The zebrafish provides a powerful animal model to study vascular morphogenesis because of their small, accessible, and transparent embryos. These unique features of zebrafish embryos permit sophisticated high-resolution live imaging of even deeply localized vessels during embryonic development and even in adult tissues. In this chapter, we summarize various methods for blood and lymphatic vessel imaging in zebrafish, including nonvital resin injection—based or dye injection—based vessel visualization, and alkaline phosphatase staining. We also provide protocols for vital imaging of vessels using microangiography or transgenic fluorescent reporter zebrafish lines.

INTRODUCTION

The vascular system is one of the first organ systems to begin functioning during vertebrate development, and its proper assembly is critical for embryonic survival. Blood vessels innervate all other tissues, supplying them with oxygen, nutrients, hormones, and cellular and humoral immune factors. The heart pumps blood through a complex network of blood vessels comprised of an inner single-cell thick endothelial epithelium surrounded by outer supporting pericyte or smooth muscle cells embedded in a fibrillar matrix. Lymphatic vessels are essential for the maintenance of fluid homeostasis, absorbing water, and macromolecules from interstitial spaces within tissues and then returning them to the circulatory system via evolutionarily conserved venous drainage connections. Lymphatics also absorb and transport dietary lipids and lipid-soluble vitamins and provide a route for immune cell trafficking. Unlike the blood vessels of the circulatory system, lymphatic vessels form a blind-ended tree rather than circulatory loops. In the initial lymphatic capillaries, discontinuous button-like junctions connect oak leaf—shaped endothelial cells, providing a "loose" endothelium that facilitates fluid absorption. The mechanisms of blood and lymphatic vessel growth and morphogenesis are a subject of intensive investigation, and a large number of genes important for blood vessel formation have been identified in recent years. This has been achieved through developmental studies in mice and other animal models. However, our understanding of how these genes work together to orchestrate the proper assembly of the intricate vascular networks in the living animal remains limited, in part because of the challenging nature of these studies. The architecture and context of blood and lymphatic vessels are difficult to reproduce in vitro, and most developing vessels in vivo are relatively inaccessible to observation and experimental manipulation. Furthermore, since a properly functioning vasculature is required for embryonic survival and major defects lead to early death and embryonic resorption in amniotes, genetic analysis of vessel formation has been largely limited to reverse-genetic approaches.

The zebrafish provides a number of advantages for in vivo analysis of vascular development. As noted elsewhere in this book, zebrafish embryos are readily accessible to observation and experimental manipulation. Genetic and experimental tools and methods are available for functional manipulation of the entire organism, vascular tissues, or even single vascular- or nonvascular cells. Two features in particular make zebrafish especially useful for studying vascular development. First, developing zebrafish are very small—a 2-dpf embryo is just 2 mm long. Their embryos are so small, in fact, that the cells and tissues of the zebrafish receive enough oxygen by passive diffusion to survive and develop in a reasonably normal fashion for the first 3–4 days of development, even in the complete absence of blood circulation. This makes it fairly straightforward to assess the cardiovascular specificity of genetic or experimental defects that affect the circulation. Second, zebrafish embryos and early larvae are virtually transparent. The embryos of zebrafish (and many other teleosts) are telolecithic—yolk is sequestered in a single large cell separate from the embryo proper. The absence of obscuring yolk proteins gives embryos and larvae a high degree of optically clarity. Genetic variants deficient in pigment cells or pigment formation are even more transparent. This remarkable transparency is probably the most valuable feature of the fish for studying blood and lymphatic vessels, facilitating high-resolution imaging in vivo.

In this chapter we review some of the methods used to image and assess the pattern and function of the zebrafish vasculature, both in developing animals and in adults. First, we briefly touch on visualizing vascular gene expression (in situ hybridization, immunohistochemistry). In the next section we detail methods for imaging vessels in fixed developing and adult zebrafish specimens (Resin and dye injection, alkaline phosphatase (AP) staining). In the final section we describe several methods for imaging blood vessels in living animals (microangiography, time-lapse imaging of transgenic zebrafish with fluorescently tagged blood vessels). Collectively, these methods provide an unprecedented capability to image vessels in developing and adult animals.

1. IMAGING VASCULAR GENE EXPRESSION

Experimental analysis of blood or lymphatic vessel formation during development requires the use of methods for visualizing the expression of particular genes within vessels and their progenitors. There are two general methods available to visualize endogenous gene expression within zebrafish embryos and larvae, in situ hybridization and immunohistochemistry. Neither of the methods is specific to the vasculature, and detailed protocols for these methods are available elsewhere (Hauptmann & Gerster, 1994; Westerfield, 2000). In situ hybridization is used routinely to assay the spatial and temporal patterns of vascular genes. A large number of published reports have described the expression of many different genes in zebrafish blood vessels and/or lymphatic vessels. We list just a few of the most commonly used probes for in situ hybridization in Table 1. The *fli1a* and *scl* genes are early markers of vascular and hematopoietic lateral mesoderm. The expression of the *fli1a* becomes restricted to endothelial cells, a subset of circulating myeloid cells, and cranial neural crest derivatives (Brown et al., 2000; Thompson et al., 1998), while *scl* expression becomes restricted to the hematopoietic lineage at later stages (Gering, Rodaway, Gottgens, Patient, & Green, 1998). The expression of *cmyb* and *runx1* marks definitive hematopoietic cells in the ventral wall of the zebrafish dorsa aorta (Burns et al., 2002; Kalev-Zylinska et al., 2002; Thompson et al., 1998). *Vessel-specific gene 1* (*vsg1*) is a novel gene whose expression is specific to the dorsal aorta, choroidal vascular plexus, and intersegmental vessels by 36 hpf (Qian et al., 2005). The *tie2* and *cdh5* genes are zebrafish orthologs of angiopoetin-1 receptor and VE-cadherin, respectively, and are expressed in a vascular-specific manner (Lyons, Bell, Stainier, & Peters, 1998; Sumanas, Jorniak, & Lin, 2005). The vascular-specific expression of the *egfl7* gene is conserved across human, mouse, and zebrafish, and loss of its function blocks vascular tubulogenesis (Parker et al., 2004). The *etv2/etsrp* gene encodes a member of the ETS family of transcription factors that is in the upstream hierarchy of angioblast development from the mesoderm and is known to drive the expression of *flia* and *scl* (Pham et al., 2007; Ren, Gomez, Zhang, & Lin, 2010). The *sox7* and *sox18* transcripts are first detected at bud stage and are located in bilateral stripes in the posterior lateral plate mesoderm. By 24 hpf, *sox7* and *sox18* are expressed in vascular endothelium of the head, trunk, and tail, while *sox7* is enriched in arteries in later stages (Cermenati et al., 2008; Herpers, van de Kamp, Duckers, & Schulte-Merker, 2008; Pendeville et al., 2008). Novel vascular-specific markers including *tll1*, *dusp2*, *admr*, and *crl* have been identified from *cloche* and *mfn* mutants (Clements et al., 2011; Connors, Trout, Ekker, & Mullins, 1999; Qian et al., 2005; Sumanas et al., 2005). In addition, a recent study using microarray analysis together with whole-mount in situ hybridization identified *she*, *yrk*, *aqp8*, *mrc1*, and *stab2* as vasculature markers from *etv2/etsrp* overexpressed zebrafish embryos (Wong, Proulx, Rost, & Sumanas, 2009). The specific role of these novel markers in vascular endothelium needs further investigation. The *kdrl* and *flt4* genes (Fouquet, Weinstein, Serluca, & Fishman, 1997; Sumoy, Keasey, Dittman, & Kimelman, 1997; Thompson et al., 1998) are zebrafish orthologs of mammalian endothelial-specific tyrosine kinase receptors for the important vascular signaling molecule *vascular endothelial growth factor* (*vegf*). They are initially expressed in hemangiogenic lateral mesoderm and then become restricted to angioblasts and endothelium. In the axial vessels of the trunk (dorsal aorta and posterior cardinal vein), *kdrl* becomes preferentially expressed in the aorta while *flt4* becomes preferentially expressed in the posterior cardinal vein (similar expression patterns of the corresponding orthologs are observed in mouse). Other genes such as *efnb2*, *grl*, *dll4*, *tbx20*, and *notch5* are useful as markers of specification of arterial rather than venous endothelium, although all of these markers also exhibit substantial expression in nonvascular tissues, particularly the nervous system (Herpers et al., 2008; Lawson et al., 2001; Szeto, Griffin, & Kimelman, 2002; Zhong, Rosenberg, Mohideen, Weinstein, & Fishman, 2000).

Table 1 Common Marker Genes Used in Zebrafish Vasculature Research

Marker Genes	Expression Pattern	References
fli1a	Pan-endothelial	Brown et al. (2000) and Thompson et al. (1998)
tie2	Pan-endothelial	Lyons et al. (1998)
cdh5	Pan-endothelial	Sumanas et al. (2005)
egfl7	Pan-endothelial	Parker et al. (2004)
etv2 (etsrp)	Pan-endothelial	Pham et al. (2007)
sox18	Pan-endothelial	Cermenati et al. (2008) and Pendeville et al. (2008)
tll1	Pan-endothelial	Clements et al. (2011) and Connors et al. (1999)
she	Pan-endothelial	Wong et al. (2009)
yrk	Pan-endothelial	Wong et al. (2009)
dusp2	Pan-endothelial	Qian et al. (2005)
admr	Pan-endothelial	Sumanas et al. (2005)
crl	Pan-endothelial	Sumanas et al. (2005)
kdrl (flk1)	Initially pan-endothelial, enriched in arteries at later stages	Sumoy et al. (1997) and Bussmann, Lawson, Zon, and Schulte-Merker (2008)
sox7	Initially pan-endothelial, enriched in arteries at later stages	Pendeville et al. (2008) and Cermenati et al. (2008)
aqp8	Artery only	Wong et al. (2009)
efnb2	Artery only	Lawson et al. (2001) and Zhong et al. (2000)
grl	Artery only	Zhong et al. (2000)
notch5	Artery only	Lawson et al. (2001)
dll4	Artery only	Herpers et al. (2008)
dlc	Artery only	Lawson et al. (2001)
tbx20	Artery only	Szeto et al. (2002)
mrc1	Vein only	Wong et al. (2009)
dab2	Vein only	Herpers et al. (2008)
ephb4	Vein only	Lawson et al. (2001)
flt4	Initially pan-endothelial. Later restricted to vein only.	Thompson et al. (1998)
stab2	Initially pan-endothelial, enriched in vein at later stages	Wong et al. (2009)
prox1	Lymphatic vessel	Yaniv et al. (2006)
lyve1	Lymphatic vessel	Flores et al. (2010)
nrp	Lymphatic vessel Artery and vein	Yaniv et al. (2006) and Martyn and Schulte-Merker (2004)
vsg1	Hematopoietic and endothelial	Qian et al. (2005)
cmyb	Hematopoietic	Thompson et al. (1998)
runx1	Hematopoietic	Burns et al. (2002) and Kalev-Zylinska et al. (2002)
scl	Hematopoietic	Gering et al. (1998)

Disabled-2 (Dab2), a cytosolic adapter—regulating endocytosis, localizes very specifically to early venous but not arterial endothelium in both *Xenopus* and *Zebrafish* embryos (Cheong, Choi, & Han, 2006; Herpers et al., 2008). The *ephb4* gene is a commonly used venous marker, and *flt4* has also been used to identify venous endothelium as noted earlier. The lymphatic endothelial genes *prox1* and *lyve1* are both conserved between mammals and fish, and both are useful zebrafish markers of lymphatic specification (Flores, Hall, Crosier, & Crosier, 2010; Yaniv et al., 2006). Prox1 functions as a key transcriptional factor for the differentiation of lymphatic endothelial cells. Lyve1 has been identified as a major receptor for HA (extracellular matrix glycosaminoglycan hyaluronan) on the lymph vessel wall, although its function in lymphangiogenesis is unclear. There are four zebrafish *nrp* genes (*nrp1a, nrp1b, nrp2a,* and *nrp2b*) that display distinct expression patterns in dorsal aorta and posterior cardinal vein during early development (Martyn & Schulte-Merker, 2004). In particular, *nrp2a* is expressed in the central nervous system and venous endothelium in early stage and also detected in the lymphatic vessels at 5 dpf (Yaniv et al., 2006).

2. NONVITAL BLOOD VESSEL AND LYMPHATIC VESSEL IMAGING

A number of methods are available for visualizing the pattern of blood or lymphatic vessels in fixed specimens. Microdye injection and microresin injection can be used to delineate the patent vasculature (lumenized or open blood vessels connected to the systemic circulation). Both of these methods rely on injection to fill blood vessels with dye or plastic resin that can be visualized in detail following the procedure. Dye injection methods are most useful in embryos and larvae up to a few weeks old. At later juvenile stages and in adults, tissue opacity and thickness interfere with dye visualization in deeper vessels, and resin injections can be more useful. Resin injections are difficult to perform on small specimens (such as embryos) but could be used to visualize vessels at almost any stage of development. While technically challenging, resin injection provides excellent visualization of the adult vasculature, since tissues surrounding the plastic resin are digested away and do not interfere with vessel observation. In addition to these two injection methods for lumenized vessels, staining for the endogenous AP activity of vascular endothelium can also be used to visualize vessels in fixed specimens. This method is useful for easy, rapid observation of vessel patterns and does not require the vessel be patent, but it cannot be used effectively prior to approximately 3 days postfertilization due to low signal and high background staining. Even at 3 dpf the method gives a relatively high background and is not particularly useful for visualizing cranial vessels. This method is also less useful at later stages due to increasing background. We describe the procedures for all of these methods in detail later.

2.1 MICRODYE AND MICRORESIN INJECTION

Since the 19th century, the dye injection method has been the most widely used tool for visualizing the developing circulatory system. Pioneering vascular embryologists such as Florence Sabin carried out their groundbreaking descriptive studies by injecting India ink into blood vessels of vertebrate embryos to reveal their patterns (for example Evans, 1910; Sabin, 1917). In the 1970s the corrosive resin casting method, previously employed to visualize larger adult blood vessels, was combined with scanning electron microscopy to permit its use for visualizing vessels on a microscopic scale, such as in the developing renal vasculature (Murakami, 1972). Although microangiography and vascular-specific transgenic fish have now become the tools of choice in most cases for visualizing vessels in living zebrafish embryos (see later discussion), these newer methods have limited usefulness in later stage larvae, juveniles, and adult fish. At these later stages the "classical" dye or resin injection methods still provide the best visualization of the majority of blood vessels (via direct injection into the dorsal aorta or caudal artery) or lymphatic vessels (via direct injection into the thoracic duct) (Yaniv et al., 2006). The resin casting method involves injection of a plastic resin that is allowed to harden in situ, followed by etching away of tissues to leave behind only the plastic cast. The cast is rotary shadowed and visualized by scanning electron microscopy. The dye injection method described later involves injection of Berlin Blue dye followed by fixation and clearing of the embryos or larvae and whole-mount microscopic visualization.

2.1.1 Resin injection method
2.1.1.1 Materials

- Paraffin bed (see Fig. 1)
- Injection apparatus for circulating saline buffer ($\times 2$; see Fig. 3)
- Injection apparatus for fixative
- Injection apparatus for resin injection (one apparatus per sample to be injected)
- Physiological saline buffer suitable for bony fish
- 2% glutaraldehyde solution in saline buffer (Sigma cat# G6403, 50% solution in water)
- Methacrylate resin components
  - Methyl methacrylate monomer (Aldrich cat# M55909 or Fluka cat# 03989)
  - Ethyl methacrylate monomer (Aldrich cat# 234893 or Fluka cat# 65852)
  - 2-Hydroxypropy methacrylate monomer (Aldrich cat# 268542 or Fluka cat# 17351)

2.1.1.2 Protocol
2.1.1.2.1 Preparation of the apparatus

1. The *paraffin bed* is made in a 9-cm glass petri dish by pouring molten paraffin wax (Fig. 1). While the wax is solidifying, tilt the dish approximately 15 degrees to create a gentle slope. A depression is made in the middle of the bed for settling a fish.

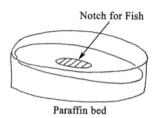

FIGURE 1

Paraffin bed used for holding adult zebrafish.

2. The *glass needles* are made from stock glass tubing (3 mm outside diameter). The tubes are cut into 10–12 cm length (Fig. 2). The needles are pulled from the tube by heating the middle of the tube with a Bunsen burner. When the color of the glass tube is changed to red and the tubes feel soft, remove the tube from the heat and pull on both ends. This should produce two injection needles with length of 5–6 cm. Let the needles cool down. Then holding thick end of the needle by hand and the sharp end of the needle by a pair of forceps, reheat the sharp end on a Bunsen burner, and pull as before. By pulling the needles twice,

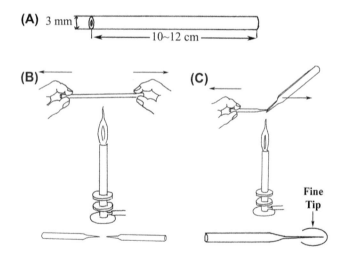

FIGURE 2

Preparation of glass needles for injection of adult zebrafish. Glass stock (A) is pulled on a Bunsen burner (B), the tips are repulled (C), and then the needles are bonded to vinyl tubing with super glue (D).

it is possible to create needles with very fine points. The tip of the needles made in this manner is closed, and the tip needs to be broken open just before use.

3. The *apparatus for injecting physiological saline buffer* is made by attaching a glass needle (see step 2) to a clear vinyl tubing (3 mm inside diameter, 20 cm in length). When plugging the needle in, a small amount of superglue is applied to reinforce the attachment. For injection of the buffer a 2- to 10-mL glass syringe is attached as shown in Fig. 3.
4. The *apparatus for injecting fixative* is prepared as described in step 3.
5. The *apparatus for injecting resin* is prepared as described in step 3, except that a 10-mL disposable syringe is attached instead of a glass syringe and heat-resistant silicone tubing is used.

Important Caution on Resin Use

As resin polymerizes, heat is generated and the viscosity of the resin increases. The heated vinyl tube can detach from the syringe suddenly, causing pressurized, viscous hot resin to splatter. The resin is harmful to skin and mucous membranes, and personal safety measures should always be taken when performing this procedure (eg, use of goggles, face masks, gloves, and other protective clothing). In addition, make sure that the end of the vinyl tubing is securely attached to the syringe and use heat-resistant silicone tubing for the resin injection apparatus.

6. *Preparing resin (methyl methacrylate and ethyl methacrylate monomer)*. Commercially available methacrylate monomers contain monomethyl ether hydroxyquinone to prevent polymerization, and it is necessary to remove this before use. Prepare 500 mL of 5% NaOH. Pour 100 mL methacrylate monomer and 50 mL 5% NaOH in a separating funnel and shake (Fig. 4A). Wait until the two solutions separate, and then remove the lower 5% NaOH layer (should be brown in color). Repeat until the NaOH solution remains clear, and then remove the NaOH by extracting the methacrylate monomer (upper layer) with distilled water. Pour 100 mL of distilled water in the separating funnel containing the methacrylate monomer, shake, and wait until the two layers separate. Remove the lower distilled water layer. Repeat three to four times. Filter methacrylate monomer using double filter paper, and incubate in a 150 mL air-tight container with sodium sulfite overnight (Fig. 4B). Place this container within a desiccator containing silica-gel, and store in a refrigerator at 4°C.

2.1.1.2.2 Experimental procedure

Steps 1, 2, 4 are to be carried out under a dissecting microscope.

FIGURE 3

Apparatus for injecting adult zebrafish with saline and fixative. For injecting resin, the glass syringe is replaced with a disposable plastic syringe.

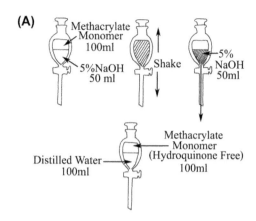

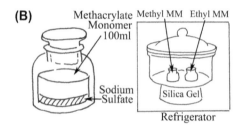

FIGURE 4

Preparation of resin for injection. Commercial resin is supplied with monomethyl ether hydroxyquinone to prevent polymerization. This must be extracted before use (A) as described in the text. After extraction, resin is stored refrigerated over sodium sulfate in a desiccator (B).

1. *Washing the circulatory system with saline buffer.* Place anesthetized adult zebrafish on the depressed part of the paraffin bed ventral side up. Use a pair of watchmaker's forceps and a pair of fine surgical or iridectomy scissors to remove the outer skin and pericardial sac surrounding the heart. Use the forceps to sever the sinus venosus to allow blood to drain. Break the tip of the needle to the size of the ventricle, and attach to a glass syringe containing saline buffer. Stab the glass needle into the ventricle in the direction of the head, and apply pressure on the syringe to flush the circulatory system with buffer (Fig. 5A). Do not stop until the system is very well flushed out; flushing should be continued well after the flow from the sinus venosus has become clear saline.
2. *Fixation with 2% glutaraldehyde.* Break a glass needle as given in step 1, attach to a syringe containing the 2% glutaraldehyde solution, and start circulating the fixative in the same manner as for the saline buffer (Fig. 5B). Fix well.
3. *Mixing the resin.* Mix together 3 mL methyl methacrylate monomer, 1.75 mL ethyl methacrylate monomer, and 5.25 mL 2-hydroxypropyl methacrylate monomer (to make 10 mL final volume) in a disposable plastic 100 mL cup

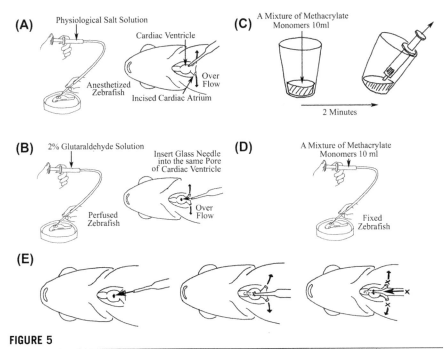

FIGURE 5
Resin injection of adult zebrafish. Anesthetized animal is thoroughly flushed with physiological salt solution (A) and then with glutaraldehyde fixative (B). Mixed resin is taken up into plastic syringe (C) which is attached to the rest of the resin injection apparatus and injection is performed (D). Injection should be stopped when resin hardens and flow ceases (E). See text for details.

(Fig. 5C). To this mixture, add 0.15 g benzoyl peroxide (catalyst), 0.15 mL N,N-dimethylaniline (polymerization agent), and Sudan III (dye), and then mix well and sonicate for 2 min.

4. *Resin injection.* Immediately after sonication, suck the resin mixture up in a 10-mL disposable plastic syringe and attach a glass needle. This time, break the needle so that it will have a slightly larger bore. Make sure the vinyl tubing is attached to the end of the syringe firmly. Push the needle in the direction of the head of the fish through the same hole used for washing and fixing (Fig. 5D). If possible, push the tip of the needle all the way into the arterial cone (back of the ventricle) before beginning the injection. At first the injection will be easy since the viscosity is low, but after 3–5 min, the viscosity will increase and the resin will start to heat up. Keep pushing down on the syringe. After 10–12 min, the resin hardens sufficiently for resin flow to stop, at which point the point the injection should be stopped (Fig. 5E). Wait about 10 min more for resin to fully harden.

5. *Digestion of tissue.* The injected adult zebrafish is digested with 10–20% KOH for a few days and gently washed with distilled water to remove tissue. The resin

cast is dissected out using watchmaker's forceps and small scissors. Sometimes sonication is used to remove bones and hard-to-remove tissues, but care is required not to destroy the cast itself. For SEM observation, each local vascular system is divided and trimmed. These procedures must be performed under water using dissecting microscope. Each block is frozen in distilled water then freeze-dried.

6. *Scanning electron microscopy.* The dried block is mounted on a metal stub and coated with osmium or platinum. Observations are performed with a scanning electron microscope using an acceleration voltage of 5–10 kV.

2.1.2 Dye injection method
2.1.2.1 Materials
- Berlin blue dye solution
- Injection apparatus for embryos and early larvae (as described for microangiography, later) *or*
- Injection apparatus for juvenile and adult fish (as described for resin injection, earlier)
- Paraffin bed (for juvenile and adult injections—see resin injection earlier)

2.1.2.2 Protocol
2.1.2.2.1 Dye injection of embryos and early larvae

1. Prepare Berlin blue dye solution by adding 0.5–0.75 g Berlin blue powder (Aldrich cat# 234125, Prussian blue) to 100 mL distilled water and dissolving thoroughly. Filter solution through double layers of Whatman 3 mm filter paper and store in an air-tight bottle.
2. Glass microneedles are prepared as for microangiography (see later discussion).
3. Agarose embeds embryos/larvae as follows (Fig. 6A): Immobilize embryo in 1× tricaine in embryo media (Westerfield, 2000). Place embryo on slide in a drop of tricaine embryo media, and then remove as much of liquid as possible with a pipette. Place a single drop of 1% molten low melting temperature agarose on the embryo, allowing it to harden and embed the embryo. Attempt to orient the embryo before the agarose hardens such that either its left or right side is facing up.
4. To allow blood drain, sever the sinus venosus using the watchmaker's forceps.
5. Remove the agarose covering the caudal half of the trunk or cranial half of the tail with the forceps, and then add a drop of 1× tricaine.
6. Break the fine tip of needle to the size of the dorsal aorta or the caudal artery, and, from the needle tip, suck enough Berlin blue solution to cannulate all the vessels thoroughly (Fig. 6B).
7. To pierce precisely the dorsal aorta or the caudal artery with the fine tip, the point just beneath the notochord must be targeted. To make sure the tip is in the correct position of these vessels, inject the dye for 0.1 s using a picopump. The blood cells move according the pumping if the tip is in right position. Inject the

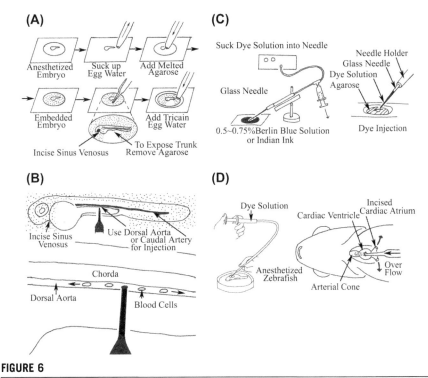

FIGURE 6

Mounting and dye injection of developing and adult zebrafish. Embryos and larvae are embedded in agarose on a glass slide for dye injection (A). Injection is performed into the dorsal aorta or caudal artery, with the sinus venosus incised to permit dye to flow through the vasculature (B). The apparatus used for dye injection is shown in (C). For dye injection of adult zebrafish, apparatus similar to that used for saline and fixative injection prior to resin injection is employed (D). See text for further details.

dye for 1–2 s, and continue the procedure until all the vessels are cannulated thoroughly (Fig. 6C). For lymphangiography, instead of the dorsal aorta, attempt to pierce the thoracic duct, which lies ventral to the dorsal aorta above the posterior cardinal vein. This vessel is more difficult to visualize and more challenging to inject than the dorsal aorta, so repeated attempts and patience will likely be needed. One can distinguish when the lymphatic rather then blood vascular system has been injected because the dye will fill the thoracic duct but will not be visible or highly diluted in blood vessels such as the cardinal vein and dorsal aorta.

8. Fix the dye injected embryos.

2.1.2.2.2 Dye injection of juvenile and adult zebrafish
Note: This method is similar to the resin injection method described earlier.

1. The injection of dye is carried out under dissecting microscope. Place anesthetized adult zebrafish on the depressed part of the paraffin bed ventral side up.
2. Use a pair of watchmaker's forceps and a pair of fine surgical or iridectomy scissors to remove the outer skin and pericardial sac surrounding the heart. Use the forceps to sever the sinus venosus to allow blood to drain during injection (Fig. 6D).
3. Break the tip of the needle to the size of the ventricle, and attach to a glass syringe containing 0.5–0.7% Berlin blue solution buffer (apparatus is very similar to that used for saline injection in the resin injection method described earlier). Stab the glass needle into the ventricle in the direction of the head, and apply pressure on the syringe to flush the circulatory system with buffer. Continue injection well after dye begins to flow from sinus venosus; make sure that the dye is thoroughly injected.
4. Fix the dye-injected sample immediately in either 4% paraformaldehyde or 10% neutral formaldehyde; store in fixative.
5. For observation, lightly wash the samples and then clear them by passing through 50%, 70%, 80%, 90%, and then 100% glycerol solution (1 solution change per day for a total of 5 days). Image samples under a dissecting microscope, dissecting away tissues as needed for observation of deeper vessels.

2.2 ALKALINE PHOSPHATASE STAINING FOR 3 DPF EMBRYOS

Zebrafish blood vessels possess endogenous AP activity. Endogenous AP activity is not detectable in 24 hpf embryos but is weakly detectable by 48 hpf and strong at 72 hpf. Staining vessels by endogenous AP activity is useful for easy and rapid visualization of the vasculature in many specimens but provides less resolution than many of the other methods. We use protocol modified from Childs, Chen, Garrity, and Fishman (2002).

2.2.1 Materials

- Fixation buffer: 10 mL 4% paraformaldehyde + 1 mL 10% Triton-X100; makes 11 mL, scale up or down as needed.
- Rinse buffer: 10 mL 10× PBS + 5 mL 10% Triton-X100 + 1 mL normal horse serum + 84 mL distilled water; makes 100 mL, scale up or down as needed.
- Staining buffer: 1 mL 5M NaCl + 2.5 mL 1M $MgCl_2$ + 5 mL 1M Tris pH 9.0–9.5 + 500 µL 10% Tween + 41 mL distilled water; makes 50 mL, scale up or down as needed.
- Staining solution: 10 mL staining buffer + 45 µL NBT + 35 µL BCIP; scale up or down as needed
- NBT: 4-Nitro Blue Tetrazolium (Boehringer-Mannheim Cat# 1-383-213), 100 mg/mL in 70% dimethylformamide.
- BCIP: X-Phosphate or 5-Bromo-4-Chloro-3-indolyl-phosphate (Boehringer-Mannheim Cat# 1-383-221), 50 mg/mL in dimethylformamide.

2.2.2 Protocol
1. Fix at room temperature (RT) for 1 h in fixation buffer.
2. Rinse 1× in rinse buffer.
3. Wash 5× 10 min at RT in rinse buffer **or** leave washing in rinse buffer at 4°C for up to several days. If doing the latter, wash again at RT for 10 min before going on to the next step.
4. Wash 2× 5 min in staining buffer.
5. Stain in 1 mL of staining solution. Color development takes about 5–30 min.
6. To stop reaction, wash 3× in rinse buffer without horse serum, then fix in 4% paraformaldehyde for 30 min, and store in fixative at 4°C.

2.2.3 Important notes
1. Avoid putting the embryos in methanol (this destroys endogenous AP activity). If embryos have been placed in methanol, some AP activity can be reconstituted by washing embryos in PBT overnight or even over a weekend before starting the staining procedure, although staining will be weaker than in nonmethanol-exposed embryos.
2. If combined AP staining and antibody staining [eg, anti-EGFP (enhanced green fluorescent protein) staining] is desired, antibody staining should be done first. After the DAB staining, do a quick postfix and go straight into the washes for the AP protocol. AP staining works very well after an antibody stain. Alternatively, stain for exogenous AP at a time point when the endogenous vascular form is not active (24 hpf).

3. VITAL IMAGING OF BLOOD AND LYMPHATIC VESSELS

While the methods described earlier are useful for visualizing vascular patterns in fixed zebrafish specimens, particularly at later developmental stages and in adults, the zebrafish is perhaps best known for its accessibility to vital imaging methods. A number of vascular imaging methods are available that take advantage of the optical clarity and experimental accessibility of zebrafish embryos and larvae. Confocal microangiography can be used for imaging blood vessels with active circulation, or patent lymphatic vessels, and to detect defects in their patterning and/or function. Confocal microangiography is performed by injecting fluorescent microspheres or quantum dots (QDs) into the blood vessels or lymphatic vessels of living embryos, and then collecting 3-dimensional image "stacks" of the fluorescently labeled vasculature in the living animal using a confocal or (preferably) multiphoton microscope. This method can be used from the initiation of circulation at approximately 1 dpf out to 10 dpf or even older larvae, although injections become progressively more technically challenging to perform after about 2 dpf. Increasing tissue depth makes high-resolution imaging of deep vessels increasingly difficult at later stages. Repeated microangiographic imaging of the same animal and microangiography on animals with impaired circulation may be difficult or impossible to perform.

Numerous transgenic zebrafish lines have been generated with endogenous fluorescent labeling of blood and/or lymphatic vessels. We list a few of these lines in Table 2. These lines facilitate high-resolution imaging of the vasculature in vivo and make possible long-term time-lapse imaging of the dynamics of vessel growth and remodeling. Unlike dye or resin injection methods, transgenic animals can be repeatedly reimaged over an extended period of time with continued normal development of the imaged vessels, particularly when multiphoton imaging is employed.

Zebrafish *fli1a*:*EGFP* (Lawson & Weinstein, 2002), *kdrl*:*EGFP* (Cross, Cook, Lin, Chen, & Rubinstein, 2003), and other transgenic zebrafish lines have already been widely used in studies of vasculogenesis and angiogenesis in the zebrafish. Since these lines permit imaging of the endothelial cells themselves, rather than vessel lumens, they can be used to image vessels that are not carrying circulation, cords of endothelial cells lacking a vascular lumen, or even isolated migrating angioblasts. The *fli1a*:*EGFP* line permits visualization of both blood and lymphatic vessels, while only blood endothelial cells are marked in the *kdrl*:*EGFP* line. The *fli1a*:*EGFP* line has already been used in a very large variety of studies, among other things to examine mechanisms of cranial (Lawson & Weinstein, 2002) and trunk (Isogai, Lawson, Torrealday, Horiguchi, & Weinstein, 2003) blood vessel formation, lymphatic vessel formation (Yaniv et al., 2006), and tumorigenesis (Stoletov, Montel, Lester, Gonias, & Klemke, 2007). The *fli1a* and *kdrl* promoters have also both been used to generate additional lines expressing other fluorescent proteins. Transgenic lines expressing nuclear-targeted EGFP (nEGFF) or membrane-targeted EGFP (EGFP-cdc42wt) have been derived. Multiphoton time-lapse imaging using $Tg(fli1a:nEGFP)^{y7}$ transgenic embryos has been used to trace the migration and lineage of individual endothelial cells and quantify their proliferation (Yaniv et al., 2006). The $Tg(fli1a:EGFP\text{-}cdc42wt)^{y48}$ line has been used to visualize the dynamics of endothelial vacuoles and their contribution to vascular lumen formation (Kamei et al., 2006). By combining microangiographic imaging of functional vascular lumens using red QDs and imaging of forming lumenal compartments using the $Tg(fli1a:EGFP\text{-}cdc42wt)^{y48}$ and $Tg(kdrl:moesin1\text{-}EGFP)$, the details of progressive formation of vascular lumens via formation and intracellular/intercellular fusion of endothelial vacuoles, and their connection to the rest of the vascular circulation, could be followed dynamically in vivo for the first time (Kamei et al., 2006; Wang et al., 2010). More recently, endothelial-specific lines and reporter constructs have been generated that permit visualization of endothelial membranes and endothelial junctions, even facilitating the identification and imaging of single endothelial cells in living animals (Lenard et al., 2015; Yu, Castranova, Pham, & Weinstein, 2015).

Many of the original transgenic reporter lines used EGFP as fluorescent reporter of choice, but more recently transgenic lines have been established employing a variety of different fluorescent proteins expressed under the control of a number of

Table 2 Zebrafish Transgenic Lines for Time-lapse Vascular Imaging

Transgenic Zebrafish Line	References
Vascular and pan-vascular	
Tg(fli1a:EGFP)[y1]	Lawson and Weinstein (2002)
Tg(fli1a:nEGFP)[y7]	Yaniv et al. (2006)
Tg(fli1a: DsRed)	Geudens et al. (2010)
Tg(fli1a:EGFP; kdrl:ras-cherry)	Hogan et al. (2009)
Tg(kdrl:G-RCFP)	Cross et al. (2003)
Tg(kdrl:memCherry)[s896]	Chi et al. (2008)
Tg(kdrl:EGFP)[s843]	Jin, Beis, Mitchell, Chen, and Stainier (2005)
Tg(flt1:YFP, kdrl:mCherryRed)	Hogan et al. (2009)
Tg(flt1_9a_cFos:GFP)[wz2]	Nicenboim et al. (2015)
Tg(flt4BAC:mCitrine)[hu7135]	van Impel et al. (2014)
Tg(etv2:EGFP)	Veldman and Lin (2012)
Tg(etv2:mCherry)	Veldman and Lin (2012)
Lymphatic and endothelial	
Tg(stabilin:YFP)[hu4453]	Hogan et al. (2009)
Tg(lyve1:egfp)[nz15]	Okuda et al. (2012)
Tg(prox1aBAC:KalTA4-4xUAS-E1b:uncTagRFP)[nim5]	van Impel et al. (2014)
Endothelial cell compartments	
Tg(fli1a:EGFP-cdc42wt)[y48]	Kamei et al. (2006)
Tg(kdrl:moesin1-EGFP)	Wang et al. (2010)
Tg(fli1:Lifeact-mCherry)	Wakayama, Fukuhara, Ando, Matsuda, and Mochizuki (2015)
Tg(kdrl:Lifeact-EGFP)[s982]	Vanhollebeke et al. (2015)
Tg(fli1a:egfp-claudin5b)[y287]	Yu et al. (2015)
Tg(UAS:EGFP-ZO1-cmlc:GFP)[ubs5-7]	Herwig et al. (2011)
Tg(UAS:VE-cadherinΔC-EGFP)[ubs12]	Lenard et al. (2013)
Vascular support	
TgBAC(pdgfrb:citrine)[s1010]	Vanhollebeke et al. (2015)
Tg(l-fabp:DBP-EGFP)	Xie, Farage, Sugimoto, and Anand-Apte (2010)
Endothelial driver	
Tg(fli1ep:GAL4FF)[ubs4]	Herwig et al. (2011)
TgBAC(cdh5:GAL4FF)[mu101]	Bussmann, Wolfe, and Siekmann (2011)
Tg(kdrl:CRE)	Bertrand et al. (2010)
Vascular lineage tracing and signaling reporters	
Tg(kdrl:Kaede)[wz3]	Nicenboim et al. (2015)
Tg(Tp1bglob:eGFP)[um14]	Parsons et al. (2009)
Tg(fli1:Gal4db-2ΔC-2A-mC)	Kashiwada et al. (2015)

Continued

Table 2 Zebrafish Transgenic Lines for Time-lapse Vascular Imaging—cont'd

Transgenic Zebrafish Line	References
Hematopoietic cells	
Tg(gata1:DsRed)^{sd2}	Traver et al. (2003)
Tg(cmyb:GFP; kdrl:memCherry)	Bertrand et al. (2010)
Tg(Runx:GFP; kdrl:RFP)	Tamplin et al. (2015)

different endothelial promoters (Table 2). Taking advantage of differences in the preferential labeling of different vascular promoters, double transgenic animals were generated for the specific observation on blood/lymph vessels, like *Tg(fli1a:EGFP; kdrl:ras-cherry)*, or artery/venous sprouts *Tg(flt1:YFP; kdrl:mCherryRed)* (Hogan et al., 2009). Lymphatic-specific transgenic lines like *Tg(lyve1:egfp)^{nz15}* have become useful for imaging lymphatics at later stages of development (Okuda et al., 2012). In addition, newly developed endothelial-specific GAL4 and CRE driver lines such as *Tg(fli1ep:GAL4FF)^{ubs4}* and *Tg(kdrl:CRE)* have greatly increased the flexibility of transgenic endothelial-specific expression in combination with a very large number of available UAS and lox reporter lines (Bertrand et al., 2010; Helker et al., 2013; Herwig et al., 2011).

In addition, double transgenic lines with both endothelial and hematopoietic promoter-driven transgenic reporter expression such as *Tg(cmyb:GFP; kdrl:memCherry)* and *Tg(Runx:GFP; kdrl:RFP)* have proven useful for studying the endothelial to hematopoietic transition and the endothelial niche for hematopoietic stem cells (Bertrand et al., 2010; Tamplin et al., 2015). Motion-based imaging of blood flow has been used to characterize of lumen formation, distinguish lymphatic vessels from blood vessels, and examine which vessels are patent and carrying blood flow in *Tg(fli1a:EGFP; gata1:DsRed)* double transgenic animals (Tong et al., 2009; Yaniv et al., 2006).

Methods for confocal microangiography and time-lapse multiphoton imaging of transgenic animals are described in more detail later. Finally, we describe some of the novel insights into in vivo vessel formation processes that have already been obtained through use of time-lapse imaging methods.

3.1 MICROANGIOGRAPHY

Confocal microangiography is useful for visualizing and assessing patent vessels that are actually carrying flow or that at least have open lumens connected to the functioning blood or lymphatic vasculature. The method facilitates detailed study of both the pattern, function, and integrity (leakiness) of vessels. The method is relatively easy to perform, particularly on younger animals, and does not require that the animal be of any particular genotype (although animals with impaired circulation may be difficult or impossible to infuse with fluorescent microspheres).

3.1.1 Materials

- 0.02–0.04 µm fluoresceinated carboxylated latex beads, available from Invitrogen. The yellow-green (cat # F8787), red-orange (cat # F8794), or dark red (cat # F8783) beads are suitable for confocal imaging using the laser lines on standard Krypton–Argon laser confocal microscopes. Other colors may be used for when multiphoton imaging is employed. QDs are fluorescent semiconductors that are especially suited for multiphoton imaging with their high quantum yield, as well as the fact that the same excitation wavelength could be used for obtaining multiple different colors depending on the QD used. For microangiography, PEG-coated nontargeted QDs are available from Invitrogen. Qtracker 565, 655, 705, and 800 (cat# Q21031MP, Q21021MP, Q21061MP, and Q21071MP, respectively) are suitable for multiphoton confocal imaging. Fluorescent bead suspension as supplied is diluted 1:1 with 2% BSA (Sigma) in deionized distilled water, sonicated approximately 25 cycles of 1″ each at maximum power on a Branson sonifier equipped with a microprobe, and subjected to centrifugation for 2 min at top speed in an Eppendorf microcentrifuge. The QDs are used as supplied.
- 1 mm OD glass capillaries (World Precision Instruments, Cat # TW100-4 without filament or TW100F-4 with an internal filament) for preparing holding and microinjection pipettes
- Two coarse micromanipulators with magnetic holders and base plates
- 30% Danieu's solution (1× Danieu's: 58 mM NaCl, 0.7 mM KCl, 0.4 mM $MgSO_4$, 0.6 mM $Ca(NO_3)_2$, 5 mM Hepes, pH 7.6)
- Holding and microinjection pipettes
- 6-cm culture dish (Falcon)
- Micromanipulator and microinjection apparatus
- Dissecting microscope equipped with epifluorescence optics

3.1.2 Protocol
3.1.2.1 Preparation of the apparatus

1. The *glass microinjection needles* are prepared from 1-mm capillaries with internal filaments using a Kopf vertical pipette puller (approximate settings: heat = 12, solenoid = 4.5; see Fig. 7A for desired shape of microneedle). Needles are broken open with a razor blade just behind their tip to give an opening of approximately 5–10 µm in width.
2. The *holding pipettes* are prepared from 1 mm capillaries without filaments by carefully partially melting one end of the capillary with a Bunsen burner, such that the opening is narrowed to approximately 0.2 mm (slightly smaller for younger embryos, slightly larger for older larvae). A photographic image of the end of the tip of a holding pipette is shown in Fig. 7B.
3. The *apparatus for microinjection* is made by attaching a glass microinjection needle (step 2) to a pipette holder (World Precision Inst Cat # MPH6912; adapter for holder and tubing to attach to picopump, WPI Cat # 5430). The

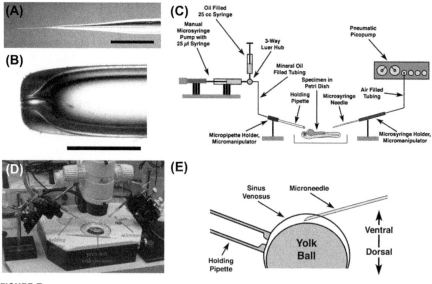

FIGURE 7

Microangiography of developing zebrafish embryos and larvae. The desired configurations for injection needles (A) and holding pipettes (B) are shown. A schematic diagram of the apparatus used is shown in (C), and a photographic image of an actual set-up is shown in (D). For injection, an embryo is held ventral side-up with suction applied through the holding pipette and injected obliquely through the sinus venosus (E). Older larvae are injected by direct intracardiac injection. See text for details. Scale bars are 3 mm (A) and 1 mm (B).

pipette holder is attached to a controlled air pressure station such as World Precision Instruments Pneumatic Picopump (catalog # PV820).

4. The *apparatus for holding embryos* is made by attaching a glass holding pipette (step 3) to a pipette holder (World Precision Inst Cat # MPH6912). The holding pipettes and their holders are attached via mineral-oil filled tubing (Stoelting Instruments, Clay-Adams cat # 427415) to a manual microsyringe pump (Stoelting Instruments cat # 51222, with 25 μL syringe).
5. Holding pipettes and microneedles and their associated holders and other equipment are arranged on either side of a stereo-dissecting microscope as diagrammed in Fig. 7C. A photographic image of a typical arrangement is shown in Fig. 7D.

3.1.2.2 Experimental procedure

1. Embryos are collected and incubated to the desired stage of development. Use of albino mutant lines or PTU treatment improves visualization of many vascular beds at later stages (see Westerfield (2000) for PTU treatment protocol).

2. A few microliters of fluorescent microsphere suspension are used to backfill a glass microneedle for injection. The tip should be broken off to the desired diameter just before use.
3. Embryos are dechorionated and anesthetized with tricaine in embryo media.
4. 1- to 3-day-old embryos and larvae are held ventral side up for injection using a holding pipette applied to the side of the yolk ball (Fig. 7E), with suction applied via a microsyringe driver. Care should be taken to not allow the holding pipette to rupture the yolk ball. 4- to 7-day-old larvae are held ventral side up for injection by embedding in 0.5% low melting temperature agarose.
5. For 1- to 3-day-old embryos, a broken glass microneedle is inserted obliquely into the sinus venosus (as diagrammed in Fig. 7E). For 4- to 7-day-old larvae, a broken glass microneedle is inserted through the pericardium directly into the ventricle.

    For lymphangiographic labeling of the lymphatic vasculature, injections are performed directly into the thoracic duct, located between the dorsal aorta and cardinal vein. Lymphangiographic injections are generally significantly more difficult than angiographic injections of comparable stage embryos.

    Labeling of the lymphatic vasculature in living animals can also be accomplished by subcutaneous or intramuscular injections after insertion of the broken glass microneedle through the tough outer periderm of the developing animal. The dyes or microspheres are preferentially taken up by and drained through the lymphatic vessels. Injections should be performed at a site relatively distant to (preferably caudal to) the area to be imaged and the animal monitored after injection to choose the optimal time for imaging. This is when the dye has filled the lymphatic vessels but has not yet diffused far from the site of injection through other tissues (which creates a high background labeling outside the lymphatics). Subcutaneous dye injection labeling of the lymphatics is most effective in older larvae (2 weeks+) in which the lymphatic vasculature is well developed. Many dyes can be used for subcutaneous injections including rhodamine dextran (Invitrogen D-7139) and Qtracker Vascular Labels (Q21061MP). Subcutaneous injections have the added benefit of allowing for the analysis of lymphatic function as well as lymphatic morphology and patterning.
6. Following microneedle insertion, many (20+) small boluses of bead suspension are delivered over the course of up to a minute. Smaller numbers of overly large boluses can cause temporary or permanent cardiac arrest. The epifluorescence attachment on the dissecting microscope can be used to monitor the success of the injection.
7. Embryos or larvae are allowed to recover from injection briefly (approximately 1 min) in tricaine-free embryo media and then rapidly mounted in 5% methyl cellulose (Sigma) or low-melt agarose (both made up in embryo media with tricaine). For short-term imaging (generally one stack of images), methylcellulose is applied to the bottom of a thick depression well slide. The rest of the well is carefully filled with 30% Danieu's solution containing 1× tricaine,

trying not to disturb the methylcellulose layer below. The injected zebrafish embryo is placed in the well, moved on top of the methylcellulose, and then gently pushed into the methylcellulose in the desired orientation to fully immobilize. Methylcellulose is only useful for short-term mounting because the embryo gradually sinks in the methylcellulose (which also loses viscosity by absorbing additional water over time). For longer term or repeated imaging, animals can be mounted in agarose, using methods such as that described later.
8. Injected, mounted animals are imaged on a confocal or multiphoton microscope using the appropriate laser lines/wavelengths. Although the fluorescent beads are initially distributed uniformly throughout the vasculature of the embryo, within minutes they began to be phagocytosed by and concentrate in selected cells lining the vessels (cf "tail reticular cells" in Westerfield (2000)). Because of this, specimens must be imaged as rapidly as possible, generally within 15 min after injection. Generally between 20 and 50 frame-averaged (5 frames) optical sections are collected with a spacing of 2–5 μm between sections, depending on the magnification (smaller spacing at higher magnifications). 3-D reconstructions can be generated using a variety of commercial packages (see later discussion).

3.2 IMAGING BLOOD AND LYMPHATIC VESSELS IN TRANSGENIC ZEBRAFISH

Confocal microangiography is a valuable tool for imaging developing blood vessels, but it has limitations. The method is well suited for delineating the luminal spaces of functional blood vessels, but those that lack circulation, vessels that have not yet formed open lumens, and isolated endothelial progenitor cells are essentially invisible. Much of the "action" of early blood vessel formation occurs prior to the initiation of circulation through the relevant vessels. The first, major axial vessels of the zebrafish trunk, the dorsal aorta and cardinal vein, coalesce as defined cords of cells at the trunk midline with distinct molecular arterial–venous identities many hours before they actually begin to carry circulation. Later-developing vessels generally form by sprouting and migration of strings of endothelial cells or even individual cells that are likewise undetectable by angiography until well after their initial growth has been completed. Furthermore, because of leakage of low molecular weight dyes, or pinocytic clearance of microspheres, injected animals can only be imaged for a short time (up to 1/2 h) after injection and repeated imaging requires reinjection of microspheres with different excitation and emission spectra. Thus, for most practical purposes, dynamic imaging of blood vessel growth using this method is not possible. What is needed is a specific and durable fluorescent "tag" for endothelial cells and their angioblast progenitors.

As already described, fluorescent proteins such as GFP (green fluorescent protein) have been used to mark a variety of tissues in transgenic zebrafish embryos and larvae. Methods for generating germ line transgenic zebrafish are now widely used, and their application and resulting lines have been thoroughly reviewed elsewhere (Lin, 2000; Udvadia & Linney, 2003). Tissue-specific expression of fluorescent (or other) proteins in germ line transgenic animals is achieved through the use of tissue-specific promoters, and a number of different promoters have been used to drive fluorescent protein expression in zebrafish vascular endothelium. Germ line transgenic zebrafish have also been generated expressing EGFP in the vasculature under the control of the zebrafish *fli1a* promoter (Lawson & Weinstein, 2002). The *fli1a* is a transcription factor expressed in the presumptive hemangioblast lineage and later restricted to vascular endothelium, cranial neural crest derivative, and a small subset of myeloid derivatives (Brown et al., 2000). These lines express abundant EGFP in the vasculature, faithfully recapitulating the expression pattern of the endogenous *fli1a* gene and permitting resolution of very fine cellular features of vascular endothelial cells in vivo. The *fli1a:EGFP* transgenic lines have become the most widely used resource for transgenic visualization of blood and lymphatic vessels and have already been used in a variety of different published studies to examining developing trunk and cranial vessels (Isogai et al., 2003; Lawson & Weinstein, 2002; Roman et al., 2002), regenerating vessels in the adult fin (Huang, Lawson, Weinstein, & Johnson, 2003), and most recently, exploring the development of lymphatic endothelium (Nicenboim et al., 2015). Transgenic zebrafish with fluorescently labeled blood vessels have also been generated by using the promoter for the *kdrl* receptor tyrosine kinase to drive EGFP expression in endothelium (Cross et al., 2003). In addition, several tools have been developed for imaging zebrafish lymphatic system. By using double transgenic lines such as *Tg(fli1a: EGFP; kdrl:ras-cherry)*, lymphatic vessels can be distinguished from blood vessels (Hogan et al., 2009). Most recently, a zebrafish transgenic line using *lyve1* promoter to drive dsRed expression has been generated (Okuda et al., 2012).

Here, we review methods for exploiting what is perhaps the most important feature of these transgenics that they permit repeated and continuous imaging of the fluorescently labeled blood vessels. This has made it possible, for the first time, to image the dynamics of blood and lymphatic vessel growth and development of vascular networks in living animals. We describe methods for mounting embryos and larvae for long-term observation and for time-lapse multiphoton microscopy of blood and lymphatic vessels within these animals. The mounting of animals for time-lapse imaging is much more difficult and in some ways more critical to the success of the experiment than the actual imaging, which is relatively straightforward to set up on most imaging systems.

3.2.1 Long-term mounting for time-lapse imaging

For time-lapse imaging of blood and lymphatic vessels in transgenics over the course of hours or even days, the animals must be carefully mounted in a way that maintains the region of the animal being imaged in a relatively fixed position, yet keeps the animal alive and developing normally throughout the course of the experiment. This task is complicated further by the fact that developing zebrafish are continuously growing and undergoing morphogenetic movements, and this must be accommodated in whatever scheme is used to hold them in place. We

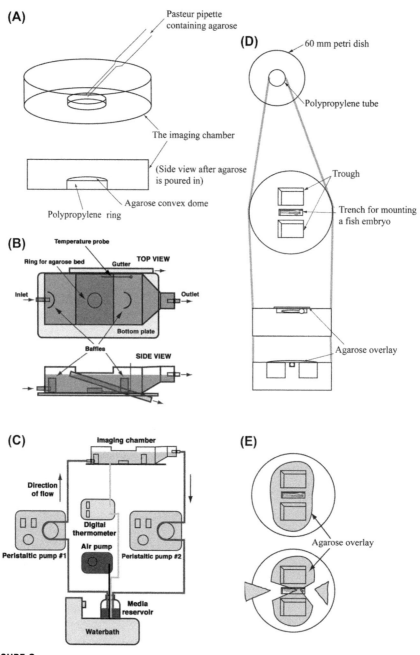

FIGURE 8

Mounting zebrafish embryos and larvae for time-lapse imaging. For shorter-term imaging, imaging chambers are prepared from a ring of polypropylene tube glued to a 60-mm petri dish (A). For long-term (greater than 1 day) time-lapse imaging, an imaging chamber is describe later a relatively simple mounting method that is adaptable to imaging different areas of embryos or larvae and holds them in place over the course of hours. For time-lapse experiments that run up to a day or more imaging chambers with buffer circulation are employed.

3.2.1.1 Materials

- Imaging vessels (see later discussion)
- 2% low melting temperature agarose made up in 30% Danieu's solution containing $1\times$ PTU (if nonalbino animals are used) and $1.25\times$ Tricaine (if nonparalyzed animals are used)
- 30% Danieu's solution with or without $1\times$ PTU and $1.25\times$ tricaine (see earlier)
- Fine forceps (Dumont #55)

3.2.1.2 Method

3.2.1.2.1 Preparation of imaging chambers

1. Imaging vessels are prepared from 6-cm polystyrene culture dishes (Falcon 3002) and 14-mL polypropylene tubes (Falcon 2059). Model cement is also required for assembly.
2. The polypropylene tube is sliced into 5-mm segments (rings) using a heated razor blade. One ring is glued to the bottom plate of the culture dish using model cement. Care should be taken to glue the slice of the polypropylene tube to the center of the dish and to avoid smearing the glue inside the polypropylene ring (to avoid obscuring the optical clarity). See Fig. 8A.
3. The glue should be allowed to dry overnight before use.
4. Just before use, the polypropylene ring in the imaging chamber should be slightly overfilled with the low melting temperature agarose to make a slightly convex dome (Fig. 8A).

---

constructed from a modified tissue culture flask (B). The approximate areas filled with water in an operating chamber are noted in blue. For details on construction, see text and Kamei and Weinstein (2005). The imaging chamber is a key part of the apparatus used for long-term time-lapse imaging of developing zebrafish, diagrammed schematically here (C). Tubing carrying water is noted in blue, wires for temperature probe are shown in green, and air line is shown in red. The inlet and outlet ports of the imaging chamber are each connected via silicone tubing through two separate peristaltic pumps to a heated, aerated reservoir of embryo buffer, forming a continuous circuit of fluid. Temperature of the imaging chamber and fluid reservoir are both monitored using separate temperature probes, and the reservoir is continuously aerated with an aquarium air pump. See text and Kamei and Weinstein (2005) for additional details. In either short- or long-term imaging chambers, the polypropylene ring in the center is filled with low-melt agarose and cavities carved out to accommodate the animal and to act as anchor points for top agarose (D). After covering the animal with top agarose much of the agarose over the animal is carved away in wedges (E). See text for further details. (See color plate)

5. Time-lapse imaging for longer duration (more than 1 day) requires a modified version of this chamber. Instead of using a 6-cm polystyrene culture dish, a T-25 tissue culture flask (Nalge Nunc cat# 163371) is used. A 1-½ inch square is cut from the upper side of the flask with a coping saw. The opening should stretch from one side of the flask to the other and be large enough to accommodate the objective lens to be used in the observation. Portions of the adjoining walls of the flask should also be cut down as shown (Fig. 8B) to allow the objective lens to enter the imaging chamber. A second opening is made in the bottom of the flask opposite the cap by drilling a 3/32″ hole using an electric drill. This second opening is to be used for the inflow port, and it should be large enough to accommodate the female Luer bulkhead (Small Parts, Inc., 3/32″ barb, cat# LCN-FB-093-25). A silicone O-ring (Small Parts, Inc., 1/4″ ID × 3/8″ OD × 1/16″ wide, cat# ORS-010-25) is inserted over the bulkhead before its attachment to the flask and then secured onto the flask with a locking nut (Small Parts, Inc., cat# LCN-LN0-25). A 5-mm-wide slice of a 14-mL Falcon tube is glued to the center of the chamber opposite the large opening using Z-Poxy resin (Small Parts, Inc., cat# EPX-PT38). Baffles are made from a lid of a 14-mL Falcon tube and also glued on inside of the chamber, one over the inlet port and another one before the outlet. Another small hole on the top of the chamber was made by drilling to allow the insertion of a temperature probe into the chamber. The cap of the flask needs to be modified so that the Luer bulkhead can be attached. The center portion of the lid is excised and discarded. The Luer female bulkhead is then screwed tightly onto the lid with an O-ring. To prevent flooding of the stage, in case accidental overflow of the chamber occurs, a safety drain is installed into the side of the chamber ("gutter" in Fig. 8B). The safety drain is constructed from a plastic drinking straw connected to a cutoff disposable pipette. The bulb and first 1 in. of the pipette are removed. A quarter of the wall of the main pipette body was also removed to form a catchment opening for chamber overflow. This piece is attached to the side of the chamber at an angle of 25–30 degrees using Z-Poxy resin. After drying, a plastic drinking straw is connected to the drain to lead the overflow to a container. The completed chamber is air-dried for overnight to allow all resin to cure and then tested for any leakage by filling it with distilled water before use.
6. The imaging chamber forms part of a loop of fluid circulation connected together with silicone tubing (Fig. 8B and C). Embryo media warmed in a bottle in a water bath is carried through silicone tubing to the chamber by a peristaltic pump. Excess media is continuously removed from the chamber and returned to the media reservoir by a second peristaltic pump. The embryo media in the reservoir is aerated by an aquarium air pump. The temperature of the water bath is calibrated to warm the media sufficiently to maintain a constant temperature (28.5°C) in the imaging chamber (an electronic temperature probe measures the temperature in the imaging chamber). The required water bath temperature should be determined empirically and depends on a number of factors including the ambient temperature of the room, the length and diameter of the tubing

between the reservoir and imaging chamber, and the flow rate. In a cool room with long tubing, the temperature of the water bath may need to be 37°C or even higher.

3.2.1.2.2 Mounting animals in imaging chambers
These procedures are done on a dissecting microscope.

1. Dechorionate and select embryos for mounting. Only a single embryo is generally mounted per imaging chamber.
2. Fill the imaging vessel with the 30% Danieu's solution to just below the rim of the Petri dish. If pigment-free albino mutant embryos are used, the PTU can be left out; if paralyzed mutant embryos are used, the tricaine can be left out of the Danieu's solution.
3. Pick an embryo for mounting, and drop it in the middle of the agarose bed (it may begin to role of because of the convex surface).
4. Using the fine forceps blades, make a shallow, narrow trench in the center of the agarose dome. This trench should be slightly wider than the dimensions of the embryo in its desired orientation for imaging, with the animal below the surface of the agarose but not too deep (Fig. 8D). It is critical that the trench be carved out carefully to make a space that holds the animal relatively motionless at rest. For imaging most portions of later stage embryos and larvae, the animal should lie on its side in the trench, for lateral view. A larger cavity should be carved out posterior to the tail to accommodate additional increases in trunk/tail length. Additional space should also be left around the head to accommodate shifting and growth, particularly in younger animals.
5. Two additional large cavities should also be carved out perpendicular to the trench on either side (Fig. 8D). These cavities will act as anchor points for the agarose layer that is overlaid on the top of the embryo.
6. Place the embryo in the trench, and slowly overlay with molten agarose. It should be warm enough to freely flow in the buffer, but not too hot to kill the embryo. We typically use glass Pasteur pipette for this since it offers more precise control. Start from one well next to the trench. Apply the agarose at steady rate, and once it filled the well, move over to the other well by moving the pipette over the embryo. Once reached over the other well, fill up this well as well (Fig. 8E). This should create an agarose bridge over the embryo and should hold the embryo down.
7. Cut excess agarose away by using the blades of a pair of fine forceps. For imaging the trunk, we slice away a triangle of agarose over the trunk and tail and over the rostral region, leaving a "bridge" of agarose over the yolk sac, posterior head, and anterior-most trunk sufficient to hold the embryo firmly in place (Fig. 8E). This ensures the optical clarity of the trunk vessels. These cuts are necessary, since the embryo is growing in anterior–posterior axis and straightening. Without removing these wedges of agarose, continued growth and straightening of the embryo/larva could not be accommodated.

3.2.2 Short-term mounting for time-lapse imaging

For relatively short period of time (up to 24 h), zebrafish embryos can be imaged by mounting them in a low percent of low melting point agarose. A bed of 1.5% agarose is set in the center of the imaging chamber. Once it is solidified, a thin layer of 0.5% low melting point agarose is poured on top of this bed. Dechorionated embryos are added to this agarose and oriented as per requirement. Low melting point agarose is allowed to set. Then the imaging chamber is mounted on the microscope stage and flooded with E3 containing $1\times$ PTU. Allow 30–40 min for agarose to equilibrate with E3 before starting image acquisition. This step allows agarose to do necessary displacement based on the differences in the salt concentrations between 1.5% agarose, low melting point agarose, and E3 and prevents embryos going out of frame within an hour after starting the time lapse.

3.2.3 Multiphoton time-lapse imaging

Once animals are properly mounted, they can be imaged by relatively straightforward time-lapse imaging methods. It is strongly recommended that a multiphoton microscopy system be used for this rather than a standard confocal microscope. The advantages of multiphoton microscopy and its use in developmental studies have been reviewed elsewhere (Denk & Svoboda, 1997; Weinstein, 2002). Multiphoton imaging reduces photodamage over the course of long imaging experiments, improves resolution of fluorescent structures deep in tissues, and improves the "three dimensionality" of resulting image reconstructions. Most imaging systems designed for or adaptable to multiphoton imaging (Leica, Zeiss, Olympus, etc.) have software interfaces that allow for simple implementation of "4-D imaging" (x, y, z, time) experiments. Below we provide a cursory experimental description with some of the important parameters and experimental considerations. The melanophores in developing embryos interfere with multiphoton imaging when 976 nm excitation (EGFP two-photon excitation wavelength) is used. This interference can be avoided either by imaging techniques (moving the field of view), by treating embryos with PTU, or by using genetic mutants lacking or with reduced melanophores (such as *albino*, *nacre*, or *casper*) (Lister, Robertson, Lepage, Johnson, & Raible, 1999; White et al., 2008). PTU should be used carefully in later-stage embryos as PTU may cause developmental defects. Use of tricaine (MS-222) for extended periods of time can also be problematic as the proper dosage for the drug becomes difficult to control, and excess tricaine can easily kill the animal imaged. For imaging experiment lasting longer than 1 day, the use of paralyzed *nic1* mutant animals (Westerfield, Liu, Kimmel, & Walker, 1990) is strongly encouraged.

1. Transfer the imaging chamber with mounted animal very carefully to the stage of a multiphoton microscope, taking care not to dislodge the animal. If the animal is very easily dislodged in transfer, then it was likely not well enough mounted and should be more securely held when remounted.

2. After locating the field to be imaged, the time-lapse parameters should be set. The following are listed some guidelines for a few of the important parameters for multiphoton transgenic blood vessel imaging:

Maximal imaging depth:	Approximately 250 μm for best image quality
Objectives:	10–100×, must pass long wavelength light
Spacing between planes:	1–5 μm, depending on magnification
Number of planes imaged:	10–60, depending on region and magnification
Interval between time points:	1–15 min (5 min is most typical)
Length of time lapse:	Up to 24 h, longer if a chamber is used with circulation of warmed buffer. With such a chamber, we have successfully imaged an animal for 5 days (Kamei & Weinstein, 2005). The chamber without flow described earlier is mainly used for shorter time-lapse experiments, and some developmental delay may be noted in longer runs.
Laser power setting:	Minimum necessary to obtain good images; if possible increase power with greater depth and use sensitive detectors to permit further decreases in required power.
Frames averaged:	Five frames averaged/plane

3. Once an imaging run has been initiated, the images being collected should be checked frequently for shifting of the field being imaged. Often some shifting occurs due to growth or morphogenetic movements of the developing animal. Some initial shifting is also sometimes seen at the beginning of a time-lapse run as the animal "settles in." The field being imaged will sometimes need to be adjusted a number of times during the course of an experiment to maintain the vessels being imaged within the field. The stage can be adjusted to reset the X and Y positions. The Z positions (bottom and top of the images stack) may also need to be reset, usually by stopping and restarting the time-lapse program. If excessive shifting occurs due to the embryo being improperly mounted, a new animal should be remounted.

3.2.4 Imaging the zebrafish vasculature using light sheet microscopy

Light sheet microscopy is another valuable tool for imaging blood vessels in zebrafish. Unlike confocal imaging, light sheet microscopes use a cylindrical lens to convert a laser beam into a sheet of light that is focused into the sample (Weber, Mickoleit, & Huisken, 2014). The light sheet excites fluorescent proteins in a single plane of the specimen, and a sensitive, high-speed camera detects the emitted light. Restricting illumination to a single plane reduces phototoxicity, and using a high-speed camera instead of scanners and detectors allows for high-speed acquisitions. Although many different types of light sheet microscopes have been developed, many use a unique mounting technique where the sample is dangled in front of the emission and detection objectives, and the movement of the sample is controlled by a four-dimensional stage. This arrangement allows the user to move the sample in X, Y, and Z, as well as to change the viewing angle by spinning the sample. Data acquired from multiple angles can be fused together giving a more encompassing data set than one acquired from a single angle using a confocal microscope (Huisken, Swoger, Del Bene, Wittbrodt, & Stelzer, 2004). By moving the Y dimension and doing multiple acquisitions, a reassembly of the entire zebrafish embryo can be done (Kaufmann, Mickoleit, Weber, & Huisken, 2012). Light sheet microscopy is also amenable to time-lapse imaging due to relatively low phototoxicity. Mounting techniques for time-lapse light sheet microscopy are nicely described by Kaufmann et al. (2012).

There are several commercial light sheet microscope options available, including offerings from Zeiss (Lightsheet Z.1) and Leica (SP8 DLS). For more adventurous laboratories, there are some cost-effective "build-it-yourself" options for light sheet microscopes (Gualda et al., 2013; Pitrone et al., 2013) including the Open-SPIM wiki, which has a parts list and step-by-step assembly instructions (http://openspim.org/(Pitrone et al., 2013)). New deconvolution algorithms developed specifically for light sheet microscopy significantly improve image quality and make light sheet microscopy an attractive option for imaging zebrafish blood vessels (Preibisch et al., 2014; Schmid & Huisken, 2015).

3.2.5 *Imaging the zebrafish vasculature using superresolution microscopy*

Standard confocal, multiphoton, and light sheet imaging methods provide adequate resolution for the vast majority of experiments focusing on the zebrafish vasculature. New imaging techniques designed to improve resolution beyond the restrictions set by the diffraction limit of light are now commercially available. Some of these techniques require fixation (STORM), large amounts of laser power (STED), or have significant depth limitations (SIM). New hardware and software developed by commercial confocal manufacturers is beginning to allow users to acquire superresolution images on more standard confocal platforms. These new imaging techniques could be used to help image and understand intraendothelial cell processes, junction formation and rearrangement, lumen formation, transmigration, and transcytosis.

CONCLUSION

Recent evidence suggests that genetic factors are critical in the formation of major vessels form during early development. Understanding the mechanisms behind the emergence of these early vascular networks will require the use of genetically and experimentally accessible model vertebrates. The accessibility and optical clarity of the zebrafish embryo and larva make it particularly useful for studies of vascular development. Studies of developing vessels are likely to have far-reaching implications for human health, since understanding mechanisms underlying the growth and morphogenesis of blood and lymphatic vessels has become critical for a number of important emerging clinical applications. Pro- and antiangiogenic therapies show great promise for treating cancer and ischemia, respectively, and a great deal of effort is currently going into uncovering and characterizing factors that can be used to promote or inhibit vessel growth in vivo. Since many of the molecules that play key roles in developing vessels carry out analogous functions during postnatal angiogenesis, it seems likely that the zebrafish will yield many important clinically applicable insights in the future.

REFERENCES

Bertrand, J. Y., Chi, N. C., Santoso, B., Teng, S., Stainier, D. Y., & Traver, D. (2010). Haematopoietic stem cells derive directly from aortic endothelium during development. *Nature, 464*, 108–111.

Brown, L. A., Rodaway, A. R., Schilling, T. F., Jowett, T., Ingham, P. W., Patient, R. K., & Sharrocks, A. D. (2000). Insights into early vasculogenesis revealed by expression of the ETS-domain transcription factor Fli-1 in wild-type and mutant zebrafish embryos. *Mechanisms of Development, 90*, 237–252.

Burns, C. E., DeBlasio, T., Zhou, Y., Zhang, J., Zon, L., & Nimer, S. D. (2002). Isolation and characterization of runxa and runxb, zebrafish members of the runt family of transcriptional regulators. *Experimental Hematology, 30*, 1381–1389.

Bussmann, J., Lawson, N., Zon, L., & Schulte-Merker, S. (2008). Zebrafish VEGF receptors: a guideline to nomenclature. *PLoS Genetics, 4*, e1000064.

Bussmann, J., Wolfe, S. A., & Siekmann, A. F. (2011). Arterial-venous network formation during brain vascularization involves hemodynamic regulation of chemokine signaling. *Development, 138*, 1717–1726.

Cermenati, S., Moleri, S., Cimbro, S., Corti, P., Del Giacco, L., Amodeo, R., ... Beltrame, M. (2008). Sox18 and Sox7 play redundant roles in vascular development. *Blood, 111*, 2657–2666.

Cheong, S. M., Choi, S. C., & Han, J. K. (2006). Xenopus Dab2 is required for embryonic angiogenesis. *BMC Developmental Biology, 6*, 63.

Chi, N. C., Shaw, R. M., De Val, S., Kang, G., Jan, L. Y., Black, B. L., & Stainier, D. Y. (2008). Foxn4 directly regulates tbx2b expression and atrioventricular canal formation. *Genes and Development, 22*, 734–739.

Childs, S., Chen, J. N., Garrity, D. M., & Fishman, M. C. (2002). Patterning of angiogenesis in the zebrafish embryo. *Development, 129*, 973–982.

Clements, W. K., Kim, A. D., Ong, K. G., Moore, J. C., Lawson, N. D., & Traver, D. (2011). A somitic Wnt16/Notch pathway specifies haematopoietic stem cells. *Nature, 474*, 220–224.

Connors, S. A., Trout, J., Ekker, M., & Mullins, M. C. (1999). The role of tolloid/mini fin in dorsoventral pattern formation of the zebrafish embryo. *Development, 126*, 3119–3130.

Cross, L. M., Cook, M. A., Lin, S., Chen, J. N., & Rubinstein, A. L. (2003). Rapid analysis of angiogenesis drugs in a live fluorescent zebrafish assay. *Arteriosclerosis, Thrombosis, and Vascular Biology, 23*, 911–912.

Denk, W., & Svoboda, K. (1997). Photon upmanship: why multiphoton imaging is more than a gimmick. *Neuron, 18*, 351–357.

Evans, H. (1910). *Manual of human embryology.* Philadelphia & London: J. B. Lippincott & Company.

Flores, M. V., Hall, C. J., Crosier, K. E., & Crosier, P. S. (2010). Visualization of embryonic lymphangiogenesis advances the use of the zebrafish model for research in cancer and lymphatic pathologies. *Developmental Dynamics: An Official Publication of the American Association of Anatomists, 239*, 2128–2135.

Fouquet, B., Weinstein, B. M., Serluca, F. C., & Fishman, M. C. (1997). Vessel patterning in the embryo of the zebrafish: guidance by notochord. *Developmental Biology, 183*, 37–48.

Gering, M., Rodaway, A. R., Gottgens, B., Patient, R. K., & Green, A. R. (1998). The SCL gene specifies haemangioblast development from early mesoderm. *EMBO Journal, 17*, 4029–4045.

Geudens, I., Herpers, R., Hermans, K., Segura, I., Ruiz de Almodovar, C., Bussmann, J., ... Dewerchin, M. (2010). Role of Dll4/Notch in the formation and wiring of the lymphatic network in zebra fish. *Arteriosclerosis, Thrombosis, and Vascular Biology, 30*, 1695–1702.

Gualda, E. J., Vale, T., Almada, P., Feijo, J. A., Martins, G. G., & Moreno, N. (2013). OpenSpinMicroscopy: an open-source integrated microscopy platform. *Nature Methods, 10*, 599–600.

Hauptmann, G., & Gerster, T. (1994). Two-color whole-mount in situ hybridization to vertebrate and *Drosophila* embryos. *Trends in Genetics, 10*, 266.

Helker, C. S., Schuermann, A., Karpanen, T., Zeuschner, D., Belting, H. G., Affolter, M., ... Herzog, W. (2013). The zebrafish common cardinal veins develop by a novel mechanism: lumen ensheathment. *Development, 140*, 2776–2786.

Herpers, R., van de Kamp, E., Duckers, H. J., & Schulte-Merker, S. (2008). Redundant roles for sox7 and sox18 in arteriovenous specification in zebrafish. *Circulation Research, 102*, 12–15.

Herwig, L., Blum, Y., Krudewig, A., Ellertsdottir, E., Lenard, A., Belting, H. G., & Affolter, M. (2011). Distinct cellular mechanisms of blood vessel fusion in the zebrafish embryo. *Current Biology: CB, 21*, 1942–1948.

Hogan, B. M., Bos, F. L., Bussmann, J., Witte, M., Chi, N. C., Duckers, H. J., & Schulte-Merker, S. (2009). Ccbe1 is required for embryonic lymphangiogenesis and venous sprouting. *Nature Genetics, 41*, 396–398.

Huang, C. C., Lawson, N. D., Weinstein, B. M., & Johnson, S. L. (2003). reg6 is required for branching morphogenesis during blood vessel regeneration in zebrafish caudal fins. *Developmental Biology, 264*, 263–274.

Huisken, J., Swoger, J., Del Bene, F., Wittbrodt, J., & Stelzer, E. H. (2004). Optical sectioning deep inside live embryos by selective plane illumination microscopy. *Science, 305*, 1007–1009.

van Impel, A., Zhao, Z., Hermkens, D. M., Roukens, M. G., Fischer, J. C., Peterson-Maduro, J., ... Schulte-Merker, S. (2014). Divergence of zebrafish and mouse lymphatic cell fate specification pathways. *Development, 141*, 1228–1238.

Isogai, S., Lawson, N. D., Torrealday, S., Horiguchi, M., & Weinstein, B. M. (2003). Angiogenic network formation in the developing vertebrate trunk. *Development, 130*, 5281–5290.

Jin, S. W., Beis, D., Mitchell, T., Chen, J. N., & Stainier, D. Y. (2005). Cellular and molecular analyses of vascular tube and lumen formation in zebrafish. *Development, 132*, 5199–5209.

Kalev-Zylinska, M. L., Horsfield, J. A., Flores, M. V., Postlethwait, J. H., Vitas, M. R., Baas, A. M., ... Crosier, K. E. (2002). Runx1 is required for zebrafish blood and vessel development and expression of a human RUNX1-CBF2T1 transgene advances a model for studies of leukemogenesis. *Development, 129*, 2015–2030.

Kamei, M., Saunders, W. B., Bayless, K. J., Dye, L., Davis, G. E., & Weinstein, B. M. (2006). Endothelial tubes assemble from intracellular vacuoles in vivo. *Nature, 442*, 453−456.

Kamei, M., & Weinstein, B. M. (2005). Long-term time-lapse fluorescence imaging of developing zebrafish. *Zebrafish, 2*, 113−123.

Kashiwada, T., Fukuhara, S., Terai, K., Tanaka, T., Wakayama, Y., Ando, K., ... Mochizuki, N. (2015). Beta-Catenin-dependent transcription is central to Bmp-mediated formation of venous vessels. *Development, 142*, 497−509.

Kaufmann, A., Mickoleit, M., Weber, M., & Huisken, J. (2012). Multilayer mounting enables long-term imaging of zebrafish development in a light sheet microscope. *Development, 139*, 3242−3247.

Lawson, N. D., Scheer, N., Pham, V. N., Kim, C. H., Chitnis, A. B., Campos-Ortega, J. A., & Weinstein, B. M. (2001). Notch signaling is required for arterial-venous differentiation during embryonic vascular development. *Development, 128*, 3675−3683.

Lawson, N. D., & Weinstein, B. M. (2002). In vivo imaging of embryonic vascular development using transgenic zebrafish. *Developmental Biology, 248*, 307−318.

Lenard, A., Daetwyler, S., Betz, C., Ellertsdottir, E., Belting, H. G., Huisken, J., & Affolter, M. (2015). Endothelial cell self-fusion during vascular pruning. *PLoS Biology, 13*, e1002126.

Lenard, A., Ellertsdottir, E., Herwig, L., Krudewig, A., Sauteur, L., Belting, H. G., & Affolter, M. (2013). In vivo analysis reveals a highly stereotypic morphogenetic pathway of vascular anastomosis. *Developmental Cell, 25*, 492−506.

Lin, S. (2000). Transgenic zebrafish. *Methods in Molecular Biology, 136*, 375−383.

Lister, J. A., Robertson, C. P., Lepage, T., Johnson, S. L., & Raible, D. W. (1999). Nacre encodes a zebrafish microphthalmia-related protein that regulates neural-crest-derived pigment cell fate. *Development, 126*, 3757−3767.

Lyons, M. S., Bell, B., Stainier, D., & Peters, K. G. (1998). Isolation of the zebrafish homologues for the tie-1 and tie-2 endothelium-specific receptor tyrosine kinases. *Developmental Dynamics: An Official Publication of the American Association of Anatomists, 212*, 133−140.

Martyn, U., & Schulte-Merker, S. (2004). Zebrafish neuropilins are differentially expressed and interact with vascular endothelial growth factor during embryonic vascular development. *Developmental Dynamics: An Official Publication of the American Association of Anatomists, 231*, 33−42.

Murakami, T. (1972). Vascular arrangement of the rat renal glomerulus: a scanning electron microscope study of corrosion casts. *Archivum Histologicum Japonicum, 34*, 87−107.

Nicenboim, J., Malkinson, G., Lupo, T., Asaf, L., Sela, Y., Mayseless, O., ... Yaniv, K. (2015). Lymphatic vessels arise from specialized angioblasts within a venous niche. *Nature, 522*, 56−61.

Okuda, K. S., Astin, J. W., Misa, J. P., Flores, M. V., Crosier, K. E., & Crosier, P. S. (2012). Lyve1 expression reveals novel lymphatic vessels and new mechanisms for lymphatic vessel development in zebrafish. *Development, 139*, 2381−2391.

Parker, L. H., Schmidt, M., Jin, S. W., Gray, A. M., Beis, D., Pham, T., ... Ye, W. (2004). The endothelial-cell-derived secreted factor Egfl7 regulates vascular tube formation. *Nature, 428*, 754−758.

Parsons, M. J., Pisharath, H., Yusuff, S., Moore, J. C., Siekmann, A. F., Lawson, N., & Leach, S. D. (2009). Notch-responsive cells initiate the secondary transition in larval zebrafish pancreas. *Mechanisms of Development, 126*, 898−912.

Pendeville, H., Winandy, M., Manfroid, I., Nivelles, O., Motte, P., Pasque, V., ... Voz, M. L. (2008). Zebrafish Sox7 and Sox18 function together to control arterial-venous identity. *Developmental Biology, 317*, 405–416.

Pham, V. N., Lawson, N. D., Mugford, J. W., Dye, L., Castranova, D., Lo, B., & Weinstein, B. M. (2007). Combinatorial function of ETS transcription factors in the developing vasculature. *Developmental Biology, 303*, 772–783.

Pitrone, P. G., Schindelin, J., Stuyvenberg, L., Preibisch, S., Weber, M., Eliceiri, K. W., ... Tomancak, P. (2013). OpenSPIM: an open-access light-sheet microscopy platform. *Nature Methods, 10*, 598–599.

Preibisch, S., Amat, F., Stamataki, E., Sarov, M., Singer, R. H., Myers, E., & Tomancak, P. (2014). Efficient Bayesian-based multiview deconvolution. *Nature Methods, 11*, 645–648.

Qian, F., Zhen, F., Ong, C., Jin, S. W., Meng Soo, H., Stainier, D. Y., ... Wen, Z. (2005). Microarray analysis of zebrafish cloche mutant using amplified cDNA and identification of potential downstream target genes. *Developmental Dynamics: An Official Publication of the American Association of Anatomists, 233*, 1163–1172.

Ren, X., Gomez, G. A., Zhang, B., & Lin, S. (2010). Scl isoforms act downstream of etsrp to specify angioblasts and definitive hematopoietic stem cells. *Blood, 115*, 5338–5346.

Roman, B. L., Pham, V. N., Lawson, N. D., Kulik, M., Childs, S., Lekven, A. C., ... Weinstein, B. M. (2002). Disruption of acvrl1 increases endothelial cell number in zebrafish cranial vessels. *Development, 129*, 3009–3019.

Sabin, F. R. (1917). Origin and development of the primitive vessels of the chick and of the pig. *Carnegie Institution of Washington publication, 266*, 61–124.

Schmid, B., & Huisken, J. (2015). Real-time multi-view deconvolution. *Bioinformatics, 31*, 3398–3400.

Stoletov, K., Montel, V., Lester, R. D., Gonias, S. L., & Klemke, R. (2007). High-resolution imaging of the dynamic tumor cell vascular interface in transparent zebrafish. *Proceedings of the National Academy of Sciences of the United States of America, 104*, 17406–17411.

Sumanas, S., Jorniak, T., & Lin, S. (2005). Identification of novel vascular endothelial-specific genes by the microarray analysis of the zebrafish cloche mutants. *Blood, 106*, 534–541.

Sumoy, L., Keasey, J. B., Dittman, T. D., & Kimelman, D. (1997). A role for notochord in axial vascular development revealed by analysis of phenotype and the expression of VEGR-2 in zebrafish flh and ntl mutant embryos. *Mechanisms of Development, 63*, 15–27.

Szeto, D. P., Griffin, K. J., & Kimelman, D. (2002). HrT is required for cardiovascular development in zebrafish. *Development, 129*, 5093–5101.

Tamplin, O. J., Durand, E. M., Carr, L. A., Childs, S. J., Hagedorn, E. J., Li, P., ... Zon, L. I. (2015). Hematopoietic stem cell arrival triggers dynamic remodeling of the perivascular niche. *Cell, 160*, 241–252.

Thompson, M. A., Ransom, D. G., Pratt, S. J., MacLennan, H., Kieran, M. W., Detrich, H. W., 3rd, ... Zon, L. I. (1998). The cloche and spadetail genes differentially affect hematopoiesis and vasculogenesis. *Developmental Biology, 197*, 248–269.

Tong, E. Y., Collins, G. C., Greene-Colozzi, A. E., Chen, J. L., Manos, P. D., Judkins, K. M., ... Levesque, T. J. (2009). Motion-based angiogenesis analysis: a simple method to quantify blood vessel growth. *Zebrafish, 6*, 239–243.

Traver, D., Paw, B. H., Poss, K. D., Penberthy, W. T., Lin, S., & Zon, L. I. (2003). Transplantation and in vivo imaging of multilineage engraftment in zebrafish bloodless mutants. *Nature Immunology, 4*, 1238–1246.

Udvadia, A. J., & Linney, E. (2003). Windows into development: historic, current, and future perspectives on transgenic zebrafish. *Developmental Biology, 256*, 1–17.

Vanhollebeke, B., Stone, O. A., Bostaille, N., Cho, C., Zhou, Y., Maquet, E., ... Stainier, D. Y. R. (2015). Tip cell-specific requirement for an atypical Gpr124- and Reck-dependent Wnt/beta-catenin pathway during brain angiogenesis. *eLife, 4*.

Veldman, M. B., & Lin, S. (2012). Etsrp/Etv2 is directly regulated by Foxc1a/b in the zebrafish angioblast. *Circulation Research, 110*, 220−229.

Wakayama, Y., Fukuhara, S., Ando, K., Matsuda, M., & Mochizuki, N. (2015). Cdc42 mediates Bmp-induced sprouting angiogenesis through Fmnl3-driven assembly of endothelial filopodia in zebrafish. *Developmental Cell, 32*, 109−122.

Wang, Y., Kaiser, M. S., Larson, J. D., Nasevicius, A., Clark, K. J., Wadman, S. A., ... Essner, J. J. (2010). Moesin1 and Ve-cadherin are required in endothelial cells during in vivo tubulogenesis. *Development, 137*, 3119−3128.

Weber, M., Mickoleit, M., & Huisken, J. (2014). Light sheet microscopy. *Methods in Cell Biology, 123*, 193−215.

Weinstein, B. (2002). Vascular cell biology in vivo: a new piscine paradigm? *Trends in Cell Biology, 12*, 439−445.

Westerfield, M. (2000). *The zebrafish book. A guide for the laboratory use of zebrafish (Danio rerio)* (4th ed.). Eugene: Univ. of Oregon Press.

Westerfield, M., Liu, D. W., Kimmel, C. B., & Walker, C. (1990). Pathfinding and synapse formation in a zebrafish mutant lacking functional acetylcholine receptors. *Neuron, 4*, 867−874.

White, R. M., Sessa, A., Burke, C., Bowman, T., LeBlanc, J., Ceol, C., ... Zon, L. I. (2008). Transparent adult zebrafish as a tool for in vivo transplantation analysis. *Cell Stem Cell, 2*, 183−189.

Wong, K. S., Proulx, K., Rost, M. S., & Sumanas, S. (2009). Identification of vasculature-specific genes by microarray analysis of Etsrp/Etv2 overexpressing zebrafish embryos. *Developmental Dynamics: An Official Publication of the American Association of Anatomists, 238*, 1836−1850.

Xie, J., Farage, E., Sugimoto, M., & Anand-Apte, B. (2010). A novel transgenic zebrafish model for blood−brain and blood−retinal barrier development. *BMC Developmental Biology, 10*, 76.

Yaniv, K., Isogai, S., Castranova, D., Dye, L., Hitomi, J., & Weinstein, B. M. (2006). Live imaging of lymphatic development in the zebrafish. *Nature Medicine, 12*, 711−716.

Yu, J. A., Castranova, D., Pham, V. N., & Weinstein, B. M. (2015). Single-cell analysis of endothelial morphogenesis in vivo. *Development, 142*, 2951−2961.

Zhong, T. P., Rosenberg, M., Mohideen, M. A., Weinstein, B., & Fishman, M. C. (2000). Gridlock, an HLH gene required for assembly of the aorta in zebrafish. *Science, 287*, 1820−1824.

CHAPTER

An eye on light-sheet microscopy

D. Kromm[a], T. Thumberger[a], J. Wittbrodt[1]

*Centre for Organismal Studies, Heidelberg University,
Heidelberg, Germany*
[1]*Corresponding author: E-mail: jochen.wittbrodt@cos.uni-heidelberg.de*

CHAPTER OUTLINE

Introduction	106
History	106
1. Principle Behind Selective Plane Illumination Microscopy	107
1.1 Speed	107
1.2 Phototoxicity	109
1.3 Resolution and Image Quality	110
1.4 Realization	110
1.5 Components of a Selective Plane Illumination Microscopy	112
2. The Microscope for Your Sample or the Sample for Your Microscope?	114
2.1 Specimen Size Versus Field of View and Movement Range of the Stages: ("Does Your Sample Fit Into the Microscope?")	114
2.2 Refractive Index and Light Attenuation of the Immersion Medium: ("Is the Culture Medium Suitable for the Optics?")	114
2.3 Desired Spatial and Temporal Resolution: ("Can I Actually Resolve the Objects I Want to Detect and Can I Image Fast/Long Enough?")	115
2.3.1 Spatial resolution	115
2.3.2 Temporal resolution	115
2.4 Mounting of the Sample: ("Is There a Reliable Routine to Mount the Specimen of Choice?")	116
3. Data Acquisition and Handling	117
4. Challenges and Perspectives	119
Acknowledgment	120
References	121

[a]These authors contributed equally.

Abstract

This chapter introduces the principles and advantages of selective plane illumination microscopy (SPIM) and compares it to commonly used epifluorescence or confocal setups. Due to the low phototoxicity, speed of imaging, high penetration depth, and spatiotemporal resolution, SPIM is predestined for in vivo imaging but can as well be used for in toto analysis of large fixed samples. Key points of light-sheet microscopy are highlighted and discussed priming the investigator to choose the best suitable system from the large collection of possible SPIM setups. Mounting of samples is shown and the demands for data acquisition, processing, handling, and visualization are discussed.

INTRODUCTION

In the last ten years, selective plane illumination microscopy (SPIM) or light-sheet fluorescence microscopy (LSFM) emerged as a powerful tool in developmental biology (Huisken, Swoger, Del Bene, Wittbrodt, & Stelzer, 2004) and today is used across a wide range of scales in biological systems ranging from the single cell (Planchon et al., 2011) to the entire organism (Huisken et al., 2004; Keller & Stelzer, 2008; Krzic, Gunther, Saunders, Streichan, & Hufnagel, 2012; Mickoleit et al., 2014; Tomer, Khairy, Amat, & Keller, 2012). There are many reasons for the recent successes of SPIM, most of which originate from the superior in vivo imaging properties in large samples over extended periods of time, but also from the way the samples are handled in these microscopes.

This chapter aims at introducing the concept of light sheet–based microscopy and its intrinsic advantages when analyzing large specimens in vivo with subcellular resolution. We will also address challenges that need to be considered prior to planning an experiment using SPIM. Further, a typical experimental procedure is laid out involving different biological samples like the fish embryo and retina. An outlook and brief overview over some notable new implementations and add-ons is given toward the end of this chapter.

HISTORY

Despite the use of SPIM mostly in biology today, particularly in developmental biology, the first publication of an optical setup resembling modern light-sheet microscopes had not been stimulated by a biological question, but was rather triggered by a physical one. More than 100 years ago, Siedentopf and Zsigmondy successfully employed a light-sheet illumination scheme to observe nanometer-sized gold particles (Siedentopf & Zsigmondy, 1902), which other techniques at the time failed to resolve. The most significant trait of modern light-sheet microscopes, ie, the illumination of a single plane, is achieved by the uncoupling of illumination and detection paths and dates back to Siedentopf's and Zsigmondy's research.

Almost a century later, Voie, Burns, & Spelman (1993) employed a cylindrical lens to create a static sheet of light and an orthogonal detection to image the inner ear cochlea of the guinea pig. Another ten years later Jan Huisken performed the first live imaging with a SPIM at EMBL in a collaboration of the Stelzer and Wittbrodt groups (Huisken et al., 2004). Driven by the biological question to follow and understand early development by following the behavior of individual cells, Philipp Keller (joint member of the Wittbrodt and Stelzer groups) (Keller, Schmidt, Wittbrodt, & Stelzer, 2008) established the digitally scanned light-sheet microscope (DSLM). The analysis of zebrafish gastrulation via the DSLM eventually settled the old debate about the gastrulation mode (involution vs ingression) by providing a holistic 4D view of zebrafish gastrulation. This allowed putting the behavior of individual cells into the organismal context and showed that the dorsal ingression behavior is gradually transiting into ventral involution (Keller et al., 2008).

Today, LSFM is usually associated with its major advantages like optical sectioning, high contrast, high imaging speed, and low phototoxicity. Before talking about the recent, more advanced SPIM setups, the following paragraphs first will shed some light on the idea behind SPIM and how these advantages arise intrinsically from the optical setup.

1. PRINCIPLE BEHIND SELECTIVE PLANE ILLUMINATION MICROSCOPY

The biggest difference between a SPIM and a conventional epifluorescence, confocal or multiphoton fluorescence microscope is the uncoupling of illumination and detection paths (Fig. 1). In LSFM typically, two objectives are placed orthogonally to each other. One of the objectives is used for illumination while the other one is employed for detection (Fig. 1B). The illumination and detection of an entire plane of the specimen is the other big difference. As the name implies, instead of illuminating the entire sample as is the case in a wide field microscope or point by point as done in a point-scanning confocal microscope (Fig. 1C), the sample is illuminated with a narrow sheet of light from the side (Fig. 1D). This way, only a section of the sample is excited and consequently the entire field of view on the detection side is illuminated simultaneously. With the use of a fast sCMOS camera ($\sim$100 frames/s) recordings of large 3D volumes can be achieved at much higher rates than with point-scanning microscopes.

The orthogonal illumination with a thin light-sheet and the use of fast cameras require different approaches; for example, sample preparation. In the following paragraphs, we will give an overview of the most prominent aspects, which distinguish light-sheet microscopy. For more (technical) details see also (Keller et al., 2008).

1.1 SPEED

The separation of illumination and detection axes facilitates a vast improvement in imaging speed. To illustrate the speed difference for 3D recordings, let's consider a zebrafish embryo, which is roughly 1 mm in diameter. For a full recording, a stack of 1 mm^3 volume has to be sampled. In a typical SPIM setup the recording of one plane

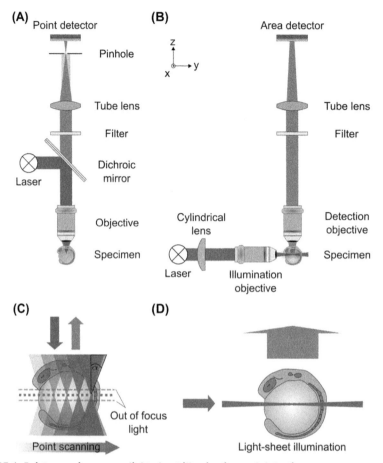

FIGURE 1 Point scanning versus light-sheet illumination and detection.

(A) In an epifluorescence or confocal microscope, the light source provides light of a desired wavelength (here blue (dark gray in print versions)), which usually is generated by a laser. The emitted light is reflected via a dichroic mirror into the direction of the objective, which focuses the light onto the specimen. Fluorophores are excited and emit light of a longer wavelength (here green (gray in print versions)), which is collected by the objective and transmitted by the dichroic mirror. The tube lens focuses the light to generate an image on the detector. In case of a confocal microscope, a pinhole is placed in front of the detector. This obstructs scattered light coming from outside the focal plane of the objective, yielding an optical sectioning. (B) Simple SPIM setup employing a cylindrical lens to generate a light-sheet in the focal plane of the detection objective. The sample is illuminated by this sheet of light, which is focused by the illumination objective, therefore restricting fluorescent excitation and emission exclusively to the focal plane. The emitted light is collected by the detection objective, spectrally separated from the excitation light with a filter, and finally focused by a tube lens onto the detector. Note that the illumination and detection arms are essentially the same as in a wide field setup but independent of each other. It is common to with a CCD sensor of 2048 by 2048 pixels takes approximately 50 ms. To sufficiently sample the entire embryo, usually 500 planes are recorded. Thus the imaging of the entire embryo would amount to 25 s. In a scanning microscope a short pixel dwell time—the time during which a pixel is excited in the sample and read out by the chip—would be 1 μs/px (usually higher values are needed). Confocal microscopes typically use detectors with 1024 by 1024 pixels or less. Under ideal imaging conditions ~1 frame/s can be achieved by confocal microscopes. Therefore, the recording of a 1 mm^3 volume will amount to at least 500 s, ie, 20 times slower compared to the SPIM setup and a factor of 4 worse in sampling on the pixel level. During early zebrafish development, cell division occurs about every 15–20 min. For robust cell tracking the recorded time points need to cover this time-span, which is hardly possible with regular confocal microscopy.

Consequently, the high imaging speed of SPIM allows following the rapid development of entire organs and organisms, approaches not suited for the low recording speeds of regular-scanning microscopes. SPIM excels at the recording of large biological samples at a high temporal resolution.

1.2 PHOTOTOXICITY

Another important fact to consider in live imaging is the energy deposition into the sample. In confocal microscopy, a bad trade-off between temporal resolution and phototoxic effects due to longer exposure time needs to be accepted. In light-sheet microscopy in contrast, the amount of energy that the sample is exposed to mostly depends on the number of recorded planes, which is due to the light's confinement to a single plane. On the contrary, in point-scanning implementations the sample is exposed to a cone of light for each recorded point. Consequently, the tissues as well as the fluorophores outside the focal plane are repeatedly and unnecessarily exposed to light. This will result in phototoxic effects and bleaching. In fact, in

---

define the detection axis as the z-axis of the system, with the y-axis being the propagation axis of the laser beam. Different tube lenses to change the magnification or other optical parameters can be used separately in both paths. To acquire images of the entire sample either the light-sheet is moved through the specimen or vice versa. (C) Sample illumination and detection in a confocal point-scanning microscope is performed point by point in an array until the image of a single plane is obtained. For each scanned point, however, the laser beam illuminates the sample also above and below the actual plane of focus. To achieve optical sectioning and a higher resolution, a pinhole is used to block out of focus light. (D) In light-sheet microscopy an entire plane is illuminated at once from the side. Instead of detecting signal in a pointlike manner, an area detector records the entire plane of focus at once. This facilitates high-speed imaging of entire volumes as well as a great reduction of excitation energy, both conditions allowing long-term and high-resolution imaging.

light-sheet microscopy the sample is exposed to ~100–1000 times less light than in confocal microscopy (Keller & Stelzer, 2008) and the effect increases with the number of optical planes imaged.

SPIM is so gentle that it allows recordings of the specimen over extended periods of hours or even several days; thus the main challenge is to maintain the sample under physiological conditions.

1.3 RESOLUTION AND IMAGE QUALITY

Most light-sheet microscopes do not require any optical elements in addition to those needed to record a regular wide field image (objective, tube lens, camera; Fig. 1B). Therefore the lateral resolution in SPIM corresponds to the regular wide field resolution, which can be approximated by Abbe's diffraction limit: $d_{lateral} = 0.61 \lambda_0/NA$.

Here, $d_{lateral}$ is the smallest resolvable size, λ_0 is the vacuum wavelength of light, and NA stands for the numerical aperture which is defined as $NA = nsin\lambda$ with n being the refractive index of the medium and λ being the opening angle of the objective.

While optical sectioning in confocal microscopy is achieved with the help of a light discriminating pinhole, in SPIM, optical sectioning is achieved solely through the illumination of the detection plane. This means that the light-sheet excites fluorophores exclusively in the focal plane of the detection objective. Typically, the thickness of the light-sheet lies in the range of the axial resolution of the detection objective, leading to a slight improvement in axial resolution compared to a normal wide field detection and can even outperform confocal microscopy for smaller numerical apertures (Engelbrecht & Stelzer, 2006).

The axial resolution in wide field detection is given by: $d_{axial} = 2n\lambda_0/NA^2$. Both wide field resolutions are by a factor of $\sqrt{2}$ worse than in confocal microscopy, but in SPIM can be improved with the help of image fusion (and deconvolution (Schmid & Huisken, 2015)) (Preibisch, Saalfeld, Schindelin, & Tomancák, 2010; Rubio-Guivernau et al., 2012; Swoger, Verveer, Greger, & Huisken, 2007). Multiview implementations and recordings enable isotropic sampling and allow gaining isotropic resolution after image fusion and image reconstruction (Chhetri et al., 2015).

Due to the fact that in SPIM more photons are detected, the signal to noise ratio and therefore the contrast is at least one order of magnitude better when compared to confocal-scanning microscopes (Keller & Stelzer, 2008; Reynaud, Krzic, Greger, & Stelzer, 2008).

1.4 REALIZATION

Currently there are just a handful of commercial light-sheet microscopes available. With a detailed parts list, and completely open blueprints and software, the OpenSPIM (Pitrone et al., 2013) platform provides the biologists with an easy entry into light-sheet microscopy. The instruments are suited for teaching as well as real imaging analyses.

To get a better feeling for how a light-sheet microscope operates, what to expect and what to be prepared for when dealing with SPIM and SPIM data, the following paragraphs will introduce some of the most important aspects of SPIM. The single plane illumination is produced either by a static light-sheet, which can be created by focusing a laser beam with a cylindrical lens (Fig. 2A) or by controllable galvanometric scan mirrors that rapidly move a focused beam up and down, generating a digitally scanned light-sheet (Fig. 2B) (Keller et al., 2008).

An example setup of a "classical" SPIM based on the use of a cylindrical lens is illustrated in Fig. 2A. The arrangement of the objectives is the same as in a confocal θ-fluorescence microscope (Stelzer & Lindek, 1994); however, a wide field detection path is used. The alignment of the focal plane of the detection objective to the plane of the thin light-sheet ensures optical sectioning. This way, only fluorophores from and nearby the focal plane are excited, which results in an image of high contrast and good focus.

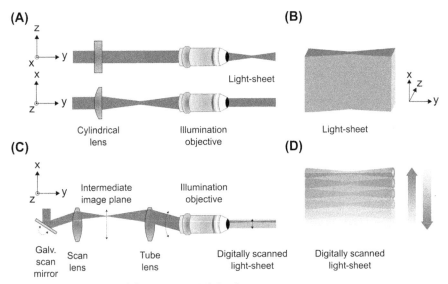

FIGURE 2 Static versus digitally scanned light-sheet.

(A–B) Generation of a static light-sheet: While the Gaussian beam stays collimated in the yz-plane, the cylindrical lens (side view up, top view down) focuses it in the xy-plane before it enters the illumination objective. This results in an extended beam in the xy-plane that is focused on the z-axis (B). (C–D) Generation of a digitally scanned light-sheet: The combination of a galvanometric scan mirror and a scan lens leads to a focused beam in the intermediate image plane between the scan lens and the tube lens. The rotation of the galvanometric scan mirror also corresponds to a rotation in front of the illumination objective, which leads to a translation along the x-axis in the intermediate image plane and therefore in the translation of the beam in the x-axis of the actual image plane. The result is a digitally scanned light-sheet (D) cf. (Keller et al., 2008).

1.5 COMPONENTS OF A SELECTIVE PLANE ILLUMINATION MICROSCOPY

In its basic implementation, only the following components are needed for a light-sheet microscope (Fig. 1):

- light source exciting the fluorophores to be analyzed (typically lasers)
- cylindrical lens, or a galvanometric mirror plus a scan lens for light-sheet generation
- illumination objective (should be used for better light-sheet quality)
- specimen chamber for immersion medium facilitating in vivo imaging
- movable specimen mount (usually motorized precision stages)
- detection objective
- optical filters
- tube lens
- area detector (eg, sCMOS/CCD camera)
- controllers for the electronic devices, a computer for software control and data recording (and processing).

Further elements can be added to expand and enhance the properties of the basic SPIM. The most common extension is the use of a second illumination objective to ensure a wider, more homogeneous illumination and to reduce possible shadow effects that can arise with the light passing through strongly absorbing media. Further, a second detection path is added to increase imaging speed and to improve image fusion capabilities (Fig. 3, MuVi-SPIM, SiM-View).

Depending on the specimen to be observed, different arrangements of the illumination and detection optics relative to the specimen have been implemented (Fig. 3, upright or inverted, diSPIM (Kumar et al., 2014)).

A very recent implementation even uses the four paths as (subsequent) illumination and detection paths such that the sample can be imaged from different angles without the need for moving it, further increasing speed and sampling quality (Chhetri et al., 2015).

On the level of the light-sheet itself, several implementations have been reported to enhance the penetration depth and resolution.

In two-photon scanned light-sheet microscopy (Truong, Supatto, Koos, Choi, & Fraser, 2011) the setup of the light-sheet microscope is combined with nonlinear excitation. Even though more energy is required for two-photon excitation, this instrument was showing excellent performance in the three-dimensional imaging of large samples in vivo, with a strongly enhanced penetration depth.

More sophisticated approaches to increase the resolution and improve penetration depth use beams and light-sheets with properties differing from the standard Gaussian beam, eg, Bessel beams (Fahrbach, Simon, & Rohrbach, 2010), Airy beams (Vettenburg et al., 2014), lattice light-sheets (Chen et al., 2014) to name only few (cf. (Höckendorf, Thumberger, & Wittbrodt, 2012)). It is also possible to use a multiphoton laser for laser manipulation experiments like ablation, cutting,

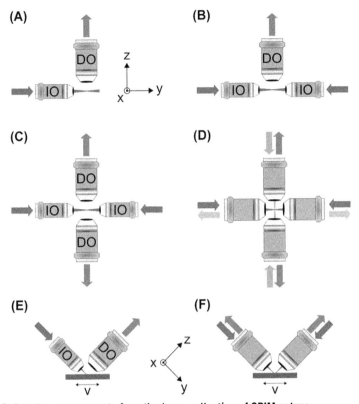

FIGURE 3 Popular arrangements from the large collection of SPIM setups.

Most SPIM setups employ orthogonal objective configurations. Depending on the sample to be imaged, the desired field of view and resolution and more criteria (see text), these setups offer different advantages over each other. Imaging from multiple sides allows during isotropical fusion of the data set during postprocessing. (A) Conventional two lens SPIM setup: The first and simplest implementation of a SPIM. (B) mSPIM: For more homogeneous illumination and the reduction of possible shadows a second illumination arm is applied (Huisken & Stainier, 2007). (C) MuVi-SPIM or SiMView: The addition of a second detection objective allows for simultaneous imaging from two sides. With a single 90 degree rotation of the sample, four views can be recorded and fused during postprocessing (Krzic et al., 2012; Tomer et al., 2012). (D) IsoView: With the addition of extra optics and two more cameras each arm can be used for illumination and detection sequentially or simultaneously. The use of different laser wavelengths and confocal line detection (http://dx.doi.org/10.1038/ncomms9881) supports even simultaneous recordings on all four arms (Chhetri et al., 2015). (E) Upright SPIM: Inverting the basic SPIM setup from (A) can be of benefit when isotropic views are not required or the sample has to be kept horizontally. In its simplest implementation only one electronically controlled translation stage is needed. The translation results in a sheared recording that requires postprocessing to correct physical dimensions (Wu et al., 2011). (F) Upright SPIM with double detection and illumination: The subsequent illumination and detection from both sides enables image fusion and therefore a better isotropic sampling of the specimen. An advanced implementation of this objective arrangement has been realized by (Kumar et al., 2014).

photoactivation, among others (optogenetic control (Arrenberg, Stainier, Baier, & Huisken, 2010), laser manipulation (Bambardekar, Clément, Blanc, Chardès, & Lenne, 2015), microsurgery (Engelbrecht et al., 2007)).

2. THE MICROSCOPE FOR YOUR SAMPLE OR THE SAMPLE FOR YOUR MICROSCOPE?

It is not surprising that many biologists are excited by the advantages of SPIM imaging over other techniques. However, choosing the setup that suits the specimen and process of interest is of outmost importance. Careful planning of the optical properties and sample handling is key to successfully imaging when designing or buying a SPIM.

In general, one has to preconceive the biological question ("What spatiotemporal level of detail needs to be examined"), the sample preparation (from whole mount in vivo specimens to fixed and stained tissue), data acquisition and processing, and finally analysis, interpretation, and visualization.

A few aspects should be carefully considered before the beginning of an experiment with special emphasis on the following key points:

2.1 SPECIMEN SIZE VERSUS FIELD OF VIEW AND MOVEMENT RANGE OF THE STAGES: ("DOES YOUR SAMPLE FIT INTO THE MICROSCOPE?")

Size becomes especially critical if large samples in the range of the field of view and above are to be imaged. The working distance of the objectives and the travel ranges of the installed stages limit the volume that can be imaged inside the specimen chamber. Any part on the outside of this volume will not be accessible for imaging. With some computational power, it is still possible to stitch the 3D data of individual stacks from a large specimen in a postprocessing routine (Preibisch, Saalfeld, & Tomancák, 2009). However, make sure to find a matching setup of specimen size and imaging objective(s).

2.2 REFRACTIVE INDEX AND LIGHT ATTENUATION OF THE IMMERSION MEDIUM: ("IS THE CULTURE MEDIUM SUITABLE FOR THE OPTICS?")

Most specimens require a certain medium, which should also be used to fill the specimen chamber. Many SPIM implementations use water immersion objectives and are therefore optimized for the refractive index of water (~ 1.33). While phosphate buffered saline and fish medium are well suited and can be used to image organisms from fly embryos to (freshwater) fish, the situation can change for different specimens or treatments, eg, sea water or oily solutions. Ideally, the chosen medium

2.3 DESIRED SPATIAL AND TEMPORAL RESOLUTION: ("CAN I ACTUALLY RESOLVE THE OBJECTS I WANT TO DETECT AND CAN I IMAGE FAST/LONG ENOUGH?")

2.3.1 Spatial resolution

While the lateral resolution of the microscope depends solely on the detection path, ie, the detection objective, the axial resolution is determined by the detection objective and the thickness of the light-sheet (Engelbrecht & Stelzer, 2006; Huisken, 2004). Most SPIM use Gaussian laser beams, which exhibit a Gaussian intensity profile perpendicular to the propagation axis. Further, the beam length depends on its smallest thickness (twice the waist size w_0) in the following manner (Saleh & Teich, 2001): $w(z) = w_0\sqrt{1 + (z/z_0)^2}$. Here, $w(z)$ is half of the thickness along the propagation axis, z the distance from the waist minimum. The *Rayleigh range* z_0 depends on the wavelength λ and the waist size $z_0 = \pi w_0^2/\lambda$. The possible resolution improvement stems from the fact that the detected signal is determined by the system's point spread function (PSF_{sys}) which is roughly speaking a multiplication of the illumination profile with the PSF of the detection objective (PSF_{det}). Therefore the PSF_{sys} can be made smaller by multiplying with a smaller illumination profile (Huisken, 2004).

It needs to be kept in mind that a Gaussian beam diverges all the more the narrower its thinnest point is. The larger the waist minimum on the other hand, the weaker the divergence, which leads to a more homogeneous illumination of the specimen. Consequently, the light-sheet should be crafted according to the needs of the user, ie, made smaller or larger depending on the needed resolution, the size of the field of view, and the optical properties of the sample. Adjusting the size of the beam thickness can be facilitated with the help of a beam expander.

http://www.nature.com/nmeth/journal/v11/n5/nmeth.2922/metrics - auth-1.

2.3.2 Temporal resolution

The intrinsic nature of the SPIM setup results in very fast imaging as the whole focal plane is illuminated and recorded in few milliseconds. The time needed to record an entire volume depends on the choice of illumination (static light-sheet via cylindrical lens or scanned beam via galvanometric mirror), the camera speed, the chosen exposure time, and volume size to be recorded (field of view and number of planes). Whether the microscope will be able to provide the desired temporal resolution strongly depends on the event to be observed, the quality of the fluorescent marker and the volume to be imaged. Today SPIM covers events spanning a temporal range from few milliseconds (Chen et al., 2014; Chhetri et al., 2015; Mickoleit et al., 2014) to many hours or even days (Keller et al., 2008; Krzic et al., 2012).

2.4 MOUNTING OF THE SAMPLE: ("IS THERE A RELIABLE ROUTINE TO MOUNT THE SPECIMEN OF CHOICE?")

The biggest transition from conventional to light-sheet microscopy relates to the mounting of the sample, which is essential for the success of the experiment and is yet a matter of the user's creativity. Especially for long-term experiments aiming at recording the development of an organ or animal, a reproducible mounting technique that ensures the stable position of the specimen, without inhibiting the development by physical, biological, or chemical constraints is a prerequisite.

Ideally, the live sample should be in a surrounding that closely resembles its physiological environment, but also permits sample displacement and unscattered light penetration. The choice of the proper medium, the density of the mounting gel (if embedded), and the temperature need to be adjusted carefully. Generally, it is recommended to establish a robust and reproducible mounting procedure before following up on advanced experiments with a SPIM microscope, especially with specimens that are imaged for the first time on a light-sheet microscope.

Assuming that the microscope of choice meets the above criteria and assuming that the sample fits the microscope, the most crucial and yet experimentally least defined step is the actual sample mounting. Compared to simple coverslip mounting for conventional microscopy, mounting for light-sheet imaging can be rather time-consuming and needs creativity and training. There are several different ways to assemble an imaging chamber, depending on the position and anchoring of the specimen holder inside the imaging chamber and the sample in question. On the one hand, a reliable positioning and alignment of the sample must be accomplished for time-lapse acquisition of stacks. On the other hand, especially for long-term imaging, this space needs to be flexible and needs to permit eg, development of the specimen. For multiview recordings or simply for imaging the sample under the desired angle, the specimen needs to be mounted and aligned in 3D, which is typically facilitated by a mounting aid, usually a glass capillary or a plastic tube. The latter two are then placed into a sample holder that can be moved and rotated inside the specimen chamber. Rather shear-resistant samples like *Drosophila melanogaster* embryos can directly be pushed into a stiff agarose gel, which subsequently is protruded out of the glass capillary to facilitate imaging. The surrounding gel allows the embryo to develop while ensuring its steady position throughout the recordings.

Fish embryos are mounted in a similar manner, ie, using the capillary as mounting-support. For this purpose a liquid agarose gel is likewise prepared, the embryo is plunged carefully into the gel, and then sucked into the capillary (Kaufmann, Mickoleit, Weber, & Huisken, 2012). Upon solidification of the gel, the embryo containing part can be pushed out of the glass capillary and imaged. While this is a viable approach for short-term imaging or for older embryos (after tail extension), early developmental stages can be affected by the gel's stiffness. Instead, lower gel concentration or best plain fish medium allow the embryo to develop freely.

To still ensure a fixed position for the latter case, a material other than glass is required to provide both the stiffness for moving and rotation of the specimen holder as well as to provide optical properties that allow distortion-free imaging. Fluorinated ethylene propylene (FEP) has proven to be a suitable material to meet these demands (Kaufmann et al., 2012). We found that attaching a thin-walled FEP tube (LUXENDO, EMBL Heidelberg) to the top of a glass capillary is the fastest and most reliable way to repeatedly image live and fixed specimens of various species and developmental stages. Sucked in and subsequently polymerized liquid low melting agarose can be used as an adjustable stabilizer of the sample position inside the FEP tube (Fig. 4A). The sample can be sucked, pipetted, or transferred by using capillary forces into the tube. A groove in the agarose stabilizer can be created to position the sample in the center. The surrounding medium in the tube as well as in the imaging chamber can be adjusted to the need of the specimen.

While one of the big advantages of SPIM imaging builds on its low phototoxicity allowing long-term whole mount imaging of dynamic processes, its high penetration depth makes it ideally suited for the imaging of large (fixed) organs. As an example, SPIM imaging can be used to reconstruct the long-term growth mode of a juvenile fish retina. For this, eyes of 3–4 weeks old medaka fish (fixed 16 days posthatching) have been analyzed. Multipotent retinal stem cells residing in the ciliary marginal zone—the stem cell niche that literally encircles the lens give rise to all descending retinal cell types (Centanin, Hoeckendorf, & Wittbrodt, 2011). To address their mode of division, sparse labeling of these cells has been performed by transplanting cells with a red fluorescent reporter construct into a host embryo of a green fluorescent reporter line (cf. (Centanin et al., 2011)). The *retinal homeobox factor 2 (rx2)* promoter drives the expression of fluorophores in retinal stem cells as well as Müller glia and photoreceptor cells of the neuroretina (Fig. 4B–B′, (Reinhardt et al., 2015)). It is the innate asymmetric mode of division of these cells to give rise to both a progenitor cell that further proliferates and develops into retinal cell types and to another stem cell to maintain the niche. This biological setup will over time—ie, development—give rise to a striped pattern (termed "arched continuous stripes," ArCoS; (Centanin et al., 2011)) along the vector of growth pointing from the central retina toward the lens with the stem cells residing at their tips (Fig. 4B–C). Sparse labeling can as well be achieved by the mosaic mode of CRISPR/Cas9-mediated targeted gene inactivation, if the coding sequence of *GFP* in eg, *wimbledon* medaka fish (ubiquitous *GFP* expression; (Centanin et al., 2011)) is targeted. Injections of a single-guide RNA and the *Cas9* mRNA into early blastomeres of *wimbledon* medaka will lead to a mosaic loss of GFP, hence to stem cells with or without proper *GFP* expression (Fig. 4C (Stemmer, Thumberger, del Sol Keyer, Wittbrodt, & Mateo, 2015)).

3. DATA ACQUISITION AND HANDLING

It is not uncommon for a single biologist to straightforwardly plan, perform, and analyze experiments on a confocal microscopy setup. While this is certainly not impossible with SPIM experiments, the task of data handling, processing, and

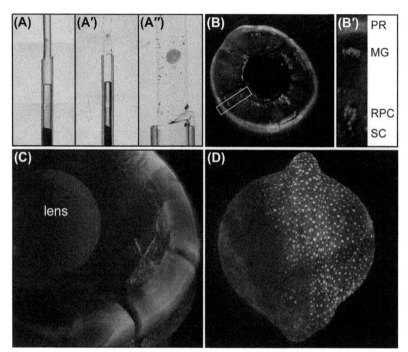

FIGURE 4 SPIM imaging of mounted samples.

(A–A″) Mounting routine of a fixed and stained enucleated juvenile medaka retina. (A) A thin-walled short fluorinated ethylene propylene (FEP) tube is inserted into a glass capillary, which provides stability for the movement of the precision stage. Sucked-in and polymerized low melting agarose serves to stabilize and fix the position of the specimen in the above small chamber. (A′) Submerging of this mounting aid demonstrates the match of the refractive indices of water and FEP ($\sim$1.33) as the FEP tube is nearly unrecognizable under water, thus not hindering imaging. (A″) Mounted juvenile medaka retina. (B–C) ArCoS imaging of medaka retinae. (B–B′) SPIM imaging of a 4-week-old medaka *rx2::GFP* line with transplanted cells from a *rx2::RFP* line. A striped pattern in the retina is detected when the donor cells are fated to become stem cells (SCs) of the ciliary marginal zone (Centanin et al., 2011). This typical ArCoS formation derives from the asymmetric mode of cell division of the neuroretinal SCs. The promoter of the *retinal homeobox factor 2* drives the expression of the fluorophores in SC, retinal progenitor cells (RPC), Müller glia (MG), and photoreceptors (PR) (Reinhardt et al., 2015). (C) ArCoS arise also due to the mosaicism of CRISPR/Cas9-mediated targeted *GFP* gene inactivation (Stemmer et al., 2015) in injected embryos from a ubiquitously expressing *GFP* line (*wimbledon*; (Centanin et al., 2011)). Four months old retina, residual GFP expression was restained in red, nuclear staining with SYTOX green (S7020, Thermo fisher). (D) Whole mount still from a developing zebrafish with green nuclei (*bact:H2B-eGFP*) and red membrane fluorescence (*bact:lyntdTomato*). (See color plate) analyzing is in many cases more demanding. Instead of dealing with a few megabytes of data per sample, a single SPIM experiment typically creates several hundred gigabytes up to several terabytes of data. The amount of data and the rate at which they are recorded places special demands on the computer hardware and the IT infrastructure as well as on the required analysis software.

To illustrate how such large amounts of data are produced, let's consider the previously mentioned volume of 1 mm^3 that has to be sampled in order to record the development of a zebrafish embryo. For a reasonable sampling of this volume ~500 planes should be recorded for each time point. The ORCA-Flash4.0 V2 Digital CMOS camera from Hamamatsu, for example, records images with 16-bit pixel depth and has a sensor size of 2048 by 2048 pixels. Therefore a single image has a size of ~8 MB. Thus, one single stack will result in ~4 GB of data. Any other view or color will add an additional factor of 2 to the data set. Recording of one view will amount to ~20 s of time which corresponds to a data rate of 200 MB/second. This exceeds most hard drives' writing speeds by several megabytes. To cope with this rate, usually RAID systems with several terabytes of space are implemented. To be able to follow and track single cells in a developing zebrafish embryo, the stacks are recorded at time intervals of 5 min (or less). Therefore, in this example, recording of 24 h of development from a single view amounts to 1.2 TB of storage. Better results will be achieved through multiview recordings, eg, with two detection arms and additional orthogonal views. As a result, close to 5 TB of storage will be required.

After the data have been acquired, they are in most cases stored on the microscopy computer and need to be moved first. For this purpose high bandwidths (>1 Gbps) and further storage should be available.

Multiview recordings need to be fused before further analysis is possible. Customized algorithms or freely available software (eg, Multi-View Fusion plugin for Fiji (Preibisch et al., 2010)) can fuse several views into one more isotropically resolved data set, enhancing the total image quality, ie, the axial resolution. For this purpose computers or workstations with high processing power, strong graphics processing units, and a large amount of memory are required.

But not only the early processing is computationally intense. Simply visualizing the data in 3D and 4D is often a time-consuming challenge. Currently, methods, such as live image fusion, image deconvolution (Schmid & Huisken, 2015), and image compression (Amat et al., 2015) aim at improving image quality and reducing data size as part of the imaging process itself. The goal for the coming years is to reduce the number of processing steps further prior to data evaluation. This way working with SPIM data eventually will become more user friendly and less time-consuming.

4. CHALLENGES AND PERSPECTIVES

The huge potential of light-sheet microscopy has triggered a rapid "evolution" and radiation of technological implementations to foster new and unprecedented experimental approaches. Technological breakthrough is often triggered by the need of the experimentalist, in the case of light-sheet microscopy by the aim of the biologist to image entire organisms alive for extended periods of time. Scaling that process will result in instruments that allow the highly parallel analyses of multiple specimens under distinct and individual conditions. Given the huge advances in genome sequencing and analyses, one could imagine light-sheet microscopy as the ultimate multidimensional phenotyping instrument allowing the association of particular phenotypic traits to specific (genetic, drug, environmental etc.,) conditions via high-throughput SPIM imaging.

On the level of the analysis of individual specimens, the functional mechanistic readout of actual biological processes is one of the big frontiers in modern biology. In particular the decoding of the nervous system can be addressed by whole mount in vivo analyses, and light-sheet microscopy has already taken the first steps along that line. The ultimate microscope will work like a scanner and will image key processes in freely swimming larva or fish.

The big challenge to be tackled in all present and future approaches is data processing, which currently takes much longer than data acquisition. On the fly approaches for image cropping, multiview live fusion, projection, and compression of data will require a rethinking of the process of data analysis and a redefinition of "raw data." As our eye is continuously processing information that we consider a representation of the reality, it will be necessary to accept on the fly preprocessing of the "big data" put out by current and future light-sheet microscopes. We can be even better than our eye by storing the "preprocessing" parameters into the actual images for critical retrospective data evaluation. There is a lot of room for new development on the level of data analysis, from highly efficient segmentation and tracking to interactive 4D analysis in virtual realities. Data analysis is the next big frontier and after the rapid evolution of the microscopy hardware, it represents the bottleneck currently preventing a rapid "radiation" of the new technology.

Ultimately several levels of sophistication of light-sheet microscopes will be appreciated by the biologist: On the one hand, there is a broad demand for a "plug and play" instrument with a complete software suite that allows the in vivo imaging and long-term studies in entire organism. Here the open SPIM community is successfully paving the way to a broader understanding and use of light-sheet microscopy (http://openspim.org).

In parallel, dedicated (likely noncommercial) high-end instruments will be tailored to addressing specific biological questions in collaborative environments providing fruitful interfaces between physics and biology.

ACKNOWLEDGMENT

Thanks to Tinatini Tavhelidse, Manuel Stemmer, Stephan Kirchmaier, and Erika Tsingos for providing the fixed and stained medaka retinae. We further thank Lars Hufnagel and Bálint

Balázs for technical support. This work was supported by the 7th framework programme of the European Union (GA 294354-ManISteC, J.W.).

REFERENCES

Amat, F., Höckendorf, B., Wan, Y., Lemon, W. C., McDole, K., & Keller, P. J. (2015). Efficient processing and analysis of large-scale light-sheet microscopy data. *Nature Protocols, 10*, 1679–1696.

Arrenberg, A. B., Stainier, D. Y. R., Baier, H., & Huisken, J. (2010). Optogenetic control of cardiac function. *Science, 330*, 971–974.

Bambardekar, K., Clément, R., Blanc, O., Chardès, C., & Lenne, P.-F. (2015). Direct laser manipulation reveals the mechanics of cell contacts in vivo. *Proceedings of the National Academy of Sciences of the United States of America, 112*, 1416–1421.

Centanin, L., Hoeckendorf, B., & Wittbrodt, J. (2011). Fate restriction and multipotency in retinal stem cells. *Stem Cell, 9*, 553–562.

Chen, B.-C., Legant, W. R., Wang, K., Shao, L., Milkie, D. E., Davidson, M. W. ... Betzig, E. (2014). Lattice light-sheet microscopy: imaging molecules to embryos at high spatiotemporal resolution. *Science, 346*, 1257998.

Chhetri, R. K., Amat, F., Wan, Y., Höckendorf, B., Lemon, W. C., & Keller, P. J. (2015). Whole-animal functional and developmental imaging with isotropic spatial resolution. *Nature Methods, 12*, 1171–1178.

Engelbrecht, C. J., Greger, K., Reynaud, E. G., Krzic, U., Colombelli, J., & Stelzer, E. H. (2007). Three-dimensional laser microsurgery in light-sheet based microscopy (SPIM). *Optics Express, 15*, 6420–6430.

Engelbrecht, C. J., & Stelzer, E. H. (2006). Resolution enhancement in a light-sheet-based microscope (SPIM). *Optics Letters, 31*(10).

Fahrbach, F. O., Simon, P., & Rohrbach, A. (2010). Microscopy with self-reconstructing beams. *Nature Photonics, 4*.

Höckendorf, B., Thumberger, T., & Wittbrodt, J. (2012). Quantitative analysis of embryogenesis: a perspective for light sheet microscopy. *Developmental Cell, 23*, 1111–1120.

Huisken, J (2004). Multi-view microscopy and multi-beam manipulation for high-resolution optical imaging. https://www.freidok.uni-freiburg.de/data/1533

Huisken, J., & Stainier, D. (2007). Even fluorescence excitation by multidirectional selective plane illumination microscopy (mSPIM). *Optics Letters, 32*.

Huisken, J., Swoger, J., Del Bene, F., Wittbrodt, J., & Stelzer, E. H. K. (2004). Optical sectioning deep inside live embryos by selective plane illumination microscopy. *Science, 305*, 1007–1009.

Kaufmann, A., Mickoleit, M., Weber, M., & Huisken, J. (2012). Multilayer mounting enables long-term imaging of zebrafish development in a light sheet microscope. *Development, 139*, 3242–3247.

Keller, P. J., Schmidt, A. D., Wittbrodt, J., & Stelzer, E. H. K. (2008). Reconstruction of zebrafish early embryonic development by scanned light sheet microscopy. *Science, 322*, 1065–1069.

Keller, P. J., & Stelzer, E. H. K. (2008). Quantitative in vivo imaging of entire embryos with digital scanned laser light sheet fluorescence microscopy. *Current Opinion in Neurobiology, 18*, 624–632.

Krzic, U., Gunther, S., Saunders, T. E., Streichan, S. J., & Hufnagel, L. (2012). Multiview light-sheet microscope for rapid in toto imaging. *Nature Methods, 9*, 730–733.

Kumar, A., Wu, Y., Christensen, R., Chandris, P., Gandler, W., McCreedy, E. ... Shroff, H. (2014). Dual-view plane illumination microscopy for rapid and spatially isotropic imaging. *Nature Protocols, 9*, 2555–2573.

Mickoleit, M., Schmid, B., Weber, M., Fahrbach, F. O., Hombach, S., Reischauer, S., & Huisken, J. (2014). High-resolution reconstruction of the beating zebrafish heart. *Nature Methods, 11*, 919–922.

Pitrone, P. G., Schindelin, J., Stuyvenberg, L., Preibisch, S., Weber, M., Eliceiri, K. W. ... Tomancák, P. (2013). OpenSPIM: an open-access light-sheet microscopy platform. *Nature Methods, 10*, 598–599.

Planchon, T. A., Gao, L., Milkie, D. E., Davidson, M. W., Galbraith, J. A., Galbraith, C. G., & Betzig, E. (2011). Rapid three-dimensional isotropic imaging of living cells using Bessel beam plane illumination. *Nature Methods, 8*, 417–423.

Preibisch, S., Saalfeld, S., Schindelin, J., & Tomancák, P. (2010). Software for bead-based registration of selective plane illumination microscopy data. *Nature Methods, 7*, 418–419.

Preibisch, S., Saalfeld, S., & Tomancák, P. (2009). Globally optimal stitching of tiled 3D microscopic image acquisitions. *Bioinformatics, 25*, 1463–1465.

Reinhardt, R., Centanin, L., Tavhelidse, T., Inoue, D., Wittbrodt, B., Concordet, J.-P. ... Wittbrodt, J. (2015). Sox2, Tlx, Gli3, and Her9 converge on Rx2 to define retinal stem cells in vivo. *The EMBO Journal, 34*, 1572–1588.

Reynaud, E. G., Krzic, U., Greger, K., & Stelzer, E. H. K. (2008). Light sheet-based fluorescence microscopy: more dimensions, more photons, and less photodamage. *HFSP Journal, 2*, 266–275.

Rubio-Guivernau, J. L., Gurchenkov, V., Luengo-Oroz, M. A., Duloquin, L., Bourgine, P., Santos, A. ... Ledesma-Carbayo, M. J. (2012). Wavelet-based image fusion in multi-view three-dimensional microscopy. *Bioinformatics, 28*, 238–245.

Saleh, B. E. A., & Teich, M. C. (2001). *Fundamentals of Photonics*.

Schmid, B., & Huisken, J. (2015). Real-time multi-view deconvolution. *Bioinformatics, 31*, 3398–3400.

Siedentopf, H., & Zsigmondy, R. (1902). Uber sichtbarmachung und größenbestimmung ultramikoskopischer teilchen, mit besonderer anwendung auf goldrubingläser. *Annalen Der Physik, 315*.

Stelzer, E., & Lindek, S. (1994). Fundamental reduction of the observation volume in far-field light microscopy by detection orthogonal to the illumination axis: confocal theta microscopy. *Optics Communications, 111*.

Stemmer, M., Thumberger, T., del Sol Keyer, M., Wittbrodt, J., & Mateo, J. L. (2015). CCTop: an intuitive, flexible and reliable CRISPR/Cas9 target prediction tool. *PLoS One, 10*, e0124633.

Swoger, J., Verveer, P., Greger, K., & Huisken, J. (2007). Multi-view image fusion improves resolution in three-dimensional microscopy. *Optics Express, 15*(13).

Tomer, R., Khairy, K., Amat, F., & Keller, P. J. (2012). Quantitative high-speed imaging of entire developing embryos with simultaneous multiview light-sheet microscopy. *Nature Methods, 9*, 755–763.

Truong, T. V., Supatto, W., Koos, D. S., Choi, J. M., & Fraser, S. E. (2011). Deep and fast live imaging with two-photon scanned light-sheet microscopy. *Nature Methods, 8*.

Vettenburg, T., Dalgarno, H. I. C., Nylk, J., Coll-Lladó, C., Ferrier, D. E. K., Čižmár, T. ... Dholakia, K. (2014). Light-sheet microscopy using an airy beam. *Nature Methods, 11*, 541–544.

Voie, A. H., Burns, D. H., & Spelman, F. A. (1993). Orthogonal-plane fluorescence optical sectioning: three-dimensional imaging of macroscopic biological specimens. *Journal of Microscopy, 170*.

Wu, Y., Ghitani, A., Christensen, R., Santella, A., Du, Z., Rondeau, G. ... Shroff, H. (2011). Inverted selective plane illumination microscopy (iSPIM) enables coupled cell identity lineaging and neurodevelopmental imaging in *Caenorhabditis elegans*. *Proceedings of the National Academy of Sciences of United States of America, 108*, 17708–17713.

CHAPTER

Single neuron morphology in vivo with confined primed conversion

6

M.A. Mohr, P. Pantazis[1]

Eidgenössische Technische Hochschule Zurich (ETH Zurich), Basel, Switzerland
[1]*Corresponding author: E-mail: periklis.pantazis@bsse.ethz.ch*

CHAPTER OUTLINE

Introduction	126
1. Photoconvertible Fluorescent Proteins	127
2. Confined Primed Conversion	128
3. Unraveling Single Neuron Morphology With Confined Primed Conversion	129
3.1 Confined Primed Conversion Enables Photoconversion of Single Cells In Vivo	129
3.2 Confined Primed Conversion of Neurons Is a Powerful Tool for Neural Morphology Analysis	132
Conclusion and Outlook	135
References	136

Abstract

Unraveling the structural organization of neurons can provide fundamental insights into brain function. However, visualizing neurite morphology in vivo remains difficult due to the high density and complexity of neural packing in the nervous system. Detailed analysis of neural morphology requires distinction of closely neighboring, highly intricate cellular structures such as neurites with high contrast. Green-to-red photoconvertible fluorescent proteins have become powerful tools to optically highlight molecular and cellular structures for developmental and cell biological studies. Yet, selective labeling of single cells of interest in vivo has been precluded due to inefficient photoconversion when using high intensity, pulsed, near-infrared laser sources that are commonly applied for achieving axially confined two-photon (2P) fluorescence excitation. Here we describe a novel optical mechanism, "confined primed conversion," which employs continuous dual-wave illumination to achieve confined green-to-red photoconversion of single cells in live zebrafish embryos. Confined primed conversion exhibits wide applicability and this chapter specifically elaborates on employing this imaging modality to analyze neural morphology of optically targeted single neurons in the developing zebrafish brain.

INTRODUCTION

Untangling the vast complexity of the brain and ultimately understanding its functionality arguably remains one of the greatest scientific challenges of our time. The brain's intricate network is thought to consist of many local computation units, so-called local neural circuits, which are connected further to execute complex functions. To acquire more insight into both the local circuitry and the global cellular brain organization maps of neural connectivity at single-cell resolution are required (Denk, Briggman, & Helmstaedter, 2012; Ji, 2014). An ideal tool to investigate neural connectivity should have the following qualities:

- the ability to resolve individual neurites and ideally single synapses,
- the means to capture a large field of view (FOV) covering long-range connections, which link different brain areas and,
- the capability to distinguish between different cells and their neurites even within densely packed, convoluted networks.

Serial electron microscopy images of small resin-embedded tissue samples can yield highly detailed reconstructions of neural processes down to individual synapses, satisfying the first requirement (Denk & Horstmann, 2004). Current studies conducted with serial electron microscopy are, however, limited to volumes much smaller than the dendritic reach of a single cell. Consequently, long neurites that connect distant brain areas are neglected. Moreover, although individual cellular structures can be distinguished through sophisticated computer reconstructions, they cannot be assigned with certainty to a specific neuron soma as they often exceed the FOV, ultimately failing to comply with the second and third requirement. It is also worth noting that serial electron microscopy cannot monitor neural dynamics, as this technique requires tissue fixation and resin embedding.

In contrast, optical imaging of fluorescently labeled neurons can provide a broad FOV covering large network areas up to whole brains, allowing imaging of long-range connections. For this purpose a host of different, sophisticated microscopy systems, optimized for different tasks have been developed—most prominently, (1) confocal laser scanning microscopy (CLSM) for high-content fluorescence imaging (Amos & White, 2003), (2) 2P Laser Scanning Microscopy (2PLSM) for deep-tissue excitation of fluorophores (Denk, Strickler, & Webb, 1990), and (3) selective plane illumination microscopy (SPIM) for very fast imaging of large brain areas (Huisken, Swoger, Del Bene, Wittbrodt, & Stelzer, 2004). These imaging means are frequently combined with noninvasive genetic labeling of neurons with different fluorescent proteins (FPs), instead of direct injection of chemical dyes into the cell soma. While genetic labeling methods render a large FOV visible for detailed optical analysis, they often lack the capability to differentially label single cells of interest. This limitation has profound effects on the ability to accurately distinguish neurites of individual neurons, which are often entangled in condensed bundles, such as in the spinal chord.

To address present shortcomings, a novel class of FPs has been developed that changes its spectral property (ie, its fluorescent color) upon light-induced photoconversion. When expressed in cells these proteins enable high-contrast sparse labeling of optically targeted cells.

1. PHOTOCONVERTIBLE FLUORESCENT PROTEINS

The discovery and evolution of FPs for optical imaging has offered a broad and colorful toolbox (Giepmans, Adams, Ellisman, & Tsien, 2006), comprising various genetically encoded fluorescent reporter molecules for noninvasive studies of biological processes. The use of tissue-specific promoters and enhancers has allowed to genetically narrow fluorescent transgene expression to certain cell types of interest, hence enabling the researcher to selectively label a distinct subset of cells in different model organisms in vivo. Yet, we still lack unambiguous genetic markers for the majority of cell types (Asakawa et al., 2008; Scott et al., 2007). Furthermore, in instances such as developmental studies and neuroscience, it is often desirable to target single cells or spatially confined subsets of cells rather than certain cell types. For example, neuroscientists would like to map the connections of the cells in a certain brain area to investigate its function without having knowledge of the specific subtypes contained within this region or of the respective specific promoters to target these cells.

Photoactivatable and photoconvertible FPs exhibit fluorescence that can be modified by light-induced chemical reactions of the chromophore. Upon photomodulation with light of a particular wavelength, photoactivatable FPs gain their fluorescent property, whereas the majority of photoconvertible FPs undergo green-to-red photoconversion. Some photoconvertible proteins take part in irreversible photoconversion reactions while for others photomodulation can be reversed using light of a specific wavelength (reviewed in Nienhaus & Nienhaus, 2014). In comparison to regular FPs, which do not change their fluorescence spectra, the capability of light-induced transitions makes these proteins ideal noninvasive molecular and cellular highlighters. This beneficial feature of such fluorescent reporters has recently been exploited for localization of fine subcellular structures and single molecules (Izeddin et al., 2014; Shcherbakova, Sengupta, Lippincott-Schwartz, & Verkhusha, 2014), for quantitative protein analysis (Müller et al., 2012; Pantazis & Bollenbach, 2012; Plachta, Bollenbach, Pease, Fraser, & Pantazis, 2011), for reversible protein activation (Zhou, Chung, Lam, & Lin, 2012), and for investigating cell dynamics during embryonic development (Dempsey, Fraser, & Pantazis, 2012; Dempsey, Qin, & Pantazis, 2014). Unlike photoactivatable FPs, photoconvertible FPs allow for visualization of the sparsely photoconverted protein population in concert with its globally labeled, nonconverted environment, making them the tool of choice when the context of the highlighted feature is of importance.

The FP Kaede—isolated from the stony coral *Trachyphyllia geoffroyi*—was the first green-to-red photoconvertible FP discovered; it exhibits a spectral shift from green to red fluorescence upon intensive ultra violet (UV) illumination (350–400 nm) (Ando, Hama, Yamamoto-Hino, Mizuno, & Miyawaki, 2002). Some of the applications of Kaede include the characterization of receptor dynamics (Schmidt et al., 2009), the tracing of cells during cell migration and tissue infiltration (Magnuson et al., 2015; Schuster & Ghysen, 2013), and the elucidation of cellular fine structure in cell clusters (Sato, Takahoko, & Okamoto, 2006). Yet, Kaede comes with inherent drawbacks: The high energy UV illumination required for efficient photoconversion of Kaede can be harmful to living cells and tissues, limiting its applicability to in vivo studies (Shaner, Patterson, & Davidson, 2007). Another major shortcoming of Kaede is its tetrameric nature, which may potentially disturb the localization and trafficking of fusion proteins, limiting its usefulness for protein fusion applications in live cell imaging (reviewed in Shaner et al., 2007). Despite these limitations, Kaede is still widely used in molecular, developmental, and cell biology. Recently potent alternative photoconvertible FPs to Kaede have been discovered and subsequently improved. This ongoing development has produced optimized monomeric proteins with bright fluorescence in both converted and unconverted states, high photoconversion efficiencies and low photobleaching, such as the two distinct Anthozoa-derived proteins mEos3 (Zhang et al., 2012) and Dendra2 (Gurskaya et al., 2006).

Given the abovementioned advantages, why is it that researches have not yet capitalized on the highly desirable features of photoconvertible FPs to study developmental or neuronal processes with high spatiotemporal resolution at the single cell level? Presumably, an important reason has been the lack of a capable method to efficiently restrict photoconversion to single cells in live tissue (Pantazis & Supatto, 2014). Traditional photoconversion using high-intensity single-photon illumination (ie, using wavelengths close to the UV range) is not confined in the axial direction and hence cannot provide highly desirable highlighting of single cells in globally labeled tissue or organ contexts. High-power, femtosecond pulsed lasers have been widely applied for spatially restricted 2P excitation. However photoconversion of single cells using these lasers is negligible (Dempsey et al., 2015), preventing the precise three dimensional (3D) confinement of photomodulation. Recently, Dempsey et al. (2015) introduced confined primed conversion as a mean to spatially restrict efficient green-to-red photoconversion to single cells. In this chapter, we will describe how this method can be used to highlight and study the morphology of individual neurons in live zebrafish larvae.

2. CONFINED PRIMED CONVERSION

Confined primed conversion uses simultaneous illumination with two different wavelengths (ie, 488 and 730 nm) to drive green-to-red photoconversion. A series of in vitro experiments led the authors to the conclusion that the 488 nm laser serves as the "priming beam" first exciting electrons of the unconverted green protein from the green ground state (S^0_{green}) to the S^2 excited state (S^2_{green})—much like regular fluorescence excitation. Without apparent fluorescence emission, the excited electrons are then further excited by absorbing photons of the 730 nm "conversion beam," which results in the transition of the protein to the red form (see Fig. 1). As a consequence of this dual-wavelength continuous-wave mechanism, photoconversion can be obtained without the use of harmful, high-power near UV-light, by instead combining two wavelengths from the visible and near-infrared spectrum (Dempsey et al., 2015).

Given the dual wavelength mechanism, restriction of photoconversion can be achieved by allowing both the priming and the converting beam to meet only at a small axially confined focal volume. Notably, spatial confinement can be achieved with a simple add-on to a conventional CLSM: a "conversion filter plate" is inserted into the light path, before the objective lens, consisting of two fused semicircular band-pass filters—each transmittable for one of the two lasers—and an opaque separator strip, separating them (Fig. 2). This device is aligned to position the separation strip parallel to the scan direction of the microscope and photoconversion is obtained by iterative simultaneous scanning of a region of interest with both priming and converting laser (Dempsey et al., 2015).

3. UNRAVELING SINGLE NEURON MORPHOLOGY WITH CONFINED PRIMED CONVERSION

Closely neighboring neurons in the nervous system extend highly intermingled processes over long distances, which renders visualization of their detailed morphology highly challenging. Singling out individual neurons by noninvasive optical means can provide a powerful tool for mapping of the structural connectome, effectively complementing current techniques such as serial electron microscopy and *trans*-synaptic or Brainbow labeling (reviewed in (Denk et al., 2012)). We demonstrate confined primed conversion of single neurons in living zebrafish larvae that express cytoplasmic Dendra2 under the control of the neural-specific HuC promoter (Kim et al., 1996; Sato et al., 2006; Fig. 4). Due to its transparency, easy breeding and broad behavioral repertoire, zebrafish has been increasingly used for neurobiological studies. Its relatively small brain, which consists of approximately 300,000 neurons, allows imaging of wide brain areas simultaneously. Though small, it shares considerable similarity with mammals at the level of fundamental components governing behavior and emotional responses, making the zebrafish an ideal model organism for connectomic studies (Portugues & Engert, 2009).

3.1 CONFINED PRIMED CONVERSION ENABLES PHOTOCONVERSION OF SINGLE CELLS IN VIVO

The zebrafish trigeminal sensory ganglion is characterized by a large central bundle of cells posterior to the eye, projecting long, subcutaneous axons across the face of

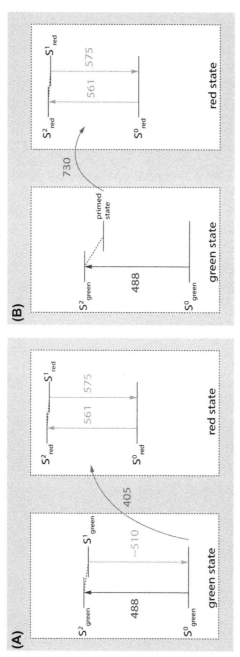

FIGURE 1 Schematic of the primed conversion mechanism.

(A) Traditional near-UV 405 nm photoconversion. In both the green and the red state, an exciting photon drives fluorescence excitation from the ground state (S^0_{green} or S^0_{red}) to an excited state (S^2_{green} or S^2_{red}). The excited electron relaxes to an energetically lower excited state (S^1_{green} S^1_{red}) before emitting a photon by relaxation back to the ground state. Absorption of 405 nm light in the green state leads to the photoconverted red state. (B) Proposed primed conversion mechanism: a 488 nm priming photon drives an electron of the chromophore from the unphotoconverted state (S^0_{green}) to an excited state (S^2_{green}), from where the primed state is reached. Subsequent absorption of a 730 nm converting photon leads to the photoconverted red state.

3. Unraveling single neuron morphology with confined primed conversion

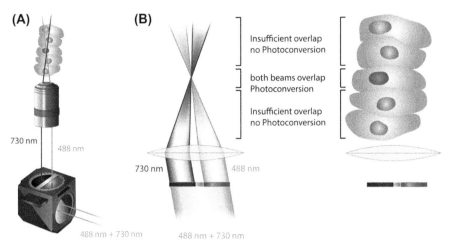

FIGURE 2 Primed conversion filter cube and scanning principle.
(A) The filter plate consisting of two semicircular optical filters transmittable for either the priming beam or the converting beam and an opaque separator strip is placed inside the light path of a confocal laser scanning microscope. The lasers (depicted as beamlets) only overlap at the focus, creating an axially defined volume of photoconversion in the sample. Cells above and below are not converted. (B) Enlarged abstracted side view.

the zebrafish (Higashijima, Hotta, & Okamoto, 2000; Knaut, Blader, Strähle, & Schier, 2005). Attempts to photoconvert individual trigeminal sensory neurons expressing cytoplasmic Dendra2 using traditional 405 nm illumination do not achieve single-cell precision and show an extended path of photoconversion in cells above and below the targeted neuron. This is due to the fact that the 405 laser beam that penetrates the tissue possesses sufficient light intensity above and below the focal volume to cause off-target photoconversion of Dendra2 expressing cells, precluding axially confined single-cell photoconversion (see Fig. 3). In contrast, confined primed conversion achieves this confinement in trigeminal ganglion neurons, expressing cytoplasmic Dendra2 (see Fig. 4). The degree of optical confinement that can be achieved with confined primed conversion is similar to the high precision observed during 2P-photoactivation of photoactivatable green fluorescent protein (paGFP) (Pantazis & González-Gaitán, 2007). However, unlike 2P-confined photoactivation, primed conversion allows for marking single cells in a globally labeled context, where many or all neurons express the unconverted FP. Furthermore, the peak-powers applied to achieve efficient primed conversion are several orders of magnitude lower than required for 2P-photoactivation (MW/cm^2 instead of TW/cm^2), greatly reducing phototoxic effects when used in live animals (Dempsey et al., 2015). Finally, in the green unphotoconverted Dendra2 channel alone, without the sparse labeling, detailed single-cell analyses are precluded by ambiguous assignment of neurites to individual cells and a lack of single-cell resolution (see Fig. 4A).

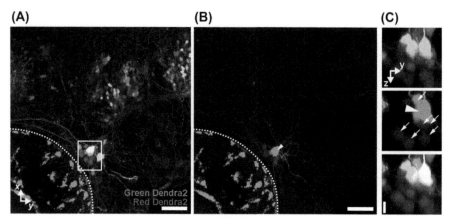

FIGURE 3 Traditional photoconversion using 405 nm light cannot specifically highlight individual neurons in tightly packed neural ganglia in vivo.

A cell surrounded by labeled neighbors within the trigeminal ganglion of a zebrafish larvae expressing cytoplasmic Dendra2 under the neural-specific HuC promoter was targeted by traditional 405 nm photoconversion. (A, B) Photoconversion of Dendra2 in a single neuron in the trigeminal ganglion at 3 days post fertilization (dpf) (maximum intensity projection (MIP), ~102 μm in depth). (C) Higher magnification of axial orientation images of the boxed region in (A, B). Many neighboring cells are seen in the green channel (green, top). In the red channel (magenta, middle), many nearby cells were photoconverted (*arrows*) along with the target cell, which can also be appreciated in the merged image (bottom). *Arrowheads* point out the target cell in the experiment. Lateral scale bars: 50 μm (A, B); axial scale bar, 10 μm (C; thickness of the bar represents 2 μm laterally). Dashed line delineates auto-fluorescent retinal pigment epithelium. (See color plate)

Adapted with permission from Dempsey, W.P., Georgieva, L., Helbling, P.M., Sonay, A.Y., Truong, T.V., Haffner, M., & Pantazis, P. (2015). In vivo single-cell labeling by confined primed conversion. Nature Methods, 12, 645–648. http://doi.org/10.1038/nmeth.3405.

3.2 CONFINED PRIMED CONVERSION OF NEURONS IS A POWERFUL TOOL FOR NEURAL MORPHOLOGY ANALYSIS

Unlike traditional photoconversion techniques using intense UV irradiation, confined primed conversion achieves photomodulation at low intensity illumination with visible and near infrared light. The ability to target single cells within a dense tissue context using low illumination powers ultimately allows for noninvasive photomodulation inside the brain of live zebrafish.

To decipher neural morphology, green cytoplasmic Dendra2 is photoconverted to its red form in the neural soma. Subsequently, the protein diffuses throughout the entire cell into the proximal neurites, allowing visualization of its entire finely branched morphology at high signal to noise ratio. Although the photoconverted red Dendra2 population is decoupled from protein production, the protein turnover is slow enough to allow for high contrast imaging and computational neurite reconstruction days after initial photoconversion (see Fig. 5).

3. Unraveling single neuron morphology with confined primed conversion 133

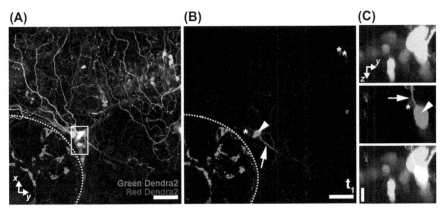

FIGURE 4 Spatially confined primed conversion enables individual neuron labeling in tightly bundled neural clusters in living zebrafish larvae.

(A, B) Visualization of a single neuron in the trigeminal ganglion directly after photoconversion of Dendra2 (time point t_1) at 3 dpf (MIP, ~82 μm in depth). (C) Higher magnification axial orientation images of the boxed region in (A). *Arrowhead* indicates the photoconverted cell, and *arrow* indicates a neurite extending from the cell body. Dashed line delineates auto-fluorescent retinal pigment epithelium. *Asterisks* in (B) and (C) indicate cells that had noticeable signal in the red Dendra2 channel even before photoconversion. (See color plate)

Adapted with permission from Dempsey, W.P., Georgieva, L., Helbling, P.M., Sonay, A.Y., Truong, T.V., Haffner, M., & Pantazis, P. (2015). In vivo single-cell labeling by confined primed conversion. Nature Methods, 12, 645–648. http://doi.org/10.1038/nmeth.3405.

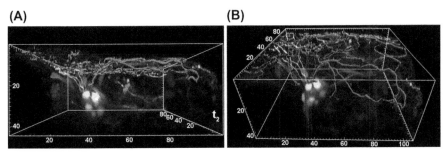

FIGURE 5 Confined primed conversion of single neurons within neural clusters in living zebrafish larvae.

Computational neurite tracing of the neuron from Fig. 4 at 24 h after photoconversion (time point t_2) of the cell soma. The side view (A) shows axial confinement of photoactivation. The traced neurites can be seen in the tilted view (B). Unconverted Dendra2 is shown in green and a computational surface reconstruction of the photoconverted Dendra2 channel is show in magenta. Major axis ticks, 20 μm. (See color plate)

Adapted with permission from Dempsey, W.P., Georgieva, L., Helbling, P.M., Sonay, A.Y., Truong, T.V., Haffner, M., & Pantazis, P. (2015). In vivo single-cell labeling by confined primed conversion. Nature Methods, 12, 645–648. http://doi.org/10.1038/nmeth.3405.

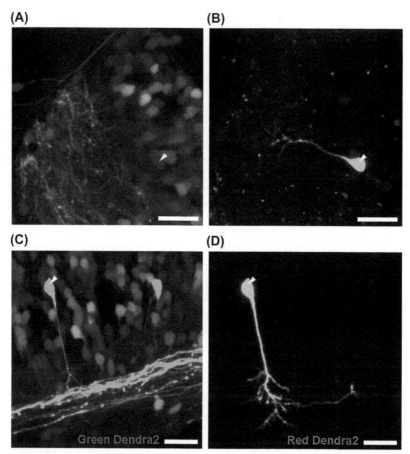

FIGURE 6 Individual labeling of various single zebrafish neurons using spatially confined primed conversion.

These panels show the individual green (left column) and red (magenta pseudocolor, right column) channels for individual neurons in the developing optic tectum (A, B) and spinal cord (C, D) within 3 dpf zebrafish expressing neural-specific Dendra2. (A, C) Individual cell projections become indiscernible within clusters of (A) synapsing neurons in the optic tectum and (C) among parallel bundled neural tracts within the developing spinal territory, which can be seen in the green Dendra2 channel. (B, D) Primed conversion was used to optically select out a cell of interest (*arrowheads*). Note that photoconverted red Dendra2 quickly diffused from the cell soma into the proximal neurites, exposing the morphology of single cells. Scale bars, 20 μm. (See color plate)

Adapted with permission from Dempsey, W.P., Georgieva, L., Helbling, P.M., Sonay, A.Y., Truong, T.V., Haffner, M., & Pantazis, P. (2015). In vivo single-cell labeling by confined primed conversion. Nature Methods, 12, 645–648. http://doi.org/10.1038/nmeth.3405.

Neural plasticity studies require morphology observation of single cells within live animals over even longer timescales. In the photoconverted target cell, green Dendra2 is constantly synthesized, which can be converted again to regain the initial high green-to-red contrast typical for primed conversion. Multiple iterations of this process could potentially allow for long-term neural plasticity studies spanning many days up to weeks.

The benefits of primed conversion for the use in single-cell neural morphology tracing in zebrafish larvae are not limited to specific brain areas. The versatility of neural tracing by primed conversion for multifacetted experiments can be also demonstrated in different areas of the zebrafish larval nervous system, including tectal cells within the central nervous system (Fig. 6 A and B) and motor neurons in the developing spinal cord (Fig. 6 C and D). Characteristic of all examples shown is the optical dissection of fine morphological structures at high contrast over long ranges, which would otherwise be difficult to achieve with current methods.

CONCLUSION AND OUTLOOK

Primed conversion is a dual-wavelength, continuous-wave mechanism, which leads to spatially confined photoconversion of photoconvertible FPs in a small volume of interest by overlapping the 488 nm priming beam and the 730 nm converting beam exclusively in the volume of interest. It is capable of spatially confining photoconversion to individual cells of interest, even within a crowded cell environment. When applied to tightly packed neurons in the zebrafish brain, this novel technique can provide structural information that can be readily integrated into a refined anatomical framework of zebrafish neural connectomics. Hence the methodology introduced here meets the aforementioned requirements for powerful neural tracing. (1) Image resolution is solely determined by the microscope being used to image the brain after photoconversion and therefore can be adjusted to the experimental needs. We have demonstrated the use of confined primed conversion in combination with CLSM. However, techniques such as SPIM, 2PLSM, and selected super-resolution microscopy techniques constitute promising approaches for postconversion fluorescence microscopy. (2) Much like image resolution, the FOV is mainly determined by the respective microscopy modality used and can be varied accordingly. (3) Finally, the capability to distinguish different cells and their neurites within complex networks is what makes primed conversion unique among all currently available techniques for neural morphology tracing. The noninvasive, sparse single-cell labeling within a dense, globally fluorescent context is highly desirable for imaging native neural morphology and linking it to the cellular surroundings.

Confined primed conversion is a noninvasive optical procedure, which requires illumination peak intensities several orders of magnitude lower than 2P-photoactivation of photoactivatable proteins that provides comparable confinement. Due to the significantly reduced phototoxicity and the relatively slow degradation of photoconverted Dendra2, it can be applied to study dynamic aspects of neural morphology in vivo.

In addition to the field of neurosciences, many areas of biological research have similar requirements for novel techniques to track cellular structures with high precision. Elucidating complex cell dynamics such as cell migration and changes in cell morphology during embryonic development requires efficient techniques to track individual cells with high fidelity in space and time (Dempsey et al., 2012). The general principle of primed conversion is a powerful approach for 3D-confined highlighting of molecular or cellular features in a complex fluorescent environment, hence uniquely satisfying these needs. Confined primed conversion can be applied to mark and visualize single cells or specific subsets of cells for various biological applications, including for (1) lineage tracing of progenitor cells during embryonic development or of stem cell populations during cell state transition, (2) analysis of cell migration and tissue penetration of immune, tumor, or stem cells, and (3) monitoring the subcellular location of fusion proteins within cells. Therefore, we predict primed conversion will impact various fields of biological research, which require 3D-confined highlighting of molecular or cellular features in vivo within an environment, where simultaneous monitoring of a global nonconverted population is desirable.

REFERENCES

Amos, W. B., & White, J. G. (2003). How the confocal laser scanning microscope entered biological research. *Biology of the Cell/Under the Auspices of the European Cell Biology Organization, 95*(6), 335–342. Retrieved from http://www.ncbi.nlm.nih.gov/pubmed/14519550.

Ando, R., Hama, H., Yamamoto-Hino, M., Mizuno, H., & Miyawaki, A. (2002). An optical marker based on the UV-induced green-to-red photoconversion of a fluorescent protein. *Proceedings of the National Academy of Sciences, 99*(20), 12651–12656. http://doi.org/10.1073/pnas.202320599.

Asakawa, K., Suster, M. L., Mizusawa, K., Nagayoshi, S., Kotani, T., Urasaki, A.... Kawakami, K. (2008). Genetic dissection of neural circuits by Tol2 transposon-mediated Gal4 gene and enhancer trapping in zebrafish. *Proceedings of the National Academy of Sciences, 105*(4), 1255–1260. http://doi.org/10.1073/pnas.0704963105.

Dempsey, W. P., Fraser, S. E., & Pantazis, P. (2012). PhOTO zebrafish: a transgenic resource for in vivo lineage tracing during development and regeneration. *PloS One, 7*(3), e32888. http://doi.org/10.1371/journal.pone.0032888.

Dempsey, W. P., Georgieva, L., Helbling, P. M., Sonay, A. Y., Truong, T. V., Haffner, M., & Pantazis, P. (2015). In vivo single-cell labeling by confined primed conversion. *Nature Methods, 12*, 645–648. http://doi.org/10.1038/nmeth.3405.

Dempsey, W. P., Qin, H., & Pantazis, P. (2014). In vivo cell tracking using PhOTO zebrafish. *Methods in Molecular Biology (Clifton, NJ), 1148*, 217–228. http://doi.org/10.1007/978-1-4939-0470-9_14.

Denk, W., Briggman, K. L., & Helmstaedter, M. (2012). Structural neurobiology: missing link to a mechanistic understanding of neural computation. *Nature Reviews. Neuroscience, 13*(5), 351–358. http://doi.org/10.1038/nrn3169.

Denk, W., & Horstmann, H. (2004). Serial block-face scanning electron microscopy to reconstruct three-dimensional tissue nanostructure. *PLoS Biology, 2*(11), e329. http://doi.org/10.1371/journal.pbio.0020329.

Denk, W., Strickler, J., & Webb, W. (1990). Two-photon laser scanning fluorescence microscopy. *Science, 248*(4951), 73–76. http://doi.org/10.1126/science.2321027.

Giepmans, B. N. G., Adams, S. R., Ellisman, M. H., & Tsien, R. Y. (2006). The fluorescent toolbox for assessing protein location and function. *Science (New York, NY), 312*(5771), 217–224. http://doi.org/10.1126/science.1124618.

Gurskaya, N. G., Verkhusha, V. V., Shcheglov, A. S., Staroverov, D. B., Chepurnykh, T. V., Fradkov, A. F. … Lukyanov, K. A. (2006). Engineering of a monomeric green-to-red photoactivatable fluorescent protein induced by blue light. *Nature Biotechnology, 24*(4), 461–465. http://doi.org/10.1038/nbt1191.

Higashijima, S., Hotta, Y., & Okamoto, H. (2000). Visualization of cranial motor neurons in live transgenic zebrafish expressing green fluorescent protein under the control of the islet-1 promoter/enhancer. *The Journal of Neuroscience: The Official Journal of the Society for Neuroscience, 20*(1), 206–218. Retrieved from http://www.jneurosci.org/content/20/1/206.abstract.

Huisken, J., Swoger, J., Del Bene, F., Wittbrodt, J., & Stelzer, E. H. K. (2004). Optical sectioning deep inside live embryos by selective plane illumination microscopy. *Science (New York, NY), 305*(5686), 1007–1009. http://doi.org/10.1126/science.1100035.

Izeddin, I., Récamier, V., Bosanac, L., Cissé, I. I., Boudarene, L., Dugast-Darzacq, C. … Darzacq, X. (2014). Single-molecule tracking in live cells reveals distinct target-search strategies of transcription factors in the nucleus. *eLife, 3*, e02230. http://doi.org/10.7554/eLife.02230.

Ji, N. (2014). The practical and fundamental limits of optical imaging in mammalian brains. *Neuron, 83*(6), 1242–1245. http://doi.org/10.1016/j.neuron.2014.08.009.

Kim, C. H., Ueshima, E., Muraoka, O., Tanaka, H., Yeo, S. Y., Huh, T. L., & Miki, N. (1996). Zebrafish elav/HuC homologue as a very early neuronal marker. *Neuroscience Letters, 216*(2), 109–112. Retrieved from http://www.ncbi.nlm.nih.gov/pubmed/8904795.

Knaut, H., Blader, P., Strähle, U., & Schier, A. F. (2005). Assembly of trigeminal sensory ganglia by chemokine signaling. *Neuron, 47*(5), 653–666. http://doi.org/10.1016/j.neuron.2005.07.014.

Magnuson, A. M., Thurber, G. M., Kohler, R. H., Weissleder, R., Mathis, D., & Benoist, C. (2015). Population dynamics of islet-infiltrating cells in autoimmune diabetes. *Proceedings of the National Academy of Sciences of the United States of America, 112*(5), 1511–1516. http://doi.org/10.1073/pnas.1423769112.

Müller, P., Rogers, K. W., Jordan, B. M., Lee, J. S., Robson, D., Ramanathan, S., & Schier, A. F. (2012). Differential diffusivity of nodal and lefty underlies a reaction-diffusion patterning system. *Science (New York, NY), 336*(6082), 721–724. http://doi.org/10.1126/science.1221920.

Nienhaus, K., & Nienhaus, G. U. (2014). Fluorescent proteins for live-cell imaging with super-resolution. *Chemical Society Reviews, 43*(4), 1088–1106. http://doi.org/10.1039/c3cs60171d.

Pantazis, P., & Bollenbach, T. (2012). Transcription factor kinetics and the emerging asymmetry in the early mammalian embryo. *Cell Cycle (Georgetown, Tex.), 11*(11), 2055–2058. http://doi.org/10.4161/cc.20118.

Pantazis, P., & González-Gaitán, M. (2007). Localized multiphoton photoactivation of paGFP in Drosophila wing imaginal discs. *Journal of Biomedical Optics, 12*(4), 044004. http://doi.org/10.1117/1.2770478.

Pantazis, P., & Supatto, W. (2014). Advances in whole-embryo imaging: a quantitative transition is underway. *Nature Reviews. Molecular Cell Biology, 15*(5), 327–339. http://doi.org/10.1038/nrm3786.

Plachta, N., Bollenbach, T., Pease, S., Fraser, S. E., & Pantazis, P. (2011). Oct4 kinetics predict cell lineage patterning in the early mammalian embryo. *Nature Cell Biology, 13*(2), 117–123. http://doi.org/10.1038/ncb2154.

Portugues, R., & Engert, F. (2009). The neural basis of visual behaviors in the larval zebrafish. *Current Opinion in Neurobiology, 19*(6), 644–647. http://doi.org/10.1016/j.conb.2009.10.007.

Sato, T., Takahoko, M., & Okamoto, H. (2006). HuC: Kaede, a useful tool to label neural morphologies in networks in vivo. *Genesis (New York, NY: 2000), 44*(3), 136–142. http://doi.org/10.1002/gene.20196.

Schmidt, A., Wiesner, B., Weißhart, K., Schulz, K., Furkert, J., Lamprecht, B. … Schülein, R. (2009). Use of Kaede fusions to visualize recycling of G protein-coupled receptors. *Traffic, 10*(1), 2–15. http://doi.org/10.1111/j.1600-0854.2008.00843.x.

Schuster, K., & Ghysen, A. (2013). Labeling defined cells or subsets of cells in zebrafish by Kaede photoconversion. *Cold Spring Harbor Protocols, 2013*(11). pdb.prot078626 http://doi.org/10.1101/pdb.prot078626.

Scott, E. K., Mason, L., Arrenberg, A. B., Ziv, L., Gosse, N. J., Xiao, T. … Baier, H. (2007). Targeting neural circuitry in zebrafish using GAL4 enhancer trapping. *Nature Methods, 4*(4), 323–326. http://doi.org/10.1038/nmeth1033.

Shaner, N. C., Patterson, G. H., & Davidson, M. W. (2007). Advances in fluorescent protein technology. *Journal of Cell Science, 120*(Pt 24), 4247–4260. http://doi.org/10.1242/jcs.005801.

Shcherbakova, D. M., Sengupta, P., Lippincott-Schwartz, J., & Verkhusha, V. V. (2014). Photocontrollable fluorescent proteins for superresolution imaging. *Annual Review of Biophysics, 43*, 303–329. http://doi.org/10.1146/annurev-biophys-051013-022836.

Zhang, M., Chang, H., Zhang, Y., Yu, J., Wu, L., Ji, W. … Xu, T. (2012). Rational design of true monomeric and bright photoactivatable fluorescent proteins. *Nature Methods, 9*(7), 727–729. http://doi.org/10.1038/nmeth.2021.

Zhou, X. X., Chung, H. K., Lam, A. J., & Lin, M. Z. (2012). Optical control of protein activity by fluorescent protein domains. *Science (New York, NY), 338*(6108), 810–814. http://doi.org/10.1126/science.1226854.

CHAPTER

Visualizing retinoic acid morphogen gradients

7

T.F. Schilling[1], J. Sosnik, Q. Nie

University of California, Irvine, CA, United States
[1]*Corresponding author: E-mail: tschilli@uci.edu*

CHAPTER OUTLINE

Introduction	140
1. Challenges for Morphogen Gradient Studies	140
2. Feedback Allows Retinoic Acid to Act as a Graded Morphogen	142
3. Cyp26s as Key Regulators of Retinoic Acid Gradient Formation	143
4. Visualizing the Retinoic Acid Gradient	146
5. Crabps and Retinoic Acid Signal Robustness	150
6. Sharpening Boundaries of Gene Expression in Response to Retinoic Acid Gradients	151
7. Noise—Both Good and Bad	156
8. Other Boundaries and Other Morphogens	157
Conclusions and Perspectives	158
Acknowledgments	159
References	159

Abstract

Morphogens were originally defined as secreted signaling molecules that diffuse from local sources to form concentration gradients, which specify multiple cell fates. More recently morphogen gradients have been shown to incorporate a range of mechanisms including short-range signal activation, transcriptional/translational feedback, and temporal windows of target gene induction. Many critical cell—cell signals implicated in both embryonic development and disease, such as Wnt, fibroblast growth factor (Fgf), hedgehog (Hh), transforming growth factor beta (TGFb), and retinoic acid (RA), are thought to act as morphogens, but key information on signal propagation and ligand distribution has been lacking for most. The zebrafish provides unique advantages for genetics and imaging to address gradients during early embryonic stages when morphogens help establish major body axes. This has been particularly informative for RA, where RA response elements (RAREs) driving fluorescent reporters as well as Fluorescence Resonance Energy Transfer (FRET) reporters of receptor binding have provided evidence for gradients, as well as regulatory mechanisms that attenuate noise and enhance gradient robustness in vivo. Here we summarize available tools in zebrafish and discuss their utility for studying dynamic regulation of RA morphogen gradients, through combined experimental and computational approaches.

INTRODUCTION

Signals that determine multiple cell fates in a concentration-dependent manner are known as morphogens. Many of the major cell-signaling pathways studied in biology—Wnt, Fgf, Tgfb, etc.—work this way in some contexts. The morphogen gradient is a fundamental concept in developmental biology, originally described by Lewis Wolpert's "French Flag" model for the developing chick limb bud, in which cells interpret different threshold concentrations of morphogen resulting in distinct fates (Fig. 1A) (Tickle, Summerbell, & Wolpert, 1975). However, the mechanisms producing morphogen gradients are probably diverse. What evidence is required to validate Wolpert's model for a given morphogen? To start with, it needs to be present as a gradient, and the perceived gradient needs to translate into gene expression activation thresholds. Furthermore for Wolpert's model to work as intended recent studies have revealed additional constraints: shaping the gradient, making its response robust, creating sharp gene expression boundaries, and dealing with biological noise (Lander, 2007; Meinhardt, 2009; Wartlick, Kicheva, & Gonzalez-Galtan, 2009). Can fields of cells really generate smooth gradients such as Wolpert envisioned or are they noisy (Fig. 1B)? If the answer is the latter, as seems likely, how do sharp boundaries of gene expression form in the face of variability in signal production, cellular architecture, and environmental fluctuations?

1. CHALLENGES FOR MORPHOGEN GRADIENT STUDIES

The problem is more complex than it appears at face value. Recent studies have revealed unexpected dynamics of both ligand and response, positive and negative feedback, and mechanisms for scaling gradients to adjust for changes in tissue size and shape (Fig. 1C,D) (Ben-Zvi, Shilo, & Barkai, 2011; Briscoe & Small, 2015; Horinaka & Morishita, 2012; Meinhardt, 2015). Gene regulatory networks that specify different cell fates based on concentration may elicit different responses depending on regulatory mechanisms (eg, feedback) within the network (Horinaka & Morishita, 2012). For example, positive feedback loops can generate bistability, where cells transition through a less stable state as the morphogen signal increases (Fig. 1C). This may help sharpen boundaries of target gene expression through switchlike responses. In contrast, negative feedback can cause oscillations due to cyclical levels of inhibition, which leads to periodic patterns (Fig. 1C). Two or more signals may also act in parallel, such as a combination of activation and inhibition, leading to responses only within a middle range of input (Horinaka & Morishita, 2012).

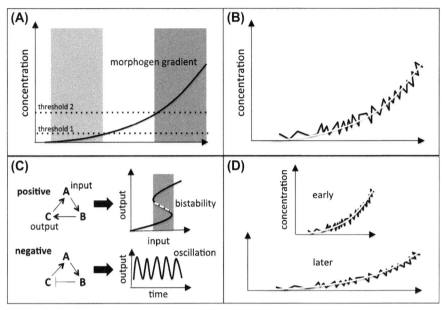

FIGURE 1 Morphogen dynamics and regulation.

(A) Standard representation of a morphogen gradient, adapted from L Wolpert's "French flag" model. The *solid line* denotes the morphogen concentration (Y axis)—highest at its source to the right of a field of responding cells (X axis). *Dotted lines* denote concentration thresholds at which cells respond differently. Blue, white, and red regions represent three distinct cell fates. (B) Hypothetical noisy morphogen gradient (*black line*) that on average matches its smooth counterpart (*white line*). (C) Examples of positive and negative feedback on signal output. Through positive feedback, a given variable input (A—X axis in graph) can result in two stable outputs (Y axis in graph), with an intervening transition state (red). Similar input driving negative feedback can result in signal oscillations over time (X axis). (D) Scaling of a morphogen gradient as the field of cells over which it acts grows (X-axis). (See color plate)

These regulatory mechanisms remain largely unknown for most putative morphogens. In fact it remains controversial if they even form gradients, and few studies have examined their spatial and temporal dynamics (Stathopoulos & Iber, 2013). The best studied is Bicoid in *Drosophila*, which clearly forms a gradient of nuclear protein along the anterior–posterior (A–P) axis in the early embryo (Driever & Nusslein-Volhard, 1988a, 1988b; Grimm, Coppey, & Wieschaus, 2010). But this is an unusual case in that Bicoid is a transcription factor, which forms a cytoplasmic gradient within a syncytium. For the more common secreted, extracellular morphogens (eg, Bmps, Fgfs, Wnts, Shh), research has relied on indirect methods to visualize the ligands or their cellular responses (eg, fluorescently tagged ligands and reporters), due to technical limitations in imaging the molecules involved directly (Alexander et al., 2011; Balasubramanian & Zhang, 2015; Bokel & Brand,

2013; Briscoe & Small, 2015; Muller et al., 2012; Ramel & Hill, 2013; Strigini & Cohen, 2000; Tuazon & Mullins, 2015). Most of these visualization methods cannot detect changes on rapid timescales. They also may miss fine cellular processes that provide direct contacts between signaling and responding cells (Prols, Sagar, & Scaal, 2015). Recent studies call into question several classic peptide morphogens, including Wg in the *Drosophila* wing disc where a membrane-tethered form can suffice for function (Alexandre, Baena-Lopez, & Vincent, 2014), and Shh in the vertebrate neural tube, where cell rearrangements rather than concentration thresholds can account for many fate outcomes (Xiong et al., 2013). There is also growing recognition that signals are noisy from embryo to embryo and from cell to cell. How do cells interpret signals in the face of stochastic fluctuations in both space and time?

One putative nonpolypeptide morphogen that has stood the test of time is the vitamin A derivative, retinoic acid (RA). RA influences the behaviors of many cell types in embryos (eg, heart, gut, somites, hindbrain, craniofacial skeleton), as well as adult stem cells (eg, neural, pancreatic), cancers (leukemia), and regenerating organs (cardiomyocytes) (Rhinn & Dolle, 2012; White & Schilling, 2008). One of the best-studied roles for RA is in anterior–posterior (A–P) patterning during vertebrate gastrulation, where it acts in parallel with Fgfs and Wnts to promote posterior development, particularly in the developing hindbrain (Kudoh, Wilson, & Dawid, 2002; Schilling, Nie, & Lander, 2012). In this context, RA fits all of the major morphogen criteria, acting at long range to determine multiple cell fates in a concentration-dependent manner. Here we summarize recent work in zebrafish combining developmental genetics, new imaging methods, and computational modeling of hindbrain development to reveal an integrated signaling network that can help explain RA's dynamics and precision as a morphogen.

2. FEEDBACK ALLOWS RETINOIC ACID TO ACT AS A GRADED MORPHOGEN

The shapes of morphogen gradients are determined by the source of the ligand, its rate of production, transport properties, and stability (Ben-Zvi & Barkai, 2010; Sample & Shvartsman, 2010; Umulis, Shimmi, O'Connor, & Othmer, 2010). Gradient shape also depends on feedback mechanisms such as self-enhanced receptor-mediated degradation, which helps make gradients robust—ie, able to compensate for variability in morphogen availability. This has been demonstrated for growth factors of the TGFb, Wg, and Hh families (Eldar, Rosin, Shilo, & Barkai, 2003; Meinhardt, 2009; Wartlick et al., 2009).

Both positive and negative feedback are critical for RA signaling (Fig. 2A) (Rhinn & Dolle, 2012; White & Schilling, 2008). Unlike polypeptide morphogens, RA is a lipophilic vitamin A derivative synthesized by aldehyde dehydrogenases (Aldhs) and degraded by cytochrome p450s (Cyp26s) within cells. Once synthesized, extracellular and intracellular binding proteins (RBPs, CRABPs) bind to, solubilize, and transport RA first into the cytoplasm and then into the nucleus where it binds nuclear hormone receptors (RARs, RXRs). RA negatively regulates its own synthesis by Aldh1a2 and positively regulates both its precursors (eg, Lrat) and receptors (RARs) (Fig. 2A). How does this influence A–P patterning of the developing hindbrain?

Morphogen gradients typically include local sources of ligand production and tightly regulated ligand degradation (Briscoe & Small, 2015; Lander, 2007; Meinhardt, 2015). For RA and A–P patterning of the hindbrain, this arrangement occurs during gastrulation. In all vertebrates, RA is synthesized posteriorly in the mesoderm (*aldh1a2* expression) and degraded anteriorly in the future forebrain/midbrain (*cyp26a1* expression), as exemplified in a zebrafish embryo at 8 h postfertilization (hpf) (Schilling et al., 2012; White, Nie, Lander, & Schilling, 2007). This suggests that RA travels from source to sink across the future hindbrain territory to establish a gradient. Most studies have focused on the potential steady-state gradients that this arrangement would produce, but as we discuss below temporal (pre-steady state) dynamics may be just as, if not more, important for hindbrain segmentation.

Evidence for an RA morphogen gradient in hindbrain development is strong. The hindbrain consists of eight segments (rhombomeres), each containing different types of interneurons and motor neurons that prefigure the cranial nerves (Fig. 2B) (Lumsden and Keynes, 1989; Trevarrow, Marks, & Kimmel, 1990). Aldh1a2 expression is restricted to the mesoderm flanking the posterior hindbrain/anterior spinal cord where it converts vitamin A into RA. This RA then diffuses or is transported anteriorly where it directly regulates transcription factors that specify different rhombomeres. Dietary depletion of vitamin A in chick embryos or loss-of-function mutations in Aldh1a2, both in zebrafish and mice, lead to a loss of posterior and expansion of anterior rhombomeres (Begemann, Schilling, Rauch, Geisler, & Ingham, 2001; Niederreither, Vermot, Schuhbaur, Chambon, & Dolle, 2000; White & Schilling, 2008). Conversely treating embryos with exogenous RA expands posterior at the expense of anterior rhombomeres. Importantly both loss- and gain-of-function approaches are dose dependent—higher doses of pharmacological inhibitors of Aldhs or RA treatments lead to progressively more severe anteriorization or posteriorization, respectively.

3. CYP26S AS KEY REGULATORS OF RETINOIC ACID GRADIENT FORMATION

How does RA degradation influence its gradient properties? Morphogen models typically require some form of tightly controlled ligand removal (Lander, 2007; Briscoe & Small, 2015). Patterns of Cyp26a1 expression suggest that it forms an anterior sink for RA at the anterior end of the hindbrain. Analyses of transgenic reporters of RA signal activation in mice (RARE-lacZ) during hindbrain segmentation have suggested that "shifting boundaries" of two other RA-degrading enzymes,

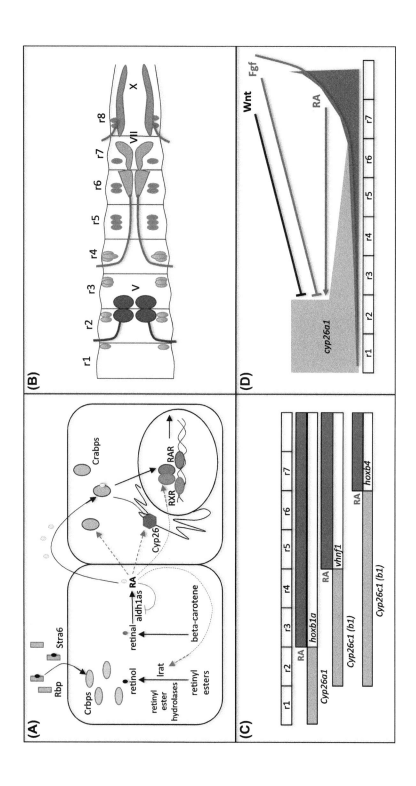

FIGURE 2 Retinoic acid (RA) as a morphogen in hindbrain patterning.

(A) Feedback in RA signaling. Signaling cell (left), responding cell (right). Vitamin A (retinol) transported by retinol-binding proteins (Rbps, *light green rectangles*) and Stra6 into cells or derived from retinyl esters via Lrat, associates with cellular retinol-binding proteins (Crbps, *light red ovals*). Retinol (black) is converted to retinal (red) and then to RA (yellow) by aldehyde dehydrogenases (Aldh1as). RA travels within cells bound to cellular RA-binding proteins (Crabps, *light blue ovals*), either to the nucleus to bind RARs (*blue ovals*) or to Cyp26s (red hexagon) associated with endoplasmic reticulum for degradation. Known positive (green, *dashed arrows*) feedback within the pathway includes Lrat, Crabps, Cyp26s, and RARs. Known negative (*red lines*) feedback includes Aldh1a2. (B) Rhombomeric organization in zebrafish. Eight rhombomeres (r1–8, anterior to the left) contain distinct sets of interneurons (blue) and motor neurons (V, trigeminal, purple; VII, facial, orange; X, vagal, green). (C) Shifting boundaries of RA degradation and hindbrain patterning based on RARE:lacZ transgenic reporters in mice. Model depicting rhombomeres at top, Cyp26s in blue, RA in red. An early Cyp26a1 domain sets the r2/3/*hoxb1a* expression boundary, a later Cyp26c1 (b1 in zebrafish) sets the r4/5/*vhnf1* expression boundary, and Cyp26c1 expands posteriorly to demarcate the r6/7/*hoxb4* boundary. (D) An integrated signaling network for hindbrain patterning. Model depicting rhombomeres at bottom, Cyp26a1 in blue, RA signaling in red, Fgf signaling in green, Wnt signaling in black. Cyp26-mediated degradation is continuously under feedback and feedforward control from Wnt, Fgf, and RA signaling, respectively, which shapes the RA gradient. (See color plate)

Adapted from White, R.J. & Schilling, T.F. (2008). How degrading: Cyp26s in hindbrain development. Developmental Dynamics, 237, 2775–2790. http://dx.doi.org/10.1002/dvdy.21695 and Schilling, T.F., Nie, Q. & Lander, A.D. (2012). Dynamics and precision in retinoic acid morphogen gradients. Current Opinion in Genetics and Development, 22, 562–569. http://dx.doi.org/10.1016/j.gde.2012.11.012.

Cyp26b1 and Cyp26c1, progressively establish more posterior rhombomeres (Fig. 2C) (Sirbu, Gresh, Barra, & Duester, 2005). Functional studies of these three enzymes in zebrafish have shown that while loss of any one Cyp26 causes mild hindbrain defects, a loss of all three transforms the entire hindbrain into an r6/7 fate (Hernandez, Putzke, Myers, Margaretha, & Moens, 2007). While these two studies confirm the importance of RA degradation in patterning, the authors also argue for a model in which domains of RA degradation, rather than a gradient per se, pattern rhombomeres. Hernandez et al (Hernandez et al., 2007) point out that a gradient model seems inconsistent with the fact that in embryos devoid of RA (ie, DEAB treated to inhibit RA synthesis), exposure to a uniform concentration of exogenous RA can restore normal patterning. This calls the morphogen model for RA in the hindbrain into question.

However, these studies (Hernandez et al., 2007; Sirbu et al., 2005) fail to take into account one critical feature of any such morphogen system, feedback. In this case the focus of feedback is at the level of degradation. Hints at this come from the observation that RA induces Cyp26a1 expression, and that the range over which RA induces target gene expression increases upon Cyp26a1 inhibition (White et al., 2007). Cyp26a1 is also expressed at lower levels throughout the hindbrain field, and inhibited by two other posteriorizing signals, Fgf and Wnt (Kudoh et al., 2002). This forms the basis for a new, modified version of the gradient model in which self-enhanced degradation of RA forms part of an integrated network of posteriorizing signals (Fig. 2D) (White et al., 2007). Computational models confirm that this integrated system can account for many of the observed results, such as the restoration of a gradient in the presence of uniform, exogenous RA. Such a system is also robust to fluctuations in RA levels and "scales," eg, it adapts as the hindbrain grows and the proximity of source and sink change (see Fig. 1D).

4. VISUALIZING THE RETINOIC ACID GRADIENT

These results beg the question of the nature of the gradient itself. Does it occur at the level of extracellular or intracellular RA? What is its shape? Unlike most putative morphogens, which are peptidic and synthesized de novo in developing embryos, RA presents additional challenges when it comes to microscopic observation. RA is a small lipophilic molecule that results from two consecutive enzymatic reactions that modify vitamin A of dietary origin (see Fig. 2A). Thus unlike Wnts or Fgfs, genetically encoded versions of fluorescently tagged RA cannot be generated. These characteristics have driven alternative strategies for visualizing RA signaling, most notably RA-response elements (RAREs) found in direct transcriptional targets of RA receptors driving lacZ or fluorescent reporters (Fig. 3). In mice a triplet of concatenated RAREs derived from the RARb receptor, along with a minimal heat shock promoter, driving lacZ has been used for decades to detect RA in the nM range (Rossant, Zirngibl, Cado, Shago, & Giguere, 1991; Sirbu et al., 2005). In zebrafish, a similar transgenic construct driving eYFP (3xRARE:eYFP) responds to RA in the same range (Fig. 3B) (Perz-Edwards, Hardison, & Linney, 2001). Analysis of 3xRARE:eYFP reveals a graded response at the hindbrain/spinal cord junction that falls off rapidly at the level of r7 (Fig. 3C) (White et al., 2007). However, it fails to detect an RA gradient further anteriorly, across the developing hindbrain field, or elsewhere that RA is known to act. This is likely due to a lack of sufficient sensitivity of the 3xRARE:eYFP reporter. More recent attempts to generate better RA reporters in zebrafish have created transgenes with more concatenated copies of RARE, promoters of other RA target genes such as Cyp26a1, or the ligand-binding domain of RAR driving another transcriptional activator and its target sequences, to try and amplify the signal (Table 1) (Huang et al., 2014; Li et al., 2012; Mandal et al., 2013; Waxman & Yelon, 2011). However, while some of these transgenics report more broadly, none show clear gradients.

Recently the Miyawaki laboratory at the RIKEN institute in Saitama, Japan, has developed Genetically Encoded reporter Probes for RA (GEPRA) (Fig. 3D) (Shimozono, Iimura, Kitaguchi, Higashijima, & Miyawaki, 2013). These fusion proteins are composed of the RA binding pocket of an RAR with blue and yellow fluorescent proteins in their C- and N- termini. These reporters function based on Fluorescence Resonance Energy Transfer (FRET) between the fluorescent proteins (CFP, YFP) that surround the binding pocket. RA binding changes the reporter conformation, which alters the resonance energy transfer. This is directly proportional to RA levels, and therefore allows visualization of the distribution of RA in vivo, revealing gradients in developing zebrafish embryos (Fig. 3E and F). Consistent with previous studies, these gradients appear during gastrulation and are eliminated by inhibiting RA synthesis. Furthermore the highest RA levels occur near the head-trunk boundary and decline both anteriorly, across the future hindbrain field, and posteriorly across the developing somitic mesoderm, after gastrulation. Interestingly depleting GEPRA-B transgenics of RA (with DEAB) and simultaneously bathing them in uniform RA reestablishes a clear RA gradient by 10– 11 hpf (three to four somites) (Fig. 3G) (Shimozono et al., 2013). These results provide strong evidence for the morphogen gradient model for RA.

However, while these GEPRA reporters have proven to be powerful tools to study RA, they have limitations. Because they rely on RA binding to the reporter, the measurements obtained are indirect—the actual microscopic observations depend on the fluorescence of the reporter and sensitivity of detection—ie, they depend on ratiometric imaging of CFP/YFP. This means that the results obtained depend on the reporter's dissociation constant (K_d; Fig. 3E). Because different reporters have different K_d, Shimozono et al. (2013) report different RA gradient shapes depending on which reporter construct is used. In addition, because they depend on K_d, the association/dissociation times of GEPRAs overlap with the temporal fluctuations of RA and render these reporters unsuitable for accurate temporal analyses. Future studies, such as Fluorescence Lifetime Microscopy to visualize RA autofluorescence, have the potential to analyze such gradient dynamics (Stringari et al., 2011).

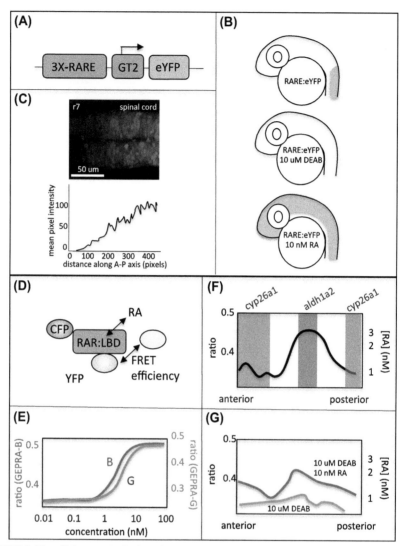

FIGURE 3 Visualizing the Retinoic acid (RA) morphogen gradient.
(A) Construct (RARE:eYFP) most commonly used to monitor RA signaling in zebrafish, containing three RA response elements (RAREs) from the mouse RARb gene, a GATA-2 basal promoter (GT2), and an enhanced yellow fluorescent protein (eYFP) (Perz-Edwards et al., 2001). (B) RARE:eYFP transgenic zebrafish embryos show expression in the spinal cord, which is lost with 10 μM DEAB treatments and induced throughout the CNS with application of 10 nM exogenous RA. (C) Confocal image of RARE-YFP fluorescence at 24 hpf (upper panel, dorsal view, anterior to the left) and quantification of YFP fluorescence at the hindbrain/spinal cord boundary (lower panel). *(Adapted from White, R.J., Nie, Q., Lander, A.D. & Schilling, T.F. (2007). Complex regulation of cyp26a1 creates a robust retinoic acid gradient in the zebrafish embryo. PLoS Biol, 5, e304.)* (D) GEPRA reporters based on the RAR ligand-binding domain (LBD) and fused to CFP and YFP. (E) Dose-response shows sensitivity between 1 and 10 nM of GEPRA-B (red) and GEPRA-G (blue) reporters. (F) Graph

Table 1 Transgenic RA Reporters in Zebrafish

Reporter	Promoter	Advantages	Disadvantages	References
3xRARE:YFP	GATA-2 (GT2)	sensitive (nM)	dim; late onset (13 hpf)	Perz-Edwards et al. (2001)
3xRARE:GFP	thymidine kinase (tk)	sensitive (nM)	dim; late onset (13 hpf); neural tube/retina only	Perz-Edwards et al. (2001)
12xRARE:eGFP	elongation factor (ef1a)	sensitive (nM)	dim; late onset (13 hpf)	Waxman and Yelon (2011)
cyp26a1:eYFP	b-actin	sensitive (nM); early onset (8 hpf)	dim; some non-RA-dependent expression	Li et al. (2012)
GDBD-RLBD; UAS:GFP	b-actin	sensitive (nM)	dim; late onset (13 hpf)	Mandal et al. (2013)
VPBD-GDBD; UAS:GFP	b-actin	hypersensitive	hypersensitive	Mandal et al. (2013)
4xRARE-cFos:QF; QUAS:GFP	c-fos	sensitive (nM)	dim; late onset (13 hpf)	Huang et al. (2014)
GEPRA-B/GEPRA-G	N/A	measure (RA) quantitatively; early onset	K_a dependence; dim	Shimozono et al. (2013)

Comparison of the composition, advantages, and disadvantages of different published RA reporters. Many use concatenated RA response elements (RAREs)—5′-ggttca(n5)agttca-3′—based on the RARb receptor in mice, with different numbers of RAREs and basal promoters (column 2). Surprisingly these all seem to have similar sensitivities in the nM range and are first detected after gastrulation, at 13 hpf. For 4RARE-cFos:QF; QUAS:GFP the RAREs were cloned upstream of a cFos minimal promoter and sequence encoding the QF transcriptional activator, and in the same transgene QF-binding upstream activating sequence drives GFP. Other reporters use the RAR ligand-binding domain (RLBD). RLBD is either fused to the Gal4 DNA-binding domain (GDBD) or a VP16-GDBD (VDBD) together with Gal4-binding upstream activating sequences (UAS) driving GFP. For Genetically Encoded Probes for RA (GEPRA) reporters the RLBD is fused to CFP and YFP to allow monitoring of RA binding using Fluorescence Resonance Energy Transfer.

based on ratiometric imaging of GEPRA fluorescence intensity measured at 12 hpf reveals graded RA levels between 0.5 and 3 nM, distributed along the anterior–posterior axis (X axis) between its source in the domain of aldh1a2 expression (red) and both anterior and posterior domains of cyp26a1 expression (blue). (G) GEPRA measurements of RA levels in embryos treated with DEAB (RA depleted, *blue line*) and simultaneously treated with 10 μM DEAB and 10 nM RA (*green line*), which partially restores the gradient. (See color plate)

Adapted from Shimozono, S., Iimura, T., Kitaguchi, T., Higashijima, S. & Miyawaki, A. (2013). Visualization of an endogenous retinoic acid gradient across embryonic development. Nature, 496, 363–366. http://dx.doi.org/10.1038/nature12037.

5. CRABPS AND RETINOIC ACID SIGNAL ROBUSTNESS

Previous studies of morphogens have largely treated cells as perfect detectors of invariant signals, but this is almost certainly never the case (Briscoe & Small, 2015; Horinaka & Morishita, 2012; Lander, 2007; Meinhardt, 2009; Wartlick et al., 2009). The concept of "robustness" in this context in embryonic development refers to the relative insensitivity of pattern formation to variability and uncertainty, such as from environmental factors (temperature, nutrition), genetic differences, or the stochastic nature of biochemical processes. One might expect this to be particularly problematic for a signal like RA, which derives from vitamin A in the diet (White & Schilling, 2008). This issue also relates to the problem of scaling. How does the RA morphogen gradient adapt to changes in the size and shape of the hindbrain?

General strategies for studying robustness and scaling involve finding mechanisms that control ligand distribution or modulate cellular responses to the signal. How are the source and rate of ligand production controlled, how is it transported, and what determines its stability (Ben-Zvi & Barkai, 2010; Sample & Shvartsman, 2010; Umulis et al., 2010)? A common feedback mechanism involving self-enhanced receptor-mediated degradation has been shown to improve robustness for many growth factors including TGFb, Wg, and Hh (Eldar et al., 2003; Meinhardt, 2009; Wartlick et al., 2009).

Prime candidates for an analogous robustness mechanism in the RA system are the Cyp26s, which degrade intracellular RA. Our experimental results suggest that the RA gradient critically depends on self-enhanced degradation for gradient maintenance (Fig. 2D) (White et al., 2007) and this depends on Cyp26a1, which is induced by RA (Fig. 4A) and limits the range of RA action (Fig. 4B). Our computational models, in which we include known parameters for RA diffusion, transport and signal transduction, show that self-enhanced degradation makes the gradient at least twofold more robust to changes in RA synthesis than the case without degradation (although this results in a much smaller change in RA gradient slope).

The other prime candidates are the RA binding proteins, which solubilize RA and transport it both extracellularly (RBPs) and intracellularly (Crabps) (see Fig. 2A) (Astrom et al., 1991; Budhu, Gillilan, & Noy, 2001; Budhu & Noy, 2002; Delva et al., 1999). Studies in mice have failed to find any functional requirements for the two Crabps, Crabp1 and Crabp2 (Lampron et al., 1995). In contrast, studies in zebrafish have shown that Crabp2a is RA inducible and required for robustness (Fig. 4C and D) (Cai et al., 2012). Similar to Cyp26a1, Crabp2a can negatively regulate RA signaling, since depleting it from zebrafish embryos makes them hypersensitive to small amounts of exogenous RA. 1 nM RA, which normally has no effect on wild-type embryos, induces RARE-YFP throughout the CNS in Crabp2a-deficient embryos. Thus like Cyp26a1, negative feedback through induction of Crabp2a dramatically improves the robustness of the response to RA.

Computational models predict that only the presence of a Crabp that acts negatively in the pathway can improve signal robustness to this extent, perhaps buffering RA in the cytoplasm away from its nuclear receptors (Fig. 4E) (Cai et al., 2012). We have used models based on known kinetics of RA interactions with RARs, Crabps, and Cyp26s to explore the effects of differing levels of RA synthesis on the resulting gradients (Fig. 4F). We calculate a robustness index (E) based on the difference in slope between the gradients generated by different levels of RA synthesis—which cross the same thresholds at different locations along the A–P axis. We derive E by calculating the normalized mean horizontal shift between the two gradients—where they cross 20% and 80% thresholds—and more robust gradients will have lower E values. This approach has the advantage that it integrates spatial information across the entire gradient as opposed to a single threshold. Crabp2a could buffer RA in the cytoplasm, preventing its nuclear localization, or it could promote RA degradation. Our models predict that Crabp2a promotes RA degradation, ie, by varying the corresponding parameters the most severe effects occur when varying the ability of Crabp2a to deliver RA to Cyp26s (Fig. 4F). Thus, the combined activities of Crabp2a and Cyp26a1 significantly improve the robustness of RA patterning in response to large variations in RA synthesis.

6. SHARPENING BOUNDARIES OF GENE EXPRESSION IN RESPONSE TO RETINOIC ACID GRADIENTS

Morphogens ultimately function to specify distinct spatial domains of gene expression and do it accurately from embryo to embryo. However, this robustness comes at a cost. Any morphogen gradient becomes shallower in slope further away from its source and more susceptible to stochastic fluctuations (ie, noise). Cells near future gene expression boundaries experience noise in morphogen concentration, ability to respond (eg, number of receptors), transcription/translation of target genes, and feedback (Elowitz, Levine, Siggia, & Swain, 2002; Kaern, Elston, Blake, & Collins, 2005; Kepler & Elston, 2001). In spatial patterning systems, noise is generally considered as detrimental.

Yet even boundaries far from the morphogen source eventually become razor sharp, leading to the question of exactly how these domains of gene expression sharpen? Our computational models predict that boundary sharpening the zebrafish hindbrain occurs via transition zones, in which cells express a mixture of genes eventually restricted to the anterior or posterior sides of the boundary (Fig. 5A) (Zhang et al., 2012). Sharpening requires large changes in gene expression in response to small changes in morphogen signal. Thus the signal needs amplification, but this potentially increases noise. Such spatial stochastic dynamics of morphogens are poorly studied in any system. How does a single morphogen specify multiple gene expression boundaries?

This problem is acute in the embryonic hindbrain, where up to seven rhombomere boundaries need to sharpen. Initially each boundary is very rough, forming

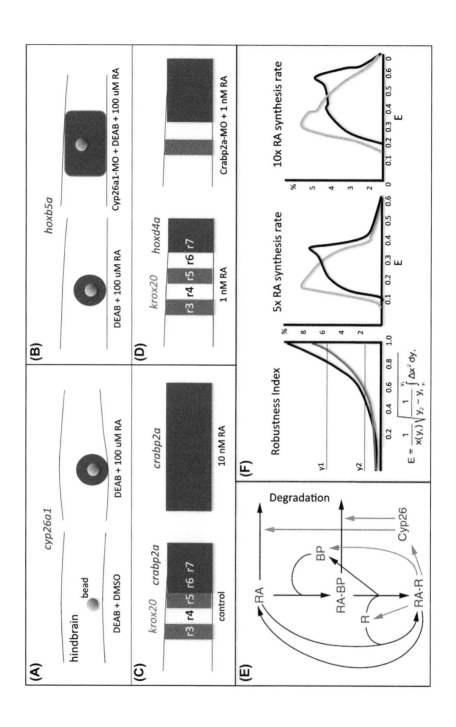

FIGURE 4 Negative feedback through Cyp26a1 and Crabp2a improve gradient robustness.

(A) Induction of *cyp26a1* expression (purple) by a bead soaked in 100 μM retinoic acid (RA) (right panel) implanted into the hindbrain region of a DEAB-treated embryo, in contrast to a control DMSO-soaked bead (left panel). Dorsal views, anterior to the left. (B) Induction of *hoxb5a* expression (purple) by an RA-soaked bead implanted into DEAB-treated embryos extends over a longer range in an embryo depleted of Cyp26a1 (injected with a Cyp26a1-MO). (C) Induction of crabp2a expression (purple) by treatment of an embryo with 10 nM RA extends throughout the hindbrain and correlates with loss of markers of anterior rhombomeres such as *krox20* (red) in r3 and r5. (D) Induction of *hoxd4a* expression (purple) by treatment with 1 nM RA extends up to the r4/5 boundary in an embryo depleted of Crabp2a (injected with a Crabp2a-MO). (E) Bound and unbound states of RA within responding cells and paths to degradation, which are included in computational models. (F) Left graph, robustness index (E—formula shown below) comparing experimental (red) and reference (black) gradients by where they cross two thresholds (y1, y2). Right graphs show two examples of probability density distributions (percentages, Y axis) of E values (X axis) for models that either include binding proteins (*blue lines*) or do not (black lines) with either a 5-fold or 10-fold increase in RA synthesis rate. (See color plate)

Adapted from White, R.J., Nie, Q., Lander, A.D. & Schilling, T.F. (2007). Complex regulation of cyp26a1 creates a robust retinoic acid gradient in the zebrafish embryo. PLoS Biology, 5, e304 and Cai, A.Q., Radtke, K., Linville, A., Lander, A.D., Nie, Q. & Schilling, T.F. (2012). Cellular retinoic acid-binding proteins are essential for hindbrain patterning and signal robustness in zebrafish. Development, 139, 2150–2155.

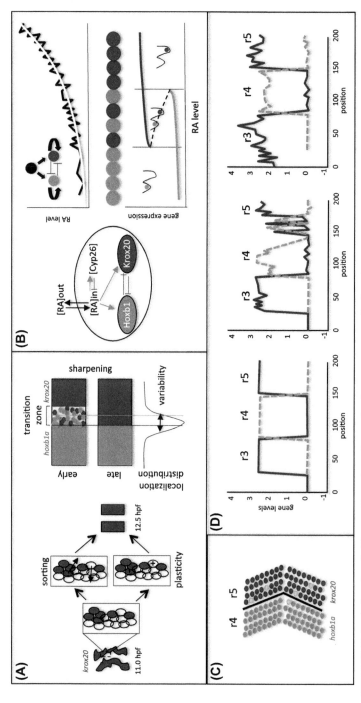

FIGURE 5 Noise-induced switching and boundary sharpening in response to retinoic acid (RA).

(A) Rhombomere boundary sharpening. Hypothetical roles of cell sorting versus plasticity in sharpening of two stripes of *krox20* expression (red) in r3 and r5 in a zebrafish embryonic hindbrain (dorsal view) between 11.0 and 12.5 hpf (left panel). Model depicting sharpening of the "transition zone" between two rhombomeres (r4, *hoxb1a*, green; r5, *krox20*, red), which normally occurs posterior to the final boundary (*dashed black line*) and contains cells expressing both genes. *Green line* indicates alternate boundary that can form if sharpening occurs at a more posterior position. (B) Noise-induced switching at the r4/5 boundary. (left panel) The model includes extracellular RA levels (RA)out, intracellular levels (RA)in, self-enhanced degradation through Cyp26a1 induction, and mutual inhibition between Hoxb1 and Krox20. (upper right panel) RA fluctuations combined with the gene regulatory network lead to fluctuations in target gene expression (green and red cells) near the boundary. (lower right panel) Noise in gene expression helps push cells into one of two stable states in the bistable region (green to red). (C) Evidence for an r4/5 transition zone. Diagram of double fluorescent in situ hybridization experiments demonstrating cells coexpressing hoxb1a (green) and krox20 (red), largely posterior to the future boundary. (D) Modeling suggests that noise-induced switching improves sharpening. Simulations using the model shown in B resolve into rhombomere-like domains of gene expression (Y axis) along a 200-μm stretch along the anterior–posterior axis (X axis) (left graph). Noise in (RA)in alone results in failure of r4/5 boundary to sharpen (middle graph). Noise in both (RA)in and in gene expression restores sharpening (right graph). (See color plate)

Adapted from Zhang, L., Radtke, K., Zheng, L., Cai, A.Q., Schilling, T.F. & Nie, Q. (2012). Noise drives sharpening of gene expression boundaries in the zebrafish hindbrain. Molecular Systems Biology, 8, 613 and Schilling, T.F., Nie, Q. & Lander, A.D. (2012). Dynamics and precision in retinoic acid morphogen gradients. Current Opinion in Genetics and Development, 22, 562–569. http://dx.doi.org/10.1016/j.gde.2012.11.012.

a transition zone of several cell diameters (Fig. 5A). Do cells within these zones sort themselves into the appropriate domains or do they show "plasticity," switching their gene expression to match their neighbors? This has been particularly well investigated for the boundaries between r3/4 and r4/5, which in zebrafish form at 10–10.7 hpf and progressively sharpen by 12.0 hpf, as evidenced by the expression of *krox20* in r3 and r5. Clearly some sorting occurs, as demonstrated recently by tracking of *krox20+* cells (Calzolari, Terriente, & Pujades, 2014; Terriente & Pujades, 2015), and previous studies have demonstrated critical roles for Ephrin/Eph signaling in repulsive interactions between cells that drive sorting (Cooke et al., 2001, Cooke, Kemp, & Moens, 2005). However, are they sufficient to sharpen the transition zones found in the developing hindbrain?

Interestingly both Hoxb1 and Krox20 activate their own transcription as well as mutually repressing one another (Bouchoucha et al., 2013), thereby forming a gene regulatory cassette, which creates three possible stable states within a cell, either one or the other gene is activated or both genes are off (Fig. 5B). This cassette is well suited for switching. If cellular plasticity utilizing this cassette is important for sharpening of the r4/5 boundary, gene expression studies should catch some cells coexpressing both *krox20* and *hoxb1a*, in the process of switching from the gene normally expressed on one side of the boundary to the other. This is the case; two-color double fluorescent in situ hybridization detects individual coexpressing cells (Fig. 5C) (Zhang et al., 2012). Most of these cells lie posterior to the future boundary, revealing the transition zone in which switching occurs.

Both *hoxb1a* and *krox20* are induced by RA (*hoxb1a* is a direct transcriptional target), as well as by other signals in the vicinity. In zebrafish, RA first induces *hoxb1a* up to the r3/4 boundary, with subsequent *krox20* expression initiated in r3, followed by expression in r5. A computational model incorporating (1) this temporal sequence of expression, (2) the *hoxb1/krox20* feedback cassette, and (3) an anteriorly declining gradient of RA, rapidly leads to distinct domains of *hoxb1a* and *krox20* expression that resemble rhombomeres (Fig. 5D). For this model to work, autoregulation of *krox20* must be stronger than *hoxb1a*, but otherwise it is remarkable how such a simple model is sufficient to generate the pattern seen in embryos.

7. NOISE—BOTH GOOD AND BAD

Noise in RA signaling and in its target genes is expected to compromise the ability of cells to interpret their positions within the morphogen gradient or to form sharp boundaries of gene expression. To test this idea computationally, we have varied noise in each component of the system individually and run model simulations to determine its effect on sharpness of the resulting r3/4 and r4/5 boundaries (Fig. 5D). Increasing noise in RA alone leads to rough boundaries that never sharpen—an initial transition/boundary zone of seven cell diameters remains broad, no less than six cell diameters. This is not surprising.

Strikingly, however, simultaneous inclusion of noise in RA and in its target genes (eg, *krox20* and *hoxb1a*) improves sharpening—an initial transition zone of seven or eight cell diameters between rhombomeres, sharpens to one cell diameter wide (Fig. 5D). How could this occur? Based on our combined experimental and computational work in zebrafish, we propose a mechanism of "noise-induced switching" for boundary sharpening (Fig. 5B) (Schilling et al., 2012; Zhang et al., 2012). In this model, stochastic fluctuations in *hoxb1a* and *krox20* expression enable cells to transition between two steady states, from *hoxb1a*+ to *krox20*+ or vice versa, by overcoming an energetic "barrier" between states. This model is counterintuitive because it argues for a positive role for noise, and suggests that the process of boundary sharpening needs noise in gene expression to work. A similar positive role for noise has been described for cells undergoing differentiation in isolation (Kuchina, Espinar, Garcia-Ojalvo, & Suel, 2011), but it has not been appreciated for gene expression boundaries and may be a general principle.

8. OTHER BOUNDARIES AND OTHER MORPHOGENS

Many of the principles revealed from zebrafish studies of RA signaling in the hindbrain have been limited to r4 and r5, due to a focus on *krox20* and the availability of transgenics for studying these segments. Less is known about other rhombomeres, particularly r1–3 where the RA gradient is predicted to be extremely shallow. In addition, the published model's effectiveness is limited to r4 and r5 (Zhang et al., 2012), suggesting that additional signals interact with RA to specify r1–3. These more anterior segments experience much smaller A–P differences in RA as the gradient declines. Fate mapping studies have shown that initially *hoxa1/b1*+ cells extend into r3 (Labalette et al., 2015). Fgf3 and Fgf8 expressed in r4 help induce *krox20* expression in neighboring rhombomeres, first in r3 and slightly later in r5, but through distinct enhancers for each segment. A Krox20 positive feedback loop subsequently maintains its own expression (Kuchina et al., 2011). Fgf3/8 also induce Sprouty4 (Spry4), which in turn inhibits downstream activation of the Fgf pathway (Labalette et al., 2011). Computational models suggest that this provides negative feedback that controls the width of r4. Similarly *Cdx* genes further posteriorly regulate the hindbrain (r7)-spinal cord boundary through interactions with RA signaling (Chang, Skromne, & Ho, 2016; Lee & Skromne, 2014; Skromne, Thorsen, Hale, Prince, & Ho, 2007).

The most thoroughly studied morphogen in zebrafish is Nodal signaling, and recent evidence hints at a dynamic system with similar principles to that of RA in terms of gradient shape, robustness, and timing of target gene expression. Graded Nodal signaling induces the formation of germ layers as well as the dorsal–ventral (D–V) axis of the embryo (Muller et al., 2012; Sampath & Robertson, 2016; Schier, 2009; Xu, Houssin, Ferri-Lagneau, Thisse, & Thisse, 2014). Of two nodal-related genes in zebrafish, Squint (Sqt) functions directly at a distance (Chen & Schier, 2001). Furthermore Sqt induces its inhibitor, Lefty1, which forms a gradient over a greater distance than the activator Nodal, as shown using GFP- or Dendra-tagged proteins, thereby forming a classic reaction-diffusion system (Meinhardt, 2009, 2015; Muller et al., 2012). Attempts to visualize the Nodal gradient have also used reporters such as Smad2:Venus as well as bimolecular fluorescence complementation to visualize the complex between Smad2 and Smad4 (Harvey & Smith, 2009). The response to a Nodal morphogen gradient in the early zebrafish embryo appears to be dictated by the timing of target gene induction, as determined by the activation of Smad2 and the expression kinetics of short- and long-range target genes (Dubrulle et al., 2015). In addition, studies using a zebrafish Nodal biosensor as well as immunofluorescence for phosphorylated Smad2 suggest that Nodal does not diffuse long distances, and that a temporal window for signal activation controls Nodal signaling domains (van Boxtel et al., 2015). Thus like RA, the Nodal morphogen gradient is shaped by feedback (self-enhanced inhibition) and rather than depending simply on morphogen levels, it depends on the kinetics of target gene induction.

CONCLUSIONS AND PERSPECTIVES

In this review, we have focused on the regulation of RA signaling and new methods for visualizing RA morphogen gradients in zebrafish. These studies highlight the fact that Wolpert's morphogen model only touches the tip of the iceberg in terms of morphogen dynamics and precision. Guided by computational models that reveal constraints in the system, experimental work with RA reporters in zebrafish has revealed that two factors, Cyp26a1 and Crabp2a, stand out as critical for the RA gradient. Self-enhanced degradation through Cyp26a1, as well as Crabp2a, help fine-tune RA levels within responding cells and binding of RA receptors. These allow the RA gradient to be surprisingly precise, robust, and able to induce sharp boundaries of target gene expression. However, these mechanisms cannot account for all of the robustness in the system, eg, self-enhanced degradation can compensate for twofold changes in RA synthesis, but zebrafish embryos are robust to at least 10-fold changes in RA concentration. Future studies are needed to identify the other mechanisms that account for this remarkable adaptability. The availability of GEPRA reporters for RA availability now make it possible to correlate these features with the spatial distribution of RA in embryos.

Computational models also reveal a surprising beneficial role for noise in boundary sharpening—noise-induced switching. While RA reporters (including GEPRAs) have kinetics that are too slow to visualize noise in RA signaling directly, new methods (eg, FLIM imaging of RA autofluorescence) on the horizon should overcome this limitation. Similarly evidence to date for noise in gene expression (eg, *hoxb1a* and *krox20*) has relied on nonquantitative methods such as in situ hybridization. Recently developed quantitative in situ methods (Choi et al., 2010) as well as live imaging methods to examine the dynamic regulation of gene expression by visualizing nascent transcripts via MS2 RNA stem loops promise to reveal the details of such dynamic expression (Bothma et al., 2014). Future studies will determine if similar noise-induced switching mechanisms control sharpening boundaries of target gene expression in response to other morphogens.

ACKNOWLEDGMENTS

We thank Pierre Le Pabic for critical comments on the manuscript. TS and QN were supported by the National Institutes of Health grants R01-GM107264, R01DE023050 and P50-GM76516. QN was also supported by National Science Foundation grant DMS1161621.

REFERENCES

Alexander, C., Zuniga, E., Blitz, I. L., Wada, N., Le Pabic, P., Javidan, Y., ... Schilling, T. F. (2011). Combinatorial roles for BMPs and endothelin 1 in patterning the dorsal-ventral axis of the craniofacial skeleton. *Development, 138*, 5135–5146. http://dx.doi.org/10.1242/dev.067801.

Alexandre, C., Baena-Lopez, A., & Vincent, J. P. (2014). Patterning and growth control by membrane-tethered Wingless. *Nature, 505*, 180–185.

Astrom, A., Tavakkol, A., Pettersson, U., Cromie, M., Elder, J. T., & Voorhees, J. J. (1991). Molecular cloning of two human cellular retinoic acid-binding proteins (CRABP). Retinoic acid-induced expression of CRABP-II but not CRABP-I in adult human skin in vivo and in skin fibroblasts in vitro. *Journal of Biological Chemistry, 266*, 17662–17666.

Balasubramanian, R., & Zhang, X. (2015). Mechanisms of FGF gradient formation during embryogenesis. *Seminars in Cell and Developmental Biology.* http://dx.doi.org/10.1016/j.semcdb.2015.10.004. pii:S1084–9521(15)00189-5.

Begemann, G., Schilling, T. F., Rauch, G. J., Geisler, R., & Ingham, P. W. (2001). The zebrafish neckless mutation reveals a requirement for raldh2 in mesodermal signals that pattern the hindbrain. *Development, 128*, 3081–3094.

Ben-Zvi, D., & Barkai, N. (2010). Scaling of morphogen gradients by an expansion-repression integral feedback control. *Proceedings of the National Academy of Sciences of the United States of America, 107*, 6924–6929. http://dx.doi.org/10.1073/pnas.0912734107.

Ben-Zvi, D., Shilo, B. Z., & Barkai, N. (2011). Scaling of morphogen gradients. *Current Opinion in Genetics and Development, 21*, 704–710. http://dx.doi.org/10.1016/j.gde.2011.07.011.

Bokel, C., & Brand, M. (2013). Generation and interpretation of FGF morphogen gradients in vertebrates. *Current Opinion in Genetics and Development, 23*, 415–422. http://dx.doi.org/10.1016/j.gde.2013.03.002.

Bothma, J. P., Garcia, H. G., Esposito, E., Schlissel, G., Gregor, T., & Levine, M. (2014). Dynamic regulation of eve stripe 2 expression reveals transcriptional bursts in living *Drosophila* embryos. *Proceedings of the National Academy of Sciences of the United States of America, 11*, 10598–10603. http://dx.doi.org/10.1073/pnas.1410022111.

Bouchoucha, Y. X., Reingruber, J., Labalette, C., Wassef, M. A., Thierion, E., Desmarquet-Trin Dinh, C., ... Charnay, P. (2013). Dissection of a Krox20 positive feedback loop driving cell fate choices in hindbrain patterning. *Molecular Systems Biology, 9*, 690. http://dx.doi.org/10.1038/msb.2013.46.

van Boxtel, A. L., Chesebro, J. E., Heliot, C., Ramel, M. C., Stone, R. K., & Hill, C. S. (2015). A temporal window for signal activation dictates the dimensions of a Nodal signaling domain. *Developmental Cell, 35*, 175—185. http://dx.doi.org/10.1016/j.devcel.2015.09.014.

Briscoe, J., & Small, S. (2015). Morphogen rules: design principles of gradient-mediated embryo patterning. *Development, 142*, 3996—4009.

Budhu, A., Gillilan, R., & Noy, N. (2001). Localization of the RAR interaction domain of cellular retinoic acid binding protein-II. *Journal of Molecular Biology, 305*, 939—949.

Budhu, A. S., & Noy, N. (2002). Direct channeling of retinoic acid between cellular retinoic acid binding protein II and retinoic acid receptor sensitizes mammary carcinoma cells to retinoic acid-induced growth arrest. *Molecular and Cellular Biology, 22*, 2632—2641.

Cai, A. Q., Radtke, K., Linville, A., Lander, A. D., Nie, Q., & Schilling, T. F. (2012). Cellular retinoic acid-binding proteins are essential for hindbrain patterning and signal robustness in zebrafish. *Development, 139*, 2150—2155.

Calzolari, S., Terriente, J., & Pujades, C. (2014). Cell segregation in the vertebrate hindbrain relies on actomyosin cables located at the interhombomeric boundaries. *EMBO Journal, 33*, 686—701. http://dx.doi.org/10.1002/embj.201386003.

Chang, J., Skromne, I., & Ho, R. K. (2016). CDX4 and retinoic acid interact to position the hindbrain-spinal cord transition. *Developmental Biology*. http://dx.doi.org/10.1016/j.ydbio.2015.12.025. pii:S0012—1606(15)30378-X.

Chen, Y., & Schier, A. F. (2001). The zebrafish Nodal signal Squint functions as a morphogen. *Nature, 411*, 607—610.

Choi, H. M., Chang, J. Y., Trinhle, A., Padilla, J. E., Fraser, S. E., & Pierce, N. A. (2010). Programmable in situ amplification for multiplexed imaging of mRNA expression. *Nature Biotechnology, 28*, 1208—1212. http://dx.doi.org/10.1038/nbt.1692.

Cooke, J., Moens, C., Roth, L., Durbin, L., Shiomi, K., Brennan, C., ... Holder, N. (2001). Eph signaling functions downstream of Val to regulate cell sorting and boundary formation in the caudal hindbrain. *Development, 128*, 571—580.

Cooke, J. E., Kemp, H. A., & Moens, C. B. (2005). EphA4 is required for cell adhesion and rhombomere boundary formation in the zebrafish. *Current Biology, 15*, 536—542.

Delva, L., Bastie, J. N., Rochette-Egly, C., Kraiba, R., Balitrand, N., Despouy, G., ... Chomienne, C. (1999). Physical and functional interactions between cellular retinoic acid binding protein II and the retinoic acid-dependent nuclear complex. *Molecular and Cellular Biology, 19*, 7158—7167.

Driever, W., & Nusslein-Volhard, C. (1988a). A gradient of bicoid protein in *Drosophila* embryos. *Cell, 54*, 83—93.

Driever, W., & Nusslein-Volhard, C. (1988b). The bicoid protein determines position in the *Drosophila* embryo in a concentration-dependent manner. *Cell, 54*, 95—104. http://dx.doi.org/10.1016/0092-8674(88)90183-3.

Dubrulle, J., Jordan, B. M., Akhmetova, L., Farrell, J. A., Kim, S. H., Solnica-Krezel, L., & Schier, A. F. (2015). Response to Nodal morphogen gradient is determined by the kinetics of target gene induction. *eLife, 4*. http://dx.doi.org/10.7554/eLife.05042.

Eldar, A., Rosin, D., Shilo, B. Z., & Barkai, N. (2003). Self-enhanced ligand degradation underlies robustness of morphogen gradients. *Developmental Cell, 5*, 635—646.

Elowitz, M. B., Levine, A. J., Siggia, E. D., & Swain, P. S. (2002). Stochastic gene expression in a single cell. *Science, 297*, 1183—1186.

Grimm, O., Coppey, M., & Wieschaus, E. (2010). Modeling the bicoid gradient. *Development, 137*, 2253—2264. http://dx.doi.org/10.1242/dev.032409.

Harvey, S. A., & Smith, J. C. (2009). Visualisation and quantification of morphogen gradient formation in the zebrafish. *PLoS Biology*, e1000101. http://dx.doi.org/10.1371/journal.pbio.1000101.

Hernandez, R. E., Putzke, A. P., Myers, J. P., Margaretha, L., & Moens, C. B. (2007). Cyp26 enzymes generate the retinoic acid response pattern necessary for hindbrain development. *Development, 134*, 177–187.

Horinaka, K., & Morishita, Y. (2012). Encoding and decoding of positional information in morphogen-dependent patterning. *Current Opinion in Genetics and Development, 22*, 553–561.

Huang, W., Wang, G., Delaspre, F., Vitery Mdel, C., Beer, R. L., & Parsons, M. J. (2014). Retinoic acid plays an evolutionarily conserved and biphasic role in pancreas development. *Developmental Biology, 394*, 83–93. http://dx.doi.org/10.1016/j.ydbio.2014.07.021.

Kaern, M., Elston, T. C., Blake, W. J., & Collins, J. J. (2005). Stochasticity in gene expression: from theories to phenotypes. *Nature Reviews Genetics, 6*, 451–464.

Kepler, T. B., & Elston, T. C. (2001). Stochasticity in transcriptional regulation: origins, consequences and mathematical representations. *Biophysical Journal, 81*, 3116–3136.

Kuchina, A., Espinar, L., Garcia-Ojalvo, J., & Suel, G. M. (2011). Reversible and noisy progression towards a commitment point enables adaptable and reliable cellular decision making. *PLoS Computational Biology, 7*, e1002273.

Kudoh, T., Wilson, S. W., & Dawid, E. B. (2002). Distinct roles for Fgf, Wnt and retinoic acid in posteriorizing the neural ectoderm. *Development, 129*, 4335–4346.

Labalette, C., Bouchoucha, Y. X., Wassef, M. A., Gongal, P. A., Le Men, J., Becker, T., ... Charnay, P. (2011). Hindbrain patterning requires fine-tuning of early krox20 transcription by Sprouty 4. *Development, 138*, 317–326.

Labalette, C., Wassef, M. A., Desmarquet-Trin Dinh, C., Bouchoucha, Y. X., Le Men, J., Charnay, P., & Gilardi-Hebenstreit, P. (2015). Molecular dissection of segment formation in the developing hindbrain. *Development, 142*, 185–195.

Lampron, C., Rochette-Egly, C., Gorry, P., Dolle, P., Mark, M., Lufkin, T., ... Chambon, P. (1995). Mice deficient in cellular retinoic acid binding protein II (CRABPII) or in both CRABPI and CRABPII are essentially normal. *Development, 121*, 539–548.

Lander, A. D. (2007). Morpheus unbound: reimagining the morphogen gradient. *Cell, 128*, 245–256.

Lee, K., & Skromne, I. (2014). Retinoic acid regulates size, pattern and alignment of tissues at the head-trunk transition. *Development, 141*, 4375–4384. http://dx.doi.org/10.1242/dev.109603.

Li, J., Hu, P., Li, K., & Zhao, Q. (2012). Identification and characterization of a novel retinoic acid response element in zebrafish cyp26a1 promoter. *Anatomical Record, 295*, 268–277. http://dx.doi.org/10.1002/ar.21520.

Lumsden, A., & Keynes, R. (1989). Segmental patterns of neuronal development in the chick hindbrain. *Nature, 337*, 424–428.

Mandal, A., Rydeen, A., Anderson, J., Sorrell, M. R., Zygmunt, T., Torres-Vazquez, J., & Waxman, J. S. (2013). Transgenic retinoic acid sensor lines in zebrafish indicate regions of available retinoic acid. *Developmental Dynamics, 242*, 989–1000. http://dx.doi.org/10.1002/dvdy.23987.

Meinhardt, H. (2009). Models for the generation and interpretation of gradients. *Cold Spring Harbor Perspectives in Biology, 1*, a001362.

Meinhardt, H. (2015). Models for patterning primary embryonic body axes: the role of space and time. *Seminars in Cell and Developmental Biology, 42*, 103. http://dx.doi.org/10.1016/j.semcdb.2015.06.005.

Muller, P., Rogers, K. W., Jordan, B. M., Lee, J. S., Robson, D., Ramanathan, S., & Schier, A. F. (2012). Differential diffusivity of Nodal and Lefty underlies a reaction-diffusion patterning system. *Science*. http://dx.doi.org/10.1126/science.1221920.

Niederreither, K., Vermot, J., Schuhbaur, B., Chambon, P., & Dolle, P. (2000). Retinoic acid synthesis and hindbrain patterning in the mouse embryo. *Development, 127*, 75–85.

Perz-Edwards, A., Hardison, N. L., & Linney, E. (2001). Retinoic acid-mediated gene expression in transgenic reporter zebrafish. *Developmental Biology, 229*, 89–101.

Prols, F., Sagar, & Scaal, M. (2015). Signaling filopodia in vertebrate embryonic development. *Cellular and Molecular Life Sciences*. http://dx.doi.org/10.1007/00018-015-2097-6.

Ramel, M. C., & Hill, C. S. (2013). The ventral to dorsal BMP activity gradient in the early zebrafish embryo is determined by graded expression of BMP ligands. *Developmental Biology, 378*, 170–182. http://dx.doi.org/10.1016/j.ydbio.2013.03.003.

Rhinn, M., & Dolle, P. (2012). Retinoic acid signaling during development. *Development, 139*, 843–858.

Rossant, J., Zirngibl, R., Cado, D., Shago, M., & Giguere, V. (1991). Expression of a retinoic acid response element-hsplacZ transgene defines specific domains of transcriptional activity during mouse embryogenesis. *Genes and Development, 5*, 1333–1344.

Sampath, K., & Robertson, E. J. (2016). Keeping a lid on nodal: transcriptional and translational repression of nodal signaling. *Open Biology*. http://dx.doi.org/10.1098/rsob.150200. pii: 150200.

Sample, C., & Shvartsman, S. Y. (2010). Multiscale modeling of diffusion in the early *Drosophila* embryo. *Proceedings of the National Academy of Sciences of the United States of America, 107*, 10092–10096. http://dx.doi.org/10.1073/pnas.1001139107.

Schier, A. F. (2009). Nodal morphogens. *Cold Spring Harbor Perspectives in Biology, 1*, a003459. http://dx.doi.org/10.1101/chsperspect.a003459.

Schilling, T. F., Nie, Q., & Lander, A. D. (2012). Dynamics and precision in retinoic acid morphogen gradients. *Current Opinion in Genetics and Development, 22*, 562–569. http://dx.doi.org/10.1016/j.gde.2012.11.012.

Shimozono, S., Iimura, T., Kitaguchi, T., Higashijima, S., & Miyawaki, A. (2013). Visualization of an endogenous retinoic acid gradient across embryonic development. *Nature, 496*, 363–366. http://dx.doi.org/10.1038/nature12037.

Sirbu, I. O., Gresh, L., Barra, J., & Duester, G. (2005). Shifting boundaries of retinoic acid activity control hindbrain segmental gene expression. *Development, 132*, 2611–2622.

Skromne, I., Thorsen, D., Hale, M., Prince, V. E., & Ho, R. K. (2007). Repression of the hindbrain developmental program by Cdx factors is required for the specification of the vertebrate spinal cord. *Development, 134*, 2147–2158.

Stathopoulos, A., & Iber, D. (2013). Studies of morphogens: keep calm and carry on. *Development, 140*, 4119–4124.

Strigini, M., & Cohen, S. M. (2000). Wingless gradient formation in the *Drosophila* wing. *Current Biology, 10*, 293–300.

Stringari, C., Cinquin, A., Cinquin, O., Digman, M. A., Donovan, P. J., & Gratton, E. (2011). Phasor approach to fluorescence lifetime microscopy distinguishes different metabolic states of germ cells in a live tissue. *Proceedings of the National Academy of Sciences of the United States of America, 108*, 13582–13587. http://dx.doi.org/10.1073/pnas.1108161108.

Terriente, J., & Pujades, C. (2015). Cell segregation in the vertebrate hindbrain: a matter of boundaries. *Cellular and Molecular Life Sciences, 72*, 3721–3730.

Tickle, C., Summerbell, D., & Wolpert, L. (1975). Positional signaling and specification of digits in chick limb morphogenesis. *Nature, 254*, 199–202.

Trevarrow, B., Marks, D. L., & Kimmel, C. B. (1990). Organization of hindbrain segments in the zebrafish embryo. *Neuron, 4*, 669–679.

Tuazon, F. B., & Mullins, M. C. (2015). Temporally coordinated signals progressively pattern the anteroposterior and dorsoventral body axes. *Seminars in Cell and Developmental Biology, 42*, 118–133.

Umulis, D. M., Shimmi, O., O'Connor, M. B., & Othmer, H. G. (2010). Organism-scale modeling of early *Drosophila* patterning via bone morphogenetic proteins. *Developmental Cell, 18*, 260–274. http://dx.doi.org/10.1016/devcel.2010.01.006.

Wartlick, O., Kicheva, A., & Gonzalez-Galtan, M. (2009). Morphogen gradient formation. *Cold Spring Harbor Perspectives in Biology, 1*, a001255.

Waxman, J. S., & Yelon, D. (2011). Zebrafish retinoic acid receptors as context dependent transcriptional activators. *Developmental Biology, 352*, 128–140.

White, R. J., Nie, Q., Lander, A. D., & Schilling, T. F. (2007). Complex regulation of cyp26a1 creates a robust retinoic acid gradient in the zebrafish embryo. *PLoS Biology, 5*, e304.

White, R. J., & Schilling, T. F. (2008). How degrading: Cyp26s in hindbrain development. *Developmental Dynamics, 237*, 2775–2790. http://dx.doi.org/10.1002/dvdy.21695.

Xiong, F., Tentner, A. R., Huang, P., Gelas, A., Mosaliganti, K. R., Souhait, L., ... Megason, S. G. (2013). Specified neural progenitors sort to form sharp domains after noisy Shh signaling. *Cell, 153*, 550–561.

Xu, P. F., Houssin, N., Ferri-Lagneau, K. F., Thisse, B., & Thisse, C. (2014). Construction of a vertebrate embryo from two opposing morphogen gradients. *Science, 344*, 87–89. http://dx.doi.org/10.1126/science.1248252.

Zhang, L., Radtke, K., Zheng, L., Cai, A. Q., Schilling, T. F., & Nie, Q. (2012). Noise drives sharpening of gene expression boundaries in the zebrafish hindbrain. *Molecular Systems Biology, 8*, 613.

CHAPTER

Using fluorescent lipids in live zebrafish larvae: from imaging whole animal physiology to subcellular lipid trafficking

J.L. Anderson, J.D. Carten, S.A. Farber[1]

Carnegie Institution for Science, Baltimore, MD, United States
[1]*Corresponding author: E-mail: farber@ciwemb.edu*

CHAPTER OUTLINE

Introduction	166
The Need for Whole Animal Studies of Lipid Metabolism	166
1. Forward Genetic Screening With Fluorescent Lipids	168
1.1 PED6	168
1.2 NBD- and BODIPY-Cholesterol	169
2. Visualizing Lipid Metabolism Using BODIPY Fatty Acid Analogs	171
2.1 BODIPY (Excitation/Emission Maxima ~503/512 nm)	171
2.2 BODIPY C2	173
2.3 BODIPY C5	174
2.4 BODIPY C12/C16	174
Summary	174
Acknowledgments	175
References	175

Abstract

Lipids serve essential functions in cells as signaling molecules, membrane components, and sources of energy. Defects in lipid metabolism are implicated in a number of pandemic human diseases, including diabetes, obesity, and hypercholesterolemia. Many aspects of how fatty acids and cholesterol are absorbed and processed by intestinal cells remain unclear and present a hurdle to developing approaches for disease prevention and treatment. Numerous studies have shown that the zebrafish is an excellent model for vertebrate lipid metabolism. In this chapter, we review commercially available fluorescent lipids that can be deployed in live zebrafish to better understand lipid signaling and metabolism. In this chapter, we present criteria one should consider when selecting specific fluorescent lipids for the study of digestive physiology or lipid metabolism in larval zebrafish.

List of Abbreviations

ACAT   Acyl CoA:cholesterol acyltransferase
FA   Fatty acid
LCFA   Long-chain fatty acid
LD   Lipid droplet
MCFA Medium-chain fatty acid
SCFA   Short-chain fatty acid
TAG   Triacylglycerol

INTRODUCTION

A significant body of work links alterations in lipid metabolism (induced through genetic mutation and/or lifestyle) with cardiovascular disease, diabetes mellitus, and obesity (Bastien, Poirier, Lemieux, & Despres, 2014; Bauer, Briss, Goodman, & Bowman, 2014; Rankinen, Sarzynski, Ghosh, & Bouchard, 2015). A better understanding of the genetic and environmental circumstances that perturb lipid uptake, trafficking, and storage in an organism is required to fully assess the impact of metabolic regulators and hormones on lipid-associated diseases. Our lab and others have made great strides in developing techniques to enable the study of lipid metabolism in vivo using zebrafish larvae (Carten, Bradford, & Farber, 2011; Clifton et al., 2010; Farber et al., 2001; Hama et al., 2009; Ho, Lorent, Pack, & Farber, 2006; Ho et al., 2004; Miyares et al., 2013; Semova et al., 2012).

THE NEED FOR WHOLE ANIMAL STUDIES OF LIPID METABOLISM

In vitro studies have laid much of the groundwork for our biochemical understanding of lipid metabolism; however, a number of caveats arise when attempting to study lipid metabolism in vitro. Such studies are often carried out in transformed cultured cells, such as liver HepG2, intestinal Caco2, and adipocyte 3LT3 cells. Such cell lines are comprised of a single cell type, which cannot duplicate the cellular heterogeneity of an entire organ, such as the intestine, that is composed of stem, enteroendocrine, immune, and goblet cells. These multiple cell types are known to influence each other through paracrine signaling that can have global effects on lipid uptake and processing. Furthermore, bile, intestinal mucus, and the gut microbiota are all known to greatly influence dietary lipid processing and absorption in the intestine (Backhed et al., 2004; Field, Dong, Beis, & Stainier, 2003; Kruit,

Groen, van Berkel, & Kuipers, 2006; Martin et al., 2008; Moschetta et al., 2005; Pack et al., 1996; Semova et al., 2012; Titus & Ahearn, 1992; Turnbaugh, Backhed, Fulton, & Gordon, 2008) and are absent in cultured cell models. For these reasons, employing whole animal in vivo strategies, in addition to cultured cell work, is vital to understanding how metabolic dysfunction arises and manifests itself in an organism.

While the zebrafish has been established as a powerful model for the study of early development, thanks largely to its accessibility, fast development, and optically clear embryos, few researchers have exploited this model organism to visualize lipid uptake and processing. In 2013, James Rothman, Randy Schekman, and Thomas Südhof received the Nobel Prize in Physiology or Medicine for their work in yeast and cultured cells to elucidate the cellular machinery that regulates intracellular vesicle traffic. Schekman stated, "Many fascinating questions remain at the molecular mechanistic level about how vesicles form and how they are directed to their target and fuse with a target membrane" (Schekman & Sudhof, 2014). Experiments using fluorescent lipids in larval zebrafish build on the pioneering studies of Rothman and Schekman to better understand critical vesicular trafficking events that move lipid and/or lipoprotein cargos in a live vertebrate. Many aspects regarding how digestive organs regulate the flow of lipids through specialized cells such as the liver hepatocyte and the intestinal enterocyte remain to be discovered, and these insights are expected to impact a host of prevalent worldwide diseases.

The ability to perform forward genetic studies in zebrafish by mutagenizing the entire genome and screening for particular phenotypes has made this vertebrate model widely used (Driever et al., 1996; Haffter et al., 1996). Mutagenesis methods commonly utilized in the zebrafish include soaking founder fish in mutagenic chemicals, such as ethylnitrosourea (ENU) to generate point mutations (Driever et al., 1996; Haffter et al., 1996), and retrovirus- or transposon-mediated gene insertions (Chen & Farese, 2002; Ivics et al., 2004). We perform ongoing screens based on ENU and gene-break transposon mutagenesis methods to search for mutations that perturb lipid processing. To find and characterize mutants, we immerse zebrafish larvae in fluorescent lipid reporters that are ingested and allow lipid processing events to be visualized in vivo (Farber et al., 2001).

Both in the wild and in the laboratory, zebrafish consume a lipid-rich diet ($\geq 10\%$ by weight) high in triacylglycerol (TAG), phospholipids, and sterols (Enzler, Smith, Lin, & Olcott, 1974; Spence, Gerlach, Lawrence, & Smith, 2008). Prior to absorption by the intestine, these lipids must be processed and solubilized by the digestive enzymes and bile that make up the intraluminal intestinal milieu. As in humans, bile is produced by hepatocytes and secreted into an extensive network of intrahepatic ducts, which drains into the gall bladder. In response to hormonal stimulation triggered by food consumption, bile is released into the intestinal lumen to emulsify dietary fat and facilitate its absorption by intestinal enterocytes. After dietary fat is emulsified, TAG and phospholipids must be broken down by luminal lipases to release free fatty acid (FA) or mono- and diacylglycerols, which can then enter the absorptive cells (enterocytes) that line the gut (Thomson, Keelan, Cheng, &

Clandinin, 1993). Zebrafish enterocytes are highly similar to mammalian enterocytes (Buhman et al., 2002), with the characteristic microvilli and basal nuclei. After food consumption, zebrafish accumulate cytoplasmic lipid droplets in their enterocytes (unpublished). Fats are burned via oxidative pathways in the mitochondria or peroxisomes or packaged into chylomicrons, which are secreted from the basolateral surface of enterocytes into lymphatic or blood vessels (Field, 2001; Levy et al., 2007). While it remains to be seen how closely the zebrafish system will model human intestinal lipoprotein metabolism, it is likely that many of the mechanisms of lipoprotein production are conserved. We can readily follow and describe these processes through the application of fluorescent lipids to live zebrafish larvae.

In this chapter, we present criteria one should consider when selecting specific fluorescent lipids for the study of digestive physiology or lipid metabolism in larval zebrafish. Determining the appropriate lipid analog or reporter to use for a given experiment will ultimately provide a better understanding of subcellular processes being examined. It is our standpoint that live-imaging studies and fluorescence-based screens are well suited for the larval zebrafish and will greatly broaden our understanding of digestive physiology in the years to come.

1. FORWARD GENETIC SCREENING WITH FLUORESCENT LIPIDS
1.1 PED6

Using lipids that alter their spectral properties when they are in specific cellular structures or are metabolized by specific enzymes can provide valuable information about where a lipid-related event occurs in the cell. One such fluorescent reporter we have used in our genetic screens is the phosphoethanolamine analog PED6, [N-((6-(2,4-dinitro-phenyl)amino)hexanoyl)-1-palmitoyl-2-BODIPY-FL-pentanoyl-*sn*-glycerol-3-phosphoethanolamine]. This reporter exhibits altered spectral characteristics upon processing by lipid-modifying enzymes (Farber et al., 2001; Hama et al., 2009; Hendrickson, Hendrickson, Johnson, & Farber, 1999) (Fig. 1A). Following ingestion by larvae, PED6 is cleaved by phospholipase A_2 (PLA_2), resulting in the release of the fluorescent BODIPY-labeled acyl chain (Farber, Olson, Clark, & Halpern, 1999) (Fig. 1B and C). When zebrafish larvae (5 dpf) are immersed in media containing PED6, bright green fluorescence is observed in the intestine, gall bladder, and liver (Fig. 1C). Because this fluorescent reporter provides a rapid readout of both lipid metabolism and digestive organ morphology in living larvae, it was used to perform the first physiological genetic screen in the zebrafish (Farber et al., 2001). This screen identified Vps51 (*fat-free*) as both a mediator of intestinal lipid processing and Golgi morphology (Ho et al., 2006). Vps51 is a component of the Golgi-associated retrograde protein (GARP) complex (Frohlich et al., 2015) and, as suggested by the *fat-free* phenotype that includes increased number of lipid droplets in liver and intestine (Liu, Lee, Tsai, & Ho, 2010), has recently been implicated in regulating cellular lipid homeostasis.

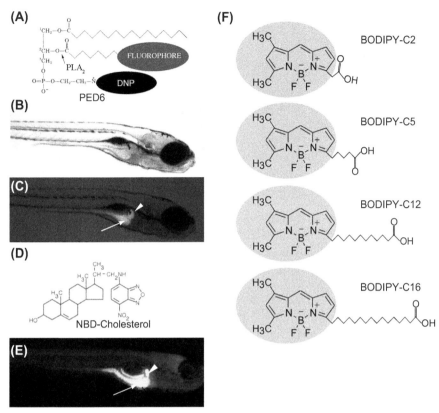

FIGURE 1 Fluorescent lipids visualize digestive organ uptake and transport in larval zebrafish.

(A) The chemical structures of PED6. The BODIPY-labeled acyl chain of PED6 is normally quenched by the dinitrophenyl group at the *sn*-3 position. Upon PLA$_2$ cleavage at the *sn*-2 position, the BODIPY-labeled acyl chain is unquenched and can fluoresce. Bright field (B) and fluorescent (C) images of 5-dpf larva following soaking in PED6 for 6 h. PED6 labeling reveals lipid processing in the gall bladder (*arrowhead*) and intestine (*arrow*). (D) The chemical structure of NBD-cholesterol. The NBD-cholesterol analog contains an NBD fluorophore where the alkyl tail at the terminal end of cholesterol would normally reside. (E) Soaking zebrafish larvae (5 dpf) in NBD-cholesterol (3 mg/mL, solubilized with fish bile) for 2 h visualizes cholesterol uptake in the gall bladder (*arrowhead*) and intestine (*arrow*). (F) The chemical structures of BODIPY fatty acids. (See color plate)

1.2 NBD- AND BODIPY-CHOLESTEROL

Continuously fluorescent lipids are equally powerful in studies performed in larval zebrafish due to the ability to follow their uptake, transport, and storage over time across multiple organs in one organism. We have utilized the sterol analog 22-NBD-cholesterol (22-[N-(7-nitrobenz-2-oxa-1,3-diazol-4-yl) amino]-23,24-bisnor-5-cholen-3-ol) (Fig. 1D) to visualize cholesterol absorption in live larvae and to screen for mutants (Fig. 1E). This reagent is different from PED6 in that it continuously fluoresces and it is more difficult to solubilize and administer via feeding. Thus, it may be more efficient to use in studies involving its direct injection into the yolk or yolk syncytial layer.

NBD-cholesterol was initially created to visualize cholesterol partitioning into membranes. While NBD-cholesterol, such as cholesterol, can be esterified by Acyl CoA:cholesterol acyltransferase (ACAT) to produce cholesterol ester (Lada et al., 2004), it was found to preferentially enter into liquid-disordered lipid domains (cholesterol enters liquid-ordered domains) (Li, Mintzer, & Bittman, 2006). This lipid packing difference makes this probe less useful as a model for all aspects of cholesterol processing. To address this limitation, a BODIPY-tagged cholesterol analog was synthesized with a modified fluorophore linker (Li & Bittman, 2007). Studies on its membrane packing found that BODIPY-cholesterol partitioned into the cholesterol-rich liquid-ordered membrane domain (Ariola, Li, Cornejo, Bittman, & Heikal, 2009) and interacted with membranes in ways similar to native sterols, making it a powerful new tool for imaging sterol trafficking in live cells (Marks, Bittman, & Pagano, 2008).

Studies done in the zebrafish have found that embryos microinjected with BODIPY-cholesterol exhibited fluorescence predominately in the yolk of developing zebrafish larvae (Holtta-Vuori et al., 2008). Ongoing work in the Farber lab involves following the path of BODIPY-cholesterol in larval intestinal enterocytes after a high-fat meal and comparing its localization to that of lipid droplets (revealed by BODIPY-labeled FA). Recent data suggest that sterol and FA initially segregate into nonoverlapping compartments (Farber, unpublished). Despite the physiologically normal lipid packing of BODIPY-cholesterol, it is an extremely poor ACAT substrate (Holtta-Vuori et al., 2008). Since this is a major pathway for dietary cholesterol processing, the lack of esterification potentially limits the degree to which this fluorophore models native cholesterol. It is our expectation that minor adjustments to BODIPY cholesterol's structure may enhance its ability to be esterified.

We have developed a number of novel feeding assays that enable us to observe fluorescently labeled dietary lipids incorporate into fat deposits within cells and track their subsequent movements between cells and digestive tissues in live zebrafish (Carten et al., 2011). Feeding fluorescent lipids together with a high-lipid meal produced striking images of lipid droplets in specific subcellular compartments and has enhanced our understanding of dietary lipid metabolism (Avraham-Davidi et al., 2012; Carten et al., 2011; Clifton et al., 2010; Delous et al., 2012; Otis & Farber, 2013; Sadler, Rawls, & Farber, 2013; Semova et al., 2012). The zebrafish is well suited to genetic manipulation because it is a genetic model organism wherein we can mutate/modify existing genes with genome engineering techniques (Bedell et al., 2012; Hwang et al., 2015; Shah, Davey, Whitebirch, Miller, & Moens, 2015) and/or introduce fluorescent proteins that can serve as transcriptional reporters (Gut et al., 2013). Similarly, we can use Tol2-mediated transgenesis (Kawakami et al., 2004) to introduce any gene of interest encoding a fluorescent fusion protein or metabolic sensor. Taken together, this approach and the unique advantages of larval zebrafish enable the visualization of specific tissues and organs in vivo in conjunction with the biochemical processing of lipid precursors within single cells and throughout the living animal.

2. VISUALIZING LIPID METABOLISM USING BODIPY FATTY ACID ANALOGS

The wide variety of fluorescent lipid analogs commercially available allows one to fully exploit the optical clarity of zebrafish larvae to study lipid metabolism. One type of analog widely used in cultured cells to visualize lipid dynamics is the BODIPY-conjugated FA. These analogs consist of an acyl chain of variable length attached to the BODIPY (4,4-difluoro-4-bora-3a, 4a-diaza-S-indacene) fluorescent moiety (Fig. 1F). First synthesized by Treibs and Kreuzer in 1968 (Treibs & Kreuzer, 1969), the BODIPY fluorophore possesses a number of advantageous qualities including high photostability, strong and narrow wavelength emission in the visible spectrum, and an overall uncharged state (Monsma et al., 1989; Pagano, Martin, Kang, & Haugland, 1991).

To administer BODIPY FA analogs to live zebrafish larvae, we developed a novel feeding assay that generates liposomes to create an emulsion of relatively hydrophobic FA analogs in embryo media (Carten et al., 2011). Following a short liposome feed, digestive organ structure and metabolic function can be assessed, as the fluorescent FAs accumulate readily throughout numerous larval organs and tissues. With this assay, we have observed that different chain length FAs (short, medium, and long) accumulate in distinct patterns throughout digestive organs and tissues, with each chain length suited to visualize particular larval organs and cellular structures. During processing of a meal delivering BODIPY-conjugated lipid, the specific transport pathway from the intestine to the liver depends on the properties of the labeled lipid. Regardless of the transport details, hepatobiliary function can be readily assayed by observing the accumulation of fluorescence in the gall bladder using a low-power (10× objective) fluorescent stereo microscope. While this approach has been covered in prior editions of this volume (Anderson, Carten, & Farber, 2011), this chapter will focus on organ morphology and subcellular lipid trafficking that can be observed with confocal microscopy (40–63× objective) of live zebrafish larvae.

2.1 BODIPY (EXCITATION/EMISSION MAXIMA ~503/512 nm)

The unconjugated BODIPY fluorophore has been primarily used to label cytoplasmic lipid droplets (LD) (Brasaemle et al., 2000; Gocze & Freeman, 1994). Administering a high-fat meal to zebrafish larvae leads to the production of copious LDs in both the intestine and liver that can be readily labeled by free BODIPY (data not shown). However, the signal-to-background noise ratio is low, and the image quality is inferior compared to data from studies based on medium- and long-chain

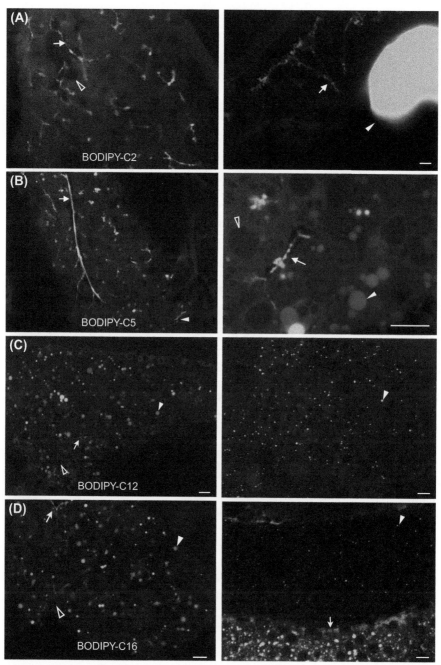

FIGURE 2 Dietary BODIPY fatty acid can be used to label digestive organs and simultaneously interrogate organ physiology.

BODIPY-conjugated FAs (described in the following section). This is most likely due to its lack of incorporation into complex lipids (Carten et al., 2011).

2.2 BODIPY C2

When comparing the chain length of a BODIPY-conjugated FA to an unlabeled FA, we estimate an additional two to four carbons account for the addition of the BODIPY fluorophore. We consider BODIPY C2 to act as a short-chain fatty acid (SCFA; a fatty acid of less than six carbon atoms) (Fig. 1F). Larvae labeled with this reagent reveal fluorescence primarily in the hepatic and pancreatic ducts (Fig. 2A), suggesting BODIPY C2 is particularly suited to illuminate the ductal networks of digestive organs. The speed of hepatic labeling (<2 h) suggests that this reagent can readily diffuse to the liver and does not require lipoproteins to transit from the intestine. One explanation is that this fluorophore engages the cellular xenobiotic efflux system, enabling its

◀──────────────────────────────

(A) BODIPY-FL C2 does not participate appreciably in lipid metabolism. Fluorescence appears primarily in the intestinal lumen of larvae (not shown). In the liver (left panel), BODIPY-FL C2 accumulates in the hepatic ducts (*arrow*) and diffusely in hepatocytes, whose nuclei are discernable (*empty arrowhead*). The ductal network of the exocrine pancreas (right panel) is also labeled (*arrow*). The gall bladder (*filled arrowhead*) is indicated. Scale bars = 10 μm. (B) BODIPY-FL C5 reveals significant subcellular details within larval digestive organs (6 dpf). In enterocytes of the anterior intestine, BODIPY-FL C5 appears in lipid droplets as well as in small endosomal-like compartments (not shown) and illuminates the intrahepatobiliary ductal network (left panel). In the right liver lobe, a single long duct (*arrow*), numerous interconnecting ducts and terminal ductules (*arrowhead*) are illuminated. Subcellular details of larval hepatocytes are also revealed (right panel). Lipid droplets (*filled arrowhead*), hepatocyte nuclei (*empty arrowhead*), and intrahepatic ducts (*arrow*) are indicated. Scale bars = 10 μm. (C) BODIPY-FL C12 is absorbed and transported in the digestive organs of zebrafish larvae. The analog appears in the intestinal lumen as well as in lipid droplets in intestinal enterocytes (not shown) and readily accumulates in the liver, visualizing large lipid droplets (*filled arrowhead*), hepatic ducts (*arrow*), and hepatocyte nuclei (*empty arrowhead*) (left panel). BODIPY-FL C12 accumulates in fluorescent foci (*arrowhead*) in the exocrine pancreas (Right panel). Scale bars = 10 μm. (D) BODIPY-FL C16 accumulates in lipid droplets in the digestive organs of live larval zebrafish. Enterocytes of the anterior intestinal bulb readily absorb BODIPY-FL C16 following an egg yolk and analog feed (not shown) and accumulates within large hepatic lipid droplets (*filled arrowhead*) and faintly in hepatic ducts (*arrow*) (left panel). Hepatocyte nuclei (*empty arrowhead*) are also visible. BODIPY-FL C16 accumulates throughout the exocrine pancreas, forming small fluorescent foci (*arrowhead*) (right panel). Beneath the exocrine pancreas, an intestinal blood vessel (*arrow*) running the length of the intestine fluoresces, indicating analog entry into the vascular system. Scale bars = 10 μm. In all images, anterior is to the right to enhance liver viewing.

Images reprinted from Carten, J.D., Bradford, M.K., & Farber, S.A. (2011). Visualizing digestive organ morphology and function using differential fatty acid metabolism in live zebrafish. Developmental Biology, *360(2), 276–285.*

rapid appearance in ductal networks. This efflux pathway is active in the renal proximal tubule, intestine, liver, and blood—brain barrier (Thiebaut et al., 1987). To address this possibility, inhibitors of xenobiotic efflux transporter P-glycoprotein could be exploited to test for suppression of BODIPY C2's labeling pattern.

2.3 BODIPY C5

Using BODIPY C5, we attain the most extensive cellular labeling. As a medium-chain fatty acid (MCFA; 6—12 carbon atoms), it can be transported directly to the liver and/or incorporated into more complex lipids and travel via the lipoprotein pathway. We found that LDs throughout a wide range of cell types were labeled, as well as ductal and arterial networks, thereby revealing the subcellular structures of multiple cell types (Fig. 2B). The ability of BODIPY C5 to label both cell membranes and TAG-rich lipid droplets is consistent with our biochemical analyses that revealed its incorporation into both phospholipids and TAG (Carten et al., 2011). The hepatic fluorescence accumulation patterns (in the cytoplasm and in lipid droplets) suggest that BODIPY C5 utilizes both chylomicron-mediated and basic secretion (chylomicron-independent) mechanisms of transport. In summary, BODIPY C5 is ideal for studies that seek widespread labeling of cellular structures and digestive organ morphology.

2.4 BODIPY C12/C16

BODIPY C12 and BODIPY C16 act as long-chain fatty acids (LCFAs; 13—21 carbon atoms) and, as such, can only be transported from the intestine via lipoproteins due to their hydrophobicity. In contrast to the shorter chain BODIPY FAs, both these fluorescent analogs appear in larval hepatic ductal networks after longer labeling times. Both fluorescent lipids readily accumulate in cytoplasmic lipid droplets (LD) in enterocytes and hepatocytes (Fig. 2C and D). Where we observe a difference between these LCFA BODIPY lipids is their ability to label cell membranes. BODIPY C12, like BODIPY C5, is incorporated into membrane phospholipids in a similar pattern to radiolabeled H^3 palmitate (C16:0) (Miyares, de Rezende, & Farber, 2014), whereas BODIPY C16 is only incorporated into TAG (Carten et al., 2011).

In summary, the key factors driving the choice of BODIPY-conjugated FA for subsequent studies are their differential ability to label ductal networks, LDs, and cell membranes.

SUMMARY

In this chapter, we have summarized the labeling patterns of BODIPY-conjugated FAs. The optical transparency of zebrafish larvae can be fully exploited by using transgenic lines, fluorescent reporters, and lipid analogs to visualize metabolic events. Furthermore, the high genetic conservation across lipid signaling and metabolic pathways allows pharmacological regents to be utilized to study the importance of lipids during early development. The high fecundity, small size, and genetic tractability of these organisms make them ideal for high-throughput screening efforts to identify genes involved in lipid metabolism and thus identify potential therapeutic targets for human diseases.

ACKNOWLEDGMENTS

This work was supported in part by the National Institute of Diabetes and Digestive and Kidney (NIDDK) F31DK091129 to J.D.C. and RO1DK093399 to S.A.F., National Institute of General Medicine (GM) RO1GM63904 to the Zebrafish Functional Genomics Consortium (Stephen Ekker and S.A.F.). The content is solely the responsibility of the authors and does not necessarily represent the official views of the National Institutes of Health (NIH). Additional support for this work was provided by the Carnegie Institution for Science endowment and the G. Harold and Leila Y. Mathers Charitable Foundation to the laboratory of S.A.F.

REFERENCES

Anderson, J. L., Carten, J. D., & Farber, S. A. (2011). Zebrafish lipid metabolism: from mediating early patterning to the metabolism of dietary fat and cholesterol. *Methods in Cell Biology, 101*, 111−141. http://dx.doi.org/10.1016/B978-0-12-387036-0.00005-0. pii: B978-0-12-387036-0.00005-0.

Ariola, F. S., Li, Z., Cornejo, C., Bittman, R., & Heikal, A. A. (2009). Membrane fluidity and lipid order in ternary giant unilamellar vesicles using a new bodipy-cholesterol derivative. *Biophysical Journal, 96*(7), 2696−2708.

Avraham-Davidi, I., Ely, Y., Pham, V. N., Castranova, D., Grunspan, M., Malkinson, G., ... Yaniv, K. (2012). ApoB-containing lipoproteins regulate angiogenesis by modulating expression of VEGF receptor 1. *Nature Medicine*. http://dx.doi.org/10.1038/nm.2759. pii:nm.2759.

Backhed, F., Ding, H., Wang, T., Hooper, L. V., Koh, G. Y., Nagy, A., ... Gordon, J. I. (2004). The gut microbiota as an environmental factor that regulates fat storage. *Proceedings of the National Academy of Sciences of the United States of America, 101*(44), 15718−15723. pii:0407076101.

Bastien, M., Poirier, P., Lemieux, I., & Despres, J. P. (2014). Overview of epidemiology and contribution of obesity to cardiovascular disease. *Progress in Cardiovascular Diseases, 56*(4), 369−381. http://dx.doi.org/10.1016/j.pcad.2013.10.016.

Bauer, U. E., Briss, P. A., Goodman, R. A., & Bowman, B. A. (2014). Prevention of chronic disease in the 21st century: elimination of the leading preventable causes of premature death and disability in the USA. *Lancet, 384*(9937), 45−52. http://dx.doi.org/10.1016/S0140-6736(14)60648-6.

Bedell, V. M., Wang, Y., Campbell, J. M., Poshusta, T. L., Starker, C. G., Krug, R. G., 2nd, ... Ekker, S. C. (2012). In vivo genome editing using a high-efficiency TALEN system. *Nature, 491*(7422), 114−118. http://dx.doi.org/10.1038/nature11537.

Brasaemle, D. L., Rubin, B., Harten, I. A., Gruia-Gray, J., Kimmel, A. R., & Londos, C. (2000). Perilipin A increases triacylglycerol storage by decreasing the rate of triacylglycerol hydrolysis. *The Journal of Biological Chemistry, 275*(49), 38486−38493. http://dx.doi.org/10.1074/jbc.M007322200.

Buhman, K. K., Smith, S. J., Stone, S. J., Repa, J. J., Wong, J. S., Knapp, F. F., Jr., ... Farese, R. V., Jr. (2002). DGAT1 is not essential for intestinal triacylglycerol absorption or chylomicron synthesis. *Journal of Biological Chemistry, 277*, 25474−25479.

Carten, J. D., Bradford, M. K., & Farber, S. A. (2011). Visualizing digestive organ morphology and function using differential fatty acid metabolism in live zebrafish. *Developmental Biology*. http://dx.doi.org/10.1016/j.ydbio.2011.09.010. pii:S0012-1606(11)01254-1.

Chen, H. C., & Farese, R. V., Jr. (2002). Fatty acids, triglycerides, and glucose metabolism: recent insights from knockout mice. *Current Opinion in Clinical Nutrition and Metabolic Care, 5*(4), 359−363.

Clifton, J. D., Lucumi, E., Myers, M. C., Napper, A., Hama, K., Farber, S. A., ... Pack, M. (2010). Identification of novel inhibitors of dietary lipid absorption using zebrafish. *PLoS One, 5*(8). http://dx.doi.org/10.1371/journal.pone.0012386.

Delous, M., Yin, C., Shin, D., Ninov, N., Debrito Carten, J., Pan, L., ... Stainier, D. Y. (2012). Sox9b is a key regulator of pancreaticobiliary ductal system development. *PLoS Genetics, 8*(6), e1002754. http://dx.doi.org/10.1371/journal.pgen.1002754. pii:PGENETICS-D-11-02406.

Driever, W., Solnica-Krezel, L., Schier, A. F., Neuhauss, S. C., Malicki, J., Stemple, D. L., ... Boggs, C. (1996). A genetic screen for mutations affecting embryogenesis in zebrafish. *Development, 123*, 37−46.

Enzler, L., Smith, V., Lin, J. S., & Olcott, H. S. (1974). The lipids of Mono Lake, California, brine shrimp (Artemia salina). *Journal of Agricultural and Food Chemistry, 22*(2), 330−331.

Farber, S. A., Olson, E. S., Clark, J. D., & Halpern, M. E. (1999). Characterization of Ca2+-dependent phospholipase A2 activity during zebrafish embryogenesis. *The Journal of Biological Chemistry, 274*(27), 19338−19346.

Farber, S. A., Pack, M., Ho, S. Y., Johnson, I. D., Wagner, D. S., Dosch, R., ... Halpern, M. E. (2001). Genetic analysis of digestive physiology using fluorescent phospholipid reporters. *Science, 292*(5520), 1385−1388.

Field, F. J. (2001). Regulation of intestinal cholesterol metabolism. In C. M. Mansbach, P. Tso, & A. Kuksis (Eds.), *Intestinal Lipid Metabolism* (pp. 235−255). New York: Kluwer Academic.

Field, H. A., Dong, P. D., Beis, D., & Stainier, D. Y. (2003). Formation of the digestive system in zebrafish. II. Pancreas morphogenesis. *Developmental Biology, 261*(1), 197−208. pii: S0012160603003087.

Frohlich, F., Petit, C., Kory, N., Christiano, R., Hannibal-Bach, H. K., Graham, M., ... Walther, T. C. (2015). The GARP complex is required for cellular sphingolipid homeostasis. *eLife, 4*. http://dx.doi.org/10.7554/eLife.08712.

Gocze, P. M., & Freeman, D. A. (1994). Factors underlying the variability of lipid droplet fluorescence in MA-10 Leydig tumor cells. *Cytometry, 17*(2), 151−158. http://dx.doi.org/10.1002/cyto.990170207.

Gut, P., Baeza-Raja, B., Andersson, O., Hasenkamp, L., Hsiao, J., Hesselson, D., ... Stainier, D. Y. (2013). Whole-organism screening for gluconeogenesis identifies activators of fasting metabolism. *Nature Chemical Biology, 9*(2), 97−104. http://dx.doi.org/10.1038/nchembio.1136.

Haffter, P., Granato, M., Brand, M., Mullins, M. C., Hammerschmidt, M., Kane, D. A., ... Nusslein-Volhard, C. (1996). The identification of genes with unique and essential functions in the development of the zebrafish, *Danio rerio*. *Development, 123*, 1−36.

Hama, K., Provost, E., Baranowski, T. C., Rubinstein, A. L., Anderson, J. L., Leach, S. D., & Farber, S. A. (2009). In vivo imaging of zebrafish digestive organ function using multiple quenched fluorescent reporters. *American Journal of Physiology. Gastrointestinal and Liver Physiology, 296*(2), G445−G453.

Hendrickson, H. S., Hendrickson, E. K., Johnson, I. D., & Farber, S. A. (1999). Intramolecularly quenched BODIPY-labeled phospholipid analogs in phospholipase A(2) and platelet-activating factor acetylhydrolase assays and in vivo fluorescence imaging. *Analytical Biochemistry, 276*(1), 27–35.

Ho, S. Y., Thorpe, J. L., Deng, Y., Santana, E., DeRose, R. A., & Farber, S. A. (2004). Lipid metabolism in zebrafish. *Methods in Cell Biology, 76*, 87–108.

Ho, S. Y., Lorent, K., Pack, M., & Farber, S. A. (2006). Zebrafish fat-free is required for intestinal lipid absorption and Golgi apparatus structure. *Cell Metabolism, 3*(4), 289–300.

Holtta-Vuori, M., Uronen, R. L., Repakova, J., Salonen, E., Vattulainen, I., Panula, P., ... Ikonen, E. (2008). BODIPY-cholesterol: a new tool to visualize sterol trafficking in living cells and organisms. *Traffic, 9*(11), 1839–1849.

Hwang, W. Y., Fu, Y., Reyon, D., Gonzales, A. P., Joung, J. K., & Yeh, J. R. (2015). Targeted mutagenesis in zebrafish using CRISPR RNA-Guided Nucleases. *Methods in Molecular Biology, 1311*, 317–334. http://dx.doi.org/10.1007/978-1-4939-2687-9_21.

Ivics, Z., Kaufman, C. D., Zayed, H., Miskey, C., Walisko, O., & Izsvak, Z. (2004). The Sleeping Beauty transposable element: evolution, regulation and genetic applications. *Current Issues in Molecular Biology, 6*(1), 43–55.

Kawakami, K., Takeda, H., Kawakami, N., Kobayashi, M., Matsuda, N., & Mishina, M. (2004). A transposon-mediated gene trap approach identifies developmentally regulated genes in zebrafish. *Developmental Cell, 7*(1), 133–144.

Kruit, J. K., Groen, A. K., van Berkel, T. J., & Kuipers, F. (2006). Emerging roles of the intestine in control of cholesterol metabolism. *World Journal of Gastroenterology, 12*(40), 6429–6439.

Lada, A. T., Davis, M., Kent, C., Chapman, J., Tomoda, H., Omura, S., & Rudel, L. L. (2004). Identification of ACAT1- and ACAT2-specific inhibitors using a novel, cell-based fluorescence assay: individual ACAT uniqueness. *Journal of Lipid Research, 45*(2), 378–386. http://dx.doi.org/10.1194/jlr.D300037-JLR200.

Levy, E., Spahis, S., Sinnett, D., Peretti, N., Maupas-Schwalm, F., Delvin, E., ... Lavoie, M. A. (2007). Intestinal cholesterol transport proteins: an update and beyond. *Current Opinion in Lipidology, 18*(3), 310–318.

Li, Z., & Bittman, R. (2007). Synthesis and spectral properties of cholesterol- and FTY720-containing boron dipyrromethene dyes. *The Journal of Organic Chemistry, 72*(22), 8376–8382. http://dx.doi.org/10.1021/jo701475q.

Li, Z., Mintzer, E., & Bittman, R. (2006). First synthesis of free cholesterol-BODIPY conjugates. *The Journal of Organic Chemistry, 71*(4), 1718–1721.

Liu, H. Y., Lee, N., Tsai, T. Y., & Ho, S. Y. (2010). Zebrafish fat-free, a novel Arf effector, regulates phospholipase D to mediate lipid and glucose metabolism. *Biochimica et Biophysica Acta, 1801*(12), 1330–1340. http://dx.doi.org/10.1016/j.bbalip.2010.08.012.

Marks, D. L., Bittman, R., & Pagano, R. E. (2008). Use of Bodipy-labeled sphingolipid and cholesterol analogs to examine membrane microdomains in cells. *Histochemistry and Cell Biology, 130*(5), 819–832.

Martin, F. P., Wang, Y., Sprenger, N., Yap, I. K., Lundstedt, T., Lek, P., ... Nicholson, J. K. (2008). Probiotic modulation of symbiotic gut microbial-host metabolic interactions in a humanized microbiome mouse model. *Molecular Systems Biology, 4*, 157.

Miyares, R. L., Stein, C., Renisch, B., Anderson, J. L., Hammerschmidt, M., & Farber, S. A. (2013). Long-chain Acyl-CoA synthetase 4A regulates Smad activity and dorsoventral patterning in the zebrafish embryo. *Developmental Cell, 27*(6), 635–647. http://dx.doi.org/10.1016/j.devcel.2013.11.011.

Miyares, R. L., de Rezende, V. B., & Farber, S. A. (2014). Zebrafish yolk lipid processing: a tractable tool for the study of vertebrate lipid transport and metabolism. *Disease Models & Mechanisms, 7*(7), 915–927. http://dx.doi.org/10.1242/dmm.015800.

Monsma, F. J., Jr., Barton, A. C., Kang, H. C., Brassard, D. L., Haugland, R. P., & Sibley, D. R. (1989). Characterization of novel fluorescent ligands with high affinity for D1 and D2 dopaminergic receptors. *Journal of Neurochemistry, 52*(5), 1641–1644.

Moschetta, A., Xu, F., Hagey, L. R., van Berge-Henegouwen, G. P., van Erpecum, K. J., Brouwers, J. F., … Hofmann, A. F. (2005). A phylogenetic survey of biliary lipids in vertebrates. *Journal of Lipid Research, 46*(10), 2221–2232.

Otis, J. P., & Farber, S. A. (2013). Imaging vertebrate digestive function and lipid metabolism. *Drug Discovery Today Disease Models, 10*(1). http://dx.doi.org/10.1016/j.ddmod.2012.02.008.

Pack, M., Solnica-Krezel, L., Malicki, J., Neuhauss, S. C., Schier, A. F., Stemple, D. L., … Fishman, M. C. (1996). Mutations affecting development of zebrafish digestive organs. *Development, 123*, 321–328.

Pagano, R. E., Martin, O. C., Kang, H. C., & Haugland, R. P. (1991). A novel fluorescent ceramide analogue for studying membrane traffic in animal cells: accumulation at the Golgi apparatus results in altered spectral properties of the sphingolipid precursor. *The Journal of Cell Biology, 113*(6), 1267–1279.

Rankinen, T., Sarzynski, M. A., Ghosh, S., & Bouchard, C. (2015). Are there genetic paths common to obesity, cardiovascular disease outcomes, and cardiovascular risk factors? *Circulation Research, 116*(5), 909–922. http://dx.doi.org/10.1161/CIRCRESAHA.116.302888.

Sadler, K. C., Rawls, J. F., & Farber, S. A. (2013). Getting the inside tract: new frontiers in zebrafish digestive system biology. *Zebrafish, 10*(2), 129–131. http://dx.doi.org/10.1089/zeb.2013.1500.

Schekman, R., & Sudhof, T. (2014). An interview with Randy Schekman and Thomas Sudhof. *Trends in Cell Biology, 24*(1), 6–8.

Semova, I., Carten, J. D., Stombaugh, J., Mackey, L. C., Knight, R., Farber, S. A., & Rawls, J. F. (2012). Microbiota regulate intestinal absorption and metabolism of fatty acids in the zebrafish. *Cell Host & Microbe, 12*(3), 277–288. http://dx.doi.org/10.1016/j.chom.2012.08.003. pii:S1931-3128(12)00274-0.

Shah, A. N., Davey, C. F., Whitebirch, A. C., Miller, A. C., & Moens, C. B. (2015). Rapid reverse genetic screening using CRISPR in zebrafish. *Nature Methods, 12*(6), 535–540. http://dx.doi.org/10.1038/nmeth.3360.

Spence, R., Gerlach, G., Lawrence, C., & Smith, C. (2008). The behaviour and ecology of the zebrafish, Danio rerio. *Biological Reviews, 83*(1), 13–34.

Thiebaut, F., Tsuruo, T., Hamada, H., Gottesman, M. M., Pastan, I., & Willingham, M. C. (1987). Cellular localization of the multidrug-resistance gene product P-glycoprotein in normal human tissues. *Proceedings of the National Academy of Sciences of the United States of America, 84*(21), 7735–7738.

Titus, E., & Ahearn, G. A. (1992). Vertebrate gastrointestinal fermentation: transport mechanisms for volatile fatty acids. *American Journal of Physiology, 262*(4 Pt 2), R547–R553.

Thomson, A. B., Keelan, M., Cheng, T., & Clandinin, M. T. (1993). Delayed effects of early nutrition with cholesterol plus saturated or polyunsaturated fatty acids on intestinal morphology and transport function in the rat. *Biochimica et Biophysica Acta, 1170*(1), 80–91. pii:0005-2760(93)90178-C.

Treibs, A., & Kreuzer, F. (1969). Difluorboryl-Komplexe von Di- und Tripyrrylmethenen. *Justus Liebigs Annalen der Chemie, 721*, 116–120.

Turnbaugh, P. J., Backhed, F., Fulton, L., & Gordon, J. I. (2008). Diet-induced obesity is linked to marked but reversible alterations in the mouse distal gut microbiome. *Cell Host & Microbe, 3*(4), 213–223. http://dx.doi.org/10.1016/j.chom.2008.02.015. pii: S1931-3128(08)00089-9.

CHAPTER

Analysis of cilia structure and function in zebrafish

E. Leventea*,[a], K. Hazime*,[a], C. Zhao*,[§], J. Malicki*,[1]

The University of Sheffield, Sheffield, United Kingdom
[§]*Ocean University of China, Qingdao, China*
[1]*Corresponding author: E-mail: j.malicki@sheffield.ac.uk*

CHAPTER OUTLINE

Introduction	180
1. Cilia in Zebrafish Organs	181
1.1 Kupffer's Vesicle	183
1.2 The Pronephros	184
1.3 Sensory Organs	185
1.3.1 Photoreceptors	185
1.3.2 Mechanosensory hair cells	187
1.3.3 Olfactory sensory neurons	189
1.4 Spinal Canal	190
2. Analytical Tools for Cilia Morphology and Motility	190
2.1 Method 1: *Detection of Ciliary Proteins Using Immunohistochemistry*	190
2.1.1 Method 1a: staining of whole embryos at 3 dpf and younger	194
2.1.2 Method 1b: staining of whole larvae at 5–7 dpf	196
2.1.3 Method 1c: staining of dissected adult tissues (ear and retina)	196
2.1.4 Method 1d: staining of cryosections (all stages)	197
2.2 Method 2: *Live Imaging of Cilia and Basal Bodies Using Transgenic Lines*	200
2.3 Method 3: *Live Imaging of Cilia Movement*	201
2.4 Method 4: *Live Imaging of Cilia Using Light Sheet Microscopy*	203
2.5 Method 5: *Analysis of Ciliary Transport Using Inducible Transgenes*	204
3. Phenotypes of Cilia Mutants in Zebrafish	208
3.1 Method 6: *Evaluation of Heart Position in Live Embryos*	209
3.2 Method 7: *Evaluation of Kidney Function*	210

[a]These authors contributed equally.

3.3 Method 8: *Analysis of Sensory Cell Morphology* 211
    3.3.1 Photoreceptor cells 211
    3.3.2 Auditory system hair cells 212
    3.3.3 Lateral line hair cells 213
    3.3.4 Olfactory sensory neurons 213
3.4 Method 9: *Staining of Neuromast Hair Cells in Live Specimen* 213
3.5 Method 10: *Labeling of Olfactory Neurons by DiI Incorporation* 214
3.6 Method 11: *Tests of Olfaction* 214
4. Future Directions 215
Acknowledgments 216
References 216

Abstract

Cilia are microtubule-based protrusions on the surface of most eukaryotic cells. They are found in most, if not all, vertebrate organs. Prominent cilia form in sensory structures, the eye, the ear, and the nose, where they are crucial for the detection of environmental stimuli, such as light and odors. Cilia are also involved in developmental processes, including left–right asymmetry formation, limb morphogenesis, and the patterning of neurons in the neural tube. Some cilia, such as those found in nephric ducts, are thought to have mechanosensory roles. Zebrafish proved very useful in genetic analysis and imaging of cilia-related processes, and in the modeling of mechanisms behind human cilia abnormalities, known as ciliopathies. A number of zebrafish defects resemble those seen in human ciliopathies. Forward and reverse genetic strategies generated a wide range of cilia mutants in zebrafish, which can be studied using sophisticated genetic and imaging approaches. In this chapter, we provide a set of protocols to examine cilia morphology, motility, and cilia-related defects in a variety of organs, focusing on the embryo and early postembryonic development.

INTRODUCTION

Cilia are microscopic microtubule-based projections of the surface on most vertebrate cells. The structure of the ciliary shaft, or the axoneme, is very well conserved among species. On cross sections, the vast majority of cilia in organisms ranging from unicellular algae to mammals feature nine evenly spaced peripheral microtubule doublets. A central pair of microtubules is also found in most motile cilia. This is known as the "9 + 2" arrangement of microtubules. This central pair is missing in most immotile cilia, which feature the "9 + 0" arrangement of microtubules (Fisch & Dupuis-Williams, 2011).

    Cilia are essential both for embryonic development and proper physiological function of fully formed organs. They play a role in a multitude of signal transduction pathways that mediate processes as diverse as embryonic patterning, vision, olfaction, and metabolism (Goetz & Anderson, 2010; Louvi & Grove, 2011; Schou, Pedersen, & Christensen, 2015). While motile cilia propagate fluid flow and propel the movement of cells, such as sperm cells or unicellular algae, immotile cilia mediate the detection and/or transduction of extracellular signals in many biological processes ranging from phototransduction to Hedgehog signaling (Green & Mykytyn, 2010; Kennedy & Malicki, 2009; Kobayashi & Takeda, 2012).

In the embryo, cilia are involved in the determination of the left—right (LR) asymmetry, limb and neural tube morphogenesis, and the differentiation of sensory neurons, such as photoreceptors in the eye (Basu & Brueckner, 2008; Haycraft et al., 2005; Tissir et al., 2010; Tsujikawa & Malicki, 2004a; Walczak-Sztulpa et al., 2010). In developmentally mature vertebrates, cilia mediate sperm cell motility, maintenance of kidney ducts, the clearing of debris from the respiratory tract, vision, olfaction, metabolism, and even perhaps some higher brain functions (Afzelius, 2004; Cardenas-Rodriguez & Badano, 2009; Guemez-Gamboa, Coufal, & Gleeson, 2014; Zimmerman & Yoder, 2015). As cilia widely contribute to biological processes that play a role in nearly every aspect of vertebrate biology, it is not surprising that they have attracted considerable attention.

The zebrafish is an excellent vertebrate model for genetic analysis and imaging of cilia-related processes (Fig. 1). This is because zebrafish embryos develop outside the mother's organism, are largely transparent, and differentiate cilia at early stages of embryogenesis. Thus, it is easy to examine the morphology and movement of cilia during development in zebrafish (Kishi, Slack, Uchiyama, & Zhdanova, 2009; Kramer-Zucker et al., 2005; Zhao & Malicki, 2007). Another important feature of the zebrafish model is that cilia mutants usually manifest a curly-body axis, a phenotype that is very easy to detect during genetic screens (Sun et al., 2004; Tsujikawa & Malicki, 2004b). Consequently, a rich collection of zebrafish ciliary mutants is available (Becker-Heck et al., 2010; Brand et al., 1996; Cao, Park, & Sun, 2010; Doerre & Malicki, 2002; Drummond et al., 1998; Malicki, Neuhauss, et al., 1996; Panizzi et al., 2012).

Although historically forward genetic strategies have been the strength of the zebrafish model, recent advances in targeted mutagenesis using TALEN and CRISPR/Cas9 nuclease systems make zebrafish reverse genetics equally attractive. Although, in general, less reliable, fast and inexpensive antisense knockdown approaches have been also used in zebrafish for a number of years (Eisen & Smith, 2008; Nasevicius & Ekker, 2000). These approaches are valuable as tools to study the genetic bases of cilia function in a living embryo. They also make it straightforward to generate zebrafish models of human disorders (Mitchison et al., 2012; Panizzi et al., 2012; Sayer et al., 2006).

1. CILIA IN ZEBRAFISH ORGANS

Below we describe the developmental timing of cilia differentiation, their morphology, and distribution in several zebrafish organs, including the Kupffer's vesicle (KV), the pronephros, the spinal canal, the eye, the ear, and the olfactory system (Fig. 1). We also discuss mutant phenotypes associated with cilia defects and methods commonly used to analyze them. As most of cilia research on zebrafish is performed on embryos and early larvae, we mostly focus on the first 7 days of development.

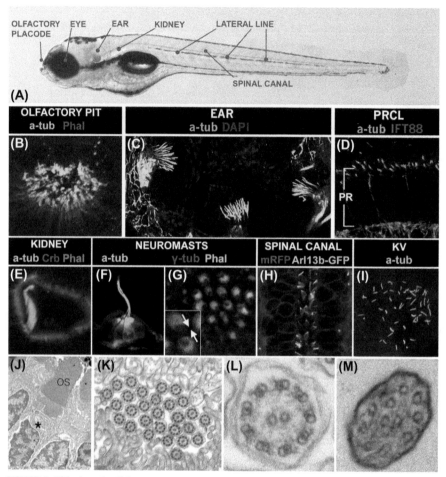

FIGURE 1 Cilia in zebrafish.

(A) The zebrafish larva at approximately 120 hpf. The locations of tissues that feature prominent cilia are indicated. (B) Confocal image of olfactory cilia stained with anti-acetylated tubulin antibody (green) at 3 dpf. F-actin is labeled with fluorophore-conjugated phalloidin (in red) to visualize the morphology of the olfactory pit. (C) A confocal image of the ear stained with anti-acetylated tubulin antibody (green) and counterstained with DAPI (blue) at 4 dpf. Z-stack of this image was processed to partly remove signal from neurites. (D) Transverse cryosection through the retina at 3 dpf. Cilia are stained with anti-acetylated tubulin (green) and anti-Ift88 (red) antibodies and imaged using confocal microscopy. The photoreceptor cell layer is indicated with a bracket. Ift88 signal is enriched at the base of cilia (green). (E) A transverse section through the pronephric duct stained with anti-acetylated tubulin antibody (green) at 4 dpf. The apical surface of the pronephric ducts differentiates actin-rich microvilli and is easy to visualize with phalloidin

1.1 KUPFFER'S VESICLE

KV, a mouse node analogue, is a cyst-like structure with the average diameter of c. 65 μm that appears in the tail bud of the zebrafish embryo around 12 hours postfertilization (hpf) (Matsui, Ishikawa, & Bessho, 2015; May-Simera et al., 2010). The inner surface of the KV is ciliated. The highest concentration of cilia in the KV is found in the anterior region of its dorsal surface (Kreiling, Prabhat, Williams, & Creton, 2007; Oteiza et al., 2010; Smith, Montenegro-Johnson, & Lopes, 2014). Both motile and immotile cilia are found in the KV. Motile cilia are responsible for the circular fluid movement inside the vesicle (Essner, Amack, Nyholm, Harris, & Yost, 2005; Yuan, Zhao, Brueckner, & Sun, 2015). Cilia motility in the KV mediates the initial symmetry-breaking process in the embryo and eventually determines organ laterality (Borovina, Superina, Voskas, & Ciruna, 2010; Kramer-Zucker et al., 2005; Kreiling et al., 2007; Okabe, Xu, & Burdine, 2008). Calcium oscillations in motile cilia have been found to be associated with this process. They are significantly more frequent on the left side of the KV, and their onset coincides with the appearance of cilia motility providing evidence that fluid flow induces their

(blue). The apical surface of ciliated cell is visualized using anti-Crumbs antibody (red). (F) Confocal image of a lateral line neuromast in whole embryo. Hair cells and their cilia are visualized with anti-acetylated tubulin antibody staining. (G) Confocal en face image of a neuromast stained with phalloidin (green) and anti-γ-tubulin antibody (red). *Arrows* in the inset indicate planar orientation of neuromast hair cells. (H) Spinal canal cilia visualized via the expression of an *Arl13-GFP* transgene (in green). To visualize cell membranes, embryos were injected with mRNA encoding membrane-bound RFP (in red). (I) Kupffer's vesicle cilia visualized using antibodies to acetylated alpha-tubulin in a whole embryo. (J) Electron micrograph showing retinal photoreceptors. Note differentiated outer segment (OS). Asterisk marks the photoreceptor cell soma. Transverse section through the central retina at 5 dpf. (K) Electron micrograph showing numerous cilia and villi in the lumen of the pronephric duct. Transverse section through the zebrafish trunk at 4 dpf. (L) A close up view of a cilium from panel (K). The 9 + 2 arrangement of microtubules is visible. (M) Ultrathin section perpendicular to the distal tip of an olfactory cilium. Microtubule singlets are visible. (See color plate)

Images in (B–G) and (I–M) courtesy of Malicki lab members, Yoshihiro Omori, Motokazu Tsujikawa, Khodor Hazime and Chengtian Zhao. In panel (D) apical is up. Image in (D) reprinted with permission from Tsujikawa, M., & Malicki, J. (2004a). Genetics of photoreceptor development and function in zebrafish. International Journal of Developmental Biology, 48(8–9), 925–934. Image in panel (H) courtesy of Brian Ciruna. Images in (I) and (M) reprinted with permission from Zhao, C., & Malicki, J. (2007). Genetic defects of pronephric cilia in zebrafish. Mechanisms of Development, 124(7–8), 605–616 and Zhao, C., Omori, Y., Brodowska, K., Kovach, P., & Malicki, J. (2012). Kinesin-2 family in vertebrate ciliogenesis. Proceedings of the National Academy of Sciences of the United States of America, 109(7), 2388–2393, respectively.

appearance (Yuan et al., 2015). These studies have to treated with caution however in light of recenty published data on ciliary calcium imaging. The KV is the earliest ciliated structure described in the zebrafish embryo so far. It can be easily imaged at 4—8 somite stage (12—14 hpf at 28.5°C) (Fig. 1I).

Defects in KV cilia lead to abnormal LR asymmetry, which first manifests itself as cardiac abnormalities. At 24 hpf, as the heart tube begins to form, the future atrial end of the heart moves to the left side. This process is referred to as cardiac jogging (Chen et al., 1997; Khodiyar, Howe, Talmud, Breckenridge, & Lovering, 2013). Leftward position of the linear heart tube ("jogging") is the first known morphological feature of zebrafish embryogenesis that can be used to evaluate LR asymmetry (See Method 6) (Fig. 4H—J). Later on, another heart phenotype, cardiac looping, is a good laterality indicator. Looping refers to the bending of the heart ventricle to the right, which occurs by 36 hpf and persists until at least 72 hpf (Fig. 4H) (Zhao & Malicki, 2007). In wild-type zebrafish, the leftward jogging precedes the rightward looping (Chen et al., 1997).

Cardiac asymmetry defects are common in mutants and morphants of zebrafish cilia genes (Jaffe et al., 2016; Mitchison et al., 2012; Panizzi et al., 2012; Zhao & Malicki, 2007). This manifests itself by the randomization of heart jogging and looping so that some embryos do not display any looping or loop their hearts in the wrong direction (Fig. 4H—J). The frequencies of these phenotypes vary (Panizzi et al., 2012; Schottenfeld, Sullivan-Brown, & Burdine, 2007; Zhao & Malicki, 2007). One also has to keep in mind that LR asymmetry defects occur with the frequency of low single digit percentages also in wild-type strains (Becker-Heck et al., 2010; Jin et al., 2014; Mitchison et al., 2012; Yuan et al., 2012). In addition to the heart, other organs are affected by LR asymmetry defects, including the liver, the gut, and the pancreas (Chen et al., 2001; Garnaas et al., 2012). Finally, central nervous system structures have been found to develop asymmetrically. An example is the diencephalic habenula in the larval zebrafish brain (Aizawa, Goto, Sato, & Okamoto, 2007). These brain structures can also be used as asymmetry indicators (Doll, Burkart, Hope, Halpern, & Gamse, 2011).

1.2 THE PRONEPHROS

The zebrafish renal system consists of two pronephroi, which originate from the intermediate mesoderm and are positioned symmetrically along both sides of the body axis. At 3 days postfertilization (dpf), each pronephros consists of the glomerulus, the pronephric tubule, and the pronephric duct. Pronephric ducts form at about 20 hpf, pronephric glomeruli appear at about 32—33 hpf, and the glomerular filtration starts between 36 and 48 hpf (Drummond et al., 1998).

Two types of ciliated cells are present in the pronephric duct, the monociliated cells and the multiciliated ones. Both monociliated and multiciliated cells are found in the anterior and the middle segments of the pronephric duct. In the posterior part of the pronephric duct, only monociliated cells are seen. Monociliated cells are observed earlier than the multiciliated ones; the former appear at about 20 hpf, while the latter at around 36 hpf. On cross sections, cilia of both types of cells show the "9 + 2" arrangement of microtubules in the ciliary axoneme (Fig. 1K, L) (Drummond et al., 1998; Liu, Pathak, Kramer-Zucker, & Drummond, 2007; Ma & Jiang, 2007).

The most common ciliary defect in the nephric system is the formation of cysts, seen both in zebrafish and mammals (Pazour et al., 2002; Sun et al., 2004; Yoder, Hou, & Guay-Woodford, 2002; Zhao & Malicki, 2007) (Fig. 4A−E). Zebrafish kidney cysts are usually located in the glomerular-tubular region and are first visible by 2 dpf. They tend to increase in size as the animal grows (Drummond et al., 1998; Kramer-Zucker et al., 2005). The pronephroi are not visible in living animals as they are transparent, but cysts can be easily seen by 3−4 dpf and later as lateral bulges slightly anterior and medial to the pectoral fin (Sun et al., 2004) (Fig. 4B and C). Kidney cysts can be studied in larvae using histological sections or transgenic lines as detailed below in this chapter (Fig. 4D and E).

1.3 SENSORY ORGANS

Vertebrate sensory neurons frequently display epithelial characteristics, and their apical surface contains sensory apparatus that frequently includes cilia (Falk, Lösl, Schröder, & Giessl, 2015). Vertebrate photoreceptors, olfactory sensory neurons, and mechanosensory hair cells all feature prominent apical cilia. The zebrafish is no exception (Fig. 1B−G; Fig. 2). Below we describe ciliated sensory cells in zebrafish visual, auditory, olfactory, and lateral line systems.

1.3.1 Photoreceptors

Similar to the mammalian eye, the zebrafish retina consists of seven major cell classes arranged in distinct layers (Schmitt & Dowling, 1999 and chapter "Analysis of the retina in the zebrafish model" by Malicki, Pooranachandran, Nikolaev, Fang, & Avanesov, 2016). Zebrafish photoreceptors form the outermost layer of the retina and display similar morphological characteristics to these in mouse and human eyes (Kennedy & Malicki, 2009). Cilia of vertebrate photoreceptors are among the most structurally complex described by biologists thus far. They consist of the so-called "outer segment" and a narrow stalk, known as the "connecting cilium," that bridges the outer segment with the cell body. The connecting cilium is currently thought to largely correspond to the ciliary transition zone (Quinlan, Tobin, & Beales, 2008; Szymanska & Johnson, 2012). Outer segments differentiate hundreds of membrane folds that harbor the visual pigment and other proteins that mediate phototransduction. Light is detected by opsin molecules tightly packed in outer segment membranes (Fotiadis et al., 2003). It is estimated that one billion visual pigment molecules are found in the photoreceptor outer segment of some species (Pugh & Lamb, 2000).

On cross sections, the photoreceptor connecting cilium features the arrangement of microtubules typical of primary cilia (Chuang, Hsu, & Sung, 2015; Kennedy & Malicki, 2009). Apical to the connecting cilium, microtubules continue to run parallel to each other but gradually lose their regular circular arrangement and distribute

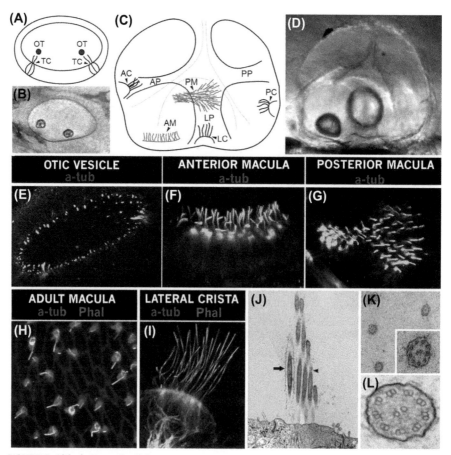

FIGURE 2 Cilia in the zebrafish ear.

(A) Schematic drawing of the otic vesicle at 24–30 hpf. *OT*, otolith; *TC*, tether cells. (B) DIC image of the otic vesicle at 30 hpf. (C) Schematic drawing of the zebrafish ear at 5 dpf. Sensory epithelia are indicated with *arrowheads*. *AC*, anterior crista; *AM*, anterior macula; *AP*, anterior pillar; *LC*, lateral crista; *LP*, lateral pillar; *PC*, posterior crista; *PM*, posterior macula; *PP*, posterior pillar. (D) DIC image of the zebrafish ear at 5 dpf. (E–G) Confocal images of the ear in whole embryos stained with anti-acetylated tubulin antibody to visualize cilia. (E) Cilia in the otic vesicle at 24 hpf. (F, G) Anterior (F) and posterior (G) maculae of a 5 dpf. (H) A confocal image of the posterior macula in the adult stained with anti-acetylated tubulin antibody (green) to visualize kinocilia. Phalloidin (red) is used to outline the apical boundaries of cells and visualize stereocilia. (I) Lateral crista at 5 dpf stained as in (H). (J) Electron micrograph of a transverse section through ear macula at 3 dpf. The kinocilium (*arrow*) and the nearby bundle of stereocilia (*arrowhead*) are indicated. (K, L) Electron micrographs of cross sections through hair cell kinocilia in the ear at 3 dpf. The 9 + 2 arrangement of microtubules is seen in kinocilia of crista (K) and macula (L). Inset in (K) shows an enlargement of a crista kinocilium. (See color plate)

Image in (I) reprinted with permission from Zhao, C., Omori, Y., Brodowska, K., Kovach, P., & Malicki, J. (2012). Kinesin-2 family in vertebrate ciliogenesis. Proceedings of the National Academy of Sciences of the United States of America, 109(7), 2388–2393.

around the circumference of a semicircular area that closely apposes outer segment discs (Insinna, Pathak, Perkins, Drummond, & Besharse, 2008; Wen, Soifer, & Wisniewski, 1982; Zhao, Omori, Brodowska, Kovach, & Malicki, 2012). Similar to other cilia, microtubule doublets transform into singlets towards the distal portion of the outer segment. In zebrafish, rudimentary outer segments become first visible on electron micrographs around 54 hpf in the ventral retina and gradually enlarge thereafter (Branchek & Bremiller, 1984; Schmitt & Dowling, 1999).

The zebrafish retina features five types of photoreceptor cells (please see chapter "Analysis of the retina in the zebrafish model" by Malicki et al., 2016). The sizes and positions of outer segments vary in different photoreceptor types (Branchek & Bremiller, 1984; Raymond, Barthel, Rounsifer, Sullivan, & Knight, 1993). The outer segments of short single cones are closest to the synaptic layer of the outer retina while rod outer segments are farthest away. In vertebrate eyes, membrane arrangement in cone outer segments differs from that in rods: rod outer segments form closed discs; whereas cones contain membrane folds that are open to the extracellular environment (Ding, Salinas, & Arshavsky, 2015; reviewed in Kennedy & Malicki, 2009). The zebrafish retina is cone rich. In the larval eye c. 80% of photoreceptors are cones (Doerre & Malicki, 2001). Mutants of intraflagellar transport (IFT) invariably lead to photoreceptor outer segment loss and cell death (Fig. 5A–E) (Doerre & Malicki, 2002; Tsujikawa & Malicki, 2004b; Zhao et al., 2012).

1.3.2 Mechanosensory hair cells

In the auditory system, mechanosensory hair cells are found in several specialized patches of cells (Fig. 2) (Platt, 1993; Bang, Sewell, & Malicki, 2001; Haddon & Lewis, 1996; Whitfield et al., 1996). These cells differentiate prominent apical cilia, known as kinocilia. The tether cells, which are thought to be hair cell precursors (Riley, Zhu, Janetopoulos, & Aufderheide, 1997; Tanimoto, Ota, Inoue, & Oda, 2011), appear at about 18–18.5 hpf and are located in two groups, one in the anterior and the other in the posterior part of the otic vesicle (Riley et al., 1997; Stooke-Vaughan, Obholzer, Baxendale, Megason, & Whitfield, 2015) (Fig. 2A and E). They are associated with otoliths, deposits of proteins and calcium salts that facilitate the detection of body movements (Fig. 2B and D) (Haddon & Lewis, 1996; Malicki, Schier, et al., 1996; Riley et al., 1997). Later during development, new cells are gradually added next to tether cells, forming patches of sensory epithelia known as the anterior (utricular) and the posterior (saccular) macula (Fig. 2C, F and G) (Bang et al., 2001; Haddon & Lewis, 1996). A third macula differentiates substantially later at around 21 to 25 dpf, in the lagena, the third sensory chamber of the ear (Bang et al., 2001).

At about 3 dpf, three additional sensory patches of hair cells, the anterior, lateral, and posterior cristae, appear in semicircular canals (Haddon & Lewis, 1996) (Fig. 1C). In adult zebrafish, areas of semicircular canals that contain cristae widen to form bulb-shaped ampullae. The kinocilia of the lateral crista can be observed in living zebrafish larvae at 5 dpf using a good quality stereomicroscope (Nicolson et al., 1998). They are unusual in that they are very long and rigid, compared to most immotile cilia (Haddon & Lewis, 1996; Zhao et al., 2012) (Fig. 2I). Despite being immotile, they feature, however, the "9 + 2" arrangement of microtubules characteristic of motile cilia (Fig. 2K). Perhaps reflecting these unusual morphological and mechanical characteristics, kinocilia of cristae also rely on distinct ciliary transport mechanisms, compared to kinocilia in other groups of hair cells (Pooranachandran & Malicki, 2016; Zhao et al., 2012).

In addition to the kinocilium, the apical surface of hair cells features so-called stereocilia. These are also finger-like protrusions of smaller diameter compared to the kinocilium. The name "stereocilia" is misleading as it suggests that these structures are related to cilia. Their cytoskeleton does not contain microtubules; rather it is actin rich, and so stereocilia are related to villi, not cilia (Tanimoto et al., 2011). Stereocilia form a bundle at the apical surface adjacent to the kinocilium (Fig. 2H and J). Their length gradually decreases with the distance from the kinocilium. They are easy to visualize using actin stains, such as phalloidin (see discussion below).

The hair cell detects mechanical stimuli, such as sound waves, through the physical displacement of the stereociliary bundle (Hudspeth, 1989). The function of the kinocilium in sound detection is not clear. In mammals, hair cell kinocilia are resorbed in the cochlea during development, indicating that they are dispensable for hearing in adult animals (Kikuchi & Hilding, 1965; Kimura, 1969). This is not the case in zebrafish, where hair cells maintain kinocilia throughout adulthood (Bang et al., 2001). Zebrafish kinocilia may contribute to mechanosensation during early stages of development (Kindt, Finch, & Nicolson, 2012). Links are observed between kinocilia and stereocilia in immature zebrafish hair cells at 3 dpf and loss of kinocilia affects mechanosensitivity of nascent hair cells. Mature zebrafish hair cells rely on stereocilia-mediated mechanotransduction (Kindt et al., 2012).

Apart from the mechanosensory patches found in the inner ear, the zebrafish and other aquatic vertebrates differentiate hair cell patches in organs called neuromasts on their body surface (Fig. 1F and G). Their role is to detect hydrodynamic movements (reviewed in Montgomery, Macdonald, Baker, & Carton, 2002). Neuromasts are found on both the head and the trunk of the fish: the former belong to the anterior lateral line system and the latter to the posterior lateral line (reviewed in Dambly-Chaudiere et al., 2003; Ledent, 2002).

The development of the posterior lateral line has been subdivided into three phases: embryonic, juvenile, and adult (Nuñez et al., 2009). The embryonic lateral line is formed by a migrating primordium that originates just posterior to the otic placode (Metcalfe, 1985; Nuñez et al., 2009). It migrates along the horizontal myoseptum between 20 and 40 hpf until it reaches the tip of the tail. While migrating, it deposits clusters of cells at intervals of five to six somites. The embryonic phase is completed by 48 hpf, at which time the lateral line consists of seven to eight neuromasts (Ledent, 2002; Nuñez et al., 2009).

The juvenile phase is distinguished by the gradual appearance of four distinct arrays of neuromasts, referred to as the ventral, lateral, dorsolateral, and dorsal lines. When the juvenile phase reaches completion at about 25 dpf, the posterior lateral line consists of 60 neuromasts. This pattern of four lines persists throughout adulthood (Nuñez et al., 2009). In older fish, each of the 60 neuromasts begins to bud off accessory ones, forming so-called "stitches" which extend along the dorsoventral axis. The number of neuromasts in stitches varies: the posterior stitches consist of six to eight neuromasts, while the anterior ones may contain up to 30 neuromasts in a 30-mm fish (Ledent, 2002; Nuñez et al., 2009). In adults, the number of neuromasts increases up to approximately 1000 (Ledent, 2002).

Hair cell ciliogenesis is closely associated with planar cell polarity. This is evident in the relative positions of the kinocilium and stereocilia, which in neighboring cells are aligned along the same axis both in the zebrafish and mammals (Bang et al., 2001; May-Simera & Kelley, 2012). In the mouse, mutations in the *ift88* gene lead to planar cell polarity defects in hair cells of the cochlea (Jones et al., 2008). Likewise, mutations in a number of other ciliary genes lead to planar polarity defects in ear epithelia (Leightner et al., 2013). Neuromasts of the lateral line are also polarized in the plane of the skin surface (Lopez-Schier & Hudspeth, 2006; Lopez-Schier, Starr, Kappler, Kollmar, & Hudspeth, 2004) (Fig. 1G). This is already visible by 3 dpf. The polarity of lateral line neuromasts makes them an attractive site for the genetic analysis of mechanisms that regulate planar cell polarity (Fig. 5J-M) (Lopez-Schier & Hudspeth, 2006; Lopez-Schier et al., 2004; McDermott et al., 2010; Raible & Kruse, 2000; Zhao & Malicki, 2011).

1.3.3 Olfactory sensory neurons

Cilia are present in the olfactory system of zebrafish and other vertebrate species already at early developmental stages (Hansen & Zeiske, 1993; Lidow & Menco, 1984; Reese, 1965). In zebrafish, olfactory placodes form by 17–18 hpf, and the nasal pit appears by 32 hpf (Hansen & Zeiske, 1993). Zebrafish differentiate three types of sensory neurons in the nasal pit: ciliated, microvillous, and crypt cells (Gayoso, Castro, Anadon, & Manso, 2012; Hansen & Zeiske, 1998). The apical surface of ciliated sensory cells differentiates a prominent olfactory knob, which contains three to seven cilia. Microvillous sensory cells, on the other hand, feature less prominent olfactory knobs and contain 10–30 microvilli. Crypt cells bear microvilli and cilia that are submerged in the apical part of the cell body (Hansen & Zeiske, 1998). Odorant specificities of these cells differ. Microvillous olfactory neurons respond to amino acids, but not to amines or bile acids, while ciliated olfactory neurons are stimulated by amines, nucleotides, and bile acids, but not by amino acids (DeMaria et al., 2013; Lipschitz & Michel, 2002). Odorant detection in ciliated and microvillous cells is achieved though olfactory and vomeronasal G-coupled protein receptors, respectively. The identity of crypt cell olfactory receptors is currently unknown (Kermen, Franco, Wyatt, & Yaksi, 2013).

The olfactory system of zebrafish contains motile and immotile cilia (Hansen & Zeiske, 1998). Structurally, zebrafish olfactory cilia display "9 + 2" microtubule configuration, they are however, immotile as they lack dynein arms (Hansen & Zeiske, 1998; Jenkins, McEwen, & Martens, 2009). Nonsensory motile multiciliated cells are also found in the olfactory epithelium (Hansen & Zeiske, 1998). These cells are thought to facilitate odorant clearance and to move the mucus in the nasal pit (Jenkins et al., 2009; Kermen et al., 2013). Similar to photoreceptor outer segments, and some invertebrate cilia, the tips of zebrafish olfactory cilia feature microtubule singlets (Fig. 1M) (Wloga et al., 2010; Zhao & Malicki, 2011). Zebrafish larval olfactory cilia can be easily visualized in whole animals.

1.4 SPINAL CANAL

The zebrafish spinal canal starts to differentiate in the neural keel at about 18 hpf (Kimmel, Ballard, Kimmel, Ullmann, & Schilling 1995). In vertebrates, ependymal epithelial cells that line the spinal canal lumen differentiate cilia (Nakayama & Loomis, 1974; Worthington & Cathcart, 1963). Cilia have been also documented in the neural tube of zebrafish embryos and early larvae (Fig. 1H) (Borovina et al., 2010; Kramer-Zucker et al., 2005). These cilia feature both the "9 + 0" and "9 + 2" arrangement of microtubules. At least some of them are motile and propagate cerebrospinal fluid flow in brain ventricles and the spinal canal (Kramer-Zucker et al., 2005; Worthington & Cathcart, 1963). Disruption of spinal canal cilia in zebrafish leads to hydrocephalus, an abnormal expansion of the brain ventricle (Kramer-Zucker et al., 2005). Also in the spinal canal, neuroepithelial cells display planar cell polarity. This is evident in the asymmetric localization of cilia to the posterior apical surface of some neuroepithelial cells (Borovina et al., 2010).

2. ANALYTICAL TOOLS FOR CILIA MORPHOLOGY AND MOTILITY

Below we provide several protocols commonly used to visualize cilia morphology, cilia motility, and the localization of ciliary proteins.

2.1 METHOD 1: *DETECTION OF CILIARY PROTEINS USING IMMUNOHISTOCHEMISTRY*

The immunolabeling of cilia in zebrafish has mostly relied on the use of antibodies that recognize microtubules and microtubule modifications. Several tubulin modifications are recognized by commercially available antibodies (Table 1). The most commonly used is the anti-acetylated α-tubulin antibody (Fig. 1B–F and H,I; Fig. 2E–I) (Duldulao, Lee, & Sun, 2009; Kishimoto, Cao, Park, & Sun, 2008; Omori & Malicki, 2006; Tsujikawa & Malicki, 2004b; Zhao et al., 2012). A disadvantage of this antibody is that it also stains cytoplasmic microtubules in many cells, neurons in particular (Fig. 1D, bottom of the panel; Fig. 2I). For example, the cytoplasm of retinal interneurons and ganglion cells contains acetylated mictrotubules which preclude the visualization of cilia in these cells (see, for example, Pooranachandran & Malicki, 2016; Zhao et al., 2012). Antibodies that recognize other tubulin modifications, such as polyglutamylation, monoglycylation, and polyglycylation are likely to improve cilia labeling in such cells (Table 1). A good alternative to anti-acetylated tubulin antibodies is GT335 which recognizes

Table 1 Markers of Cilia and Ciliated Cells in Zebrafish

Name/Specificity	Marker type/Concentration	Reported Expression in Tissues and Cells	References/Sources
Ciliary axoneme			
Arl13b Tg(βact: arl13b-GFP)	Transgene, also effective as mRNA	CNS, PND, KV	Borovina et al. (2010), Duldulao et al. (2009), and Lu et al. (2015)
Arl13b (Scorpion)	Rabbit poly-Ab; 1:200–1:2000	CNS, PND, KV	Duldulao et al. (2009)
Tubulin, polyglycylated	Rabbit poly-Ab; 1:300	PND multiciliated cells; olfactory placode, spinal canal	Pathak et al. (2011)
Tubulin, polyglycylated	Mouse mono-Ab; 1:2000	Nose, hair cells, medial section of PND	Callen et al. (1994), Wloga et al. (2010), Sigma, clone AXO49
Tubulin, α acetylated	Mouse mono-Ab; 1:500–1:1000	Ubiquitous	Duldulao et al. (2009), Pathak, Austin-Tse, Liu, Vasilyev, and Drummond (2014), Tsujikawa and Malicki (2004b), Wloga et al. (2010), Zhai et al. (2014), Sigma, clone 6-11B-1
Tubulin, α acetylated	Rabbit mono-Ab; 1:500–1:800	Ubiquitous	Malicki lab (unpublished), Cell Signaling; (D20G3) XP#5335
Tubulin, α & β monoglycylated	Mouse mono-Ab; 1:200–1:2000	CNS, PND, hair cells	Callen et al. (1994), McIntyre et al. (2012), Wloga et al. (2010), Sigma, TAP952
Basal bodies			
Centrin Tg(actb2:cetn2-GFP)	Transgene	Retina	Ramsey and Perkins (2013) and Randlett, Poggi, Zolessi, and Harris (2011)
Centrin xcentrin-GFP	mRNA for injections	CNS	Borovina et al. (2010) and Zolessi et al. (2006)
Tubulin, γ	Mouse mono-Ab; 1:200	Ubiquitous	Duldulao et al. (2009), Ramsey and Perkins (2013), Tsujikawa and Malicki (2004b), Sigma, clone GTU88

Continued

Table 1 Markers of Cilia and Ciliated Cells in Zebrafish—cont'd

Name/Specificity	Marker type/Concentration	Reported Expression in Tissues and Cells	References/Sources
Photoreceptors			
4C12	Ab (mono); 1:100	Rods	Morris et al. (2005) and Saade, Alvarez-Delfin, and Fadool (2013)
Opsin, blue (opn1sw2)	Ab (poly); 1:200–1:1000	Blue cones ($\leq$3 dpf)	Doerre and Malicki (2001) and Vihtelic, Doro, and Hyde (1999)
Opsin, blue Tg(3.5opn1sw2:mCherry)	Transgene	Blue cones	Duval, Chung, Lehmann, and Allison (2013)
Opsin, green (opn1mw1-4)	Ab (poly); 1:200–1:1000	Green cones ($\leq$3 dpf)	Doerre and Malicki (2001) and Vihtelic et al. (1999)
Opsin, green Tg(RH2-3/GFP-RH2-4/RFP)	Transgene (PAC)	Green cones (RH2-3 and RH2-4 opsin)	Tsujimura et al. (2015)
Opsin, red (opn1lw1-2)	Ab (poly); 1:200–1:1000	Red cones ($\leq$3 dpf)	Doerre and Malicki (2001) and Vihtelic et al. (1999)
Opsin, red Tg(Rho:EGFP)	Transgene	Rods	Yin et al. (2012)
Opsin, UV (opn1sw1)	Ab (poly)	UV cones ($\leq$3 dpf)	Doerre and Malicki (2001) and Vihtelic et al. (1999)
Transducin α Tg(gnat2:EGFP)	Transgene	Cones	Lagman, Callado-Pérez, Franzén, Larhammar and Abalo (2015) and Smyth, Di Lorenzo, and Kennedy (2008)
Transducin α (Gnat1)	RNA probe	Rods	Lagman et al. (2015)

Transducin α (Gnat2)	RNA probe	Cones	Brockerhoff et al. (2003) and Lagman et al. (2015)
Zpr-1 (Fret 43)	Ab (mono)	Double cones (48 hpf)	Larison and Bremiller (1990), ZIRC
Zpr-3 (FRet 11)	Ab (mono)	Rods (50 hpf)	Schmitt and Dowling (1996), ZIRC
Mechanosensory hair cells			
Brn3c Tg(brn3c:GFP)	Transgene	Neuromasts, ear, also subset of retinal ganglion cells	Carrillo et al. (2016), Sweet, Vemaraju, and Riley (2011), and Xiao et al. (2005)
HCS-1	Mouse mono-Ab; 1:250–1:500	Hair cell stoma of neuromasts and ear	Gale, Meyers, and Corwin (2000), Goodyear et al. (2010), Lopez-Schier and Hudspeth (2006) and Schibler and Malicki (2007)
SqET4 Tg(SqET4:gfp)	Transgene (enhancer trap)	Mature hair cells of the lateral line	Loh et al. (2014) and Lush and Piotrowski (2014)
Tubulin, α acetylated	Mouse mono-Ab; 1:500–1:1000	Apical termini of hair cell somata	Duldulao et al. (2009), Pathak et al. (2014), Tsujikawa and Malicki (2004b), Wloga et al. (2010), Zhai et al. (2014), Sigma, clone 6-11B-1

KV, Kupffer's vesicle; *PND*, pronephric duct.

monoglutamylated and polyglutamylated tubulin (Wolff et al., 1992) and has been successfully applied in zebrafish (Wloga et al., 2009; Zhao & Malicki, 2011). Finally, in double staining experiments, it is frequently desirable to visualize ciliary microtubules and other ciliary proteins at the same time. A frequent difficulty in such experiments is that almost all commercially available anti-acetylated tubulin antibodies are mouse monoclonals. A good rabbit monoclonal antibody to acetylated tubulin has, however, recently become available (Cell Signaling, Table 1). Although not as specific as mouse monoclonal antibodies, it is very useful in double labeling experiments.

The protocols below provide an overview of antibody staining for cilia and cilia-related proteins. We will mostly present protocols developed to stain the zebrafish ear and the retina. These protocols can be, however, applied to the analysis of other ciliated organs, such as the pronephros, the spinal canal, and the nasal pit.

2.1.1 Method 1a: staining of whole embryos at 3 dpf and younger

Materials and equipment

- Agarose, 1% in distilled water and low-melting point agarose, 1.5% in water.
- Antibodies, such as mouse anti-acetylated tubulin or mouse anti-HCS-1 primary antibodies (Table 1) and secondary antibodies.
- Blocking Solution (BS): 0.5% Triton X-100 + 10% serum in 1x PBS. Use serum that matches the species in which secondary antibodies were generated.
- Embryo medium as defined in the "Zebrafish Book" (Westerfield, 2007).
- Eppendorf tubes (1.5 mL)
- Fixative: 4% paraformaldehyde (PFA) (Sigma, P6148) in 1x PBS. Dissolve PFA in 1x PBS in a 60°C water bath. After PFA dissolves, allow the PFA solution to cool to room temperature (RT), and adjust pH to 7.4 with 1N NaOH. Sterile filter with Sartorius Minisart single-use 20 μm filter. Preferably use fresh.
- Mixer (nutating), such as Sigma—Aldrich Z674567 (use for all washing steps at RT)
- Molds to imprint the surface of agarose. We use molds that produce wells with a 1 × 1 mm opening and ~30 degrees tilted bottom. Molds can be made by local machine shop facilities or purchased from Adaptive Science Tools, Cat. Number CNF2 and PT1.
- Optional: Alexa Fluor 546 Phalloidin: Invitrogen, A22283
- PBSTr (1x PBS + 0.1% Triton X-100)
- Petri dishes: disposable large (90 mm) and small (35 mm)
- Shaker (orbital), Stuart Scientific; Cat. Number: SO3 (use for permeabilization, blocking, and antibody incubation steps)
- Tricaine (Sigma, E10521); Stock Solution: 4 mg/mL, add 1.5 mL into 20 mL of embryo medium to obtain working concentration. See the "Zebrafish Book" for additional comments on Tricaine use (Westerfield, 2007).
- Water dipping objectives. For example, Olympus LUMPlan 60x, NA 0.9, working distance 1.5 mm or LUMPlan 40x, NA 0.8, working distance 3.3 mm.

Protocol

Fixing and staining

1. Euthanize embryos at the desired stage in Tricaine and immediately transfer into fixative.
2. Fix fish in fresh 4% PFA for 4 h at RT or overnight at 4°C without shaking, which may cause curling of some cilia in the ear. Place the tube horizontally to allow the PFA to diffuse to all the fish. Subsequent steps are at RT unless stated otherwise.
3. Wash 3x in PBSTr, 5 min each.
4. Incubate embryos in 1% Triton X-100 in PBSTr for 8 h at RT. This treatment dissolves otoliths. Alternatively, otoliths can be dissolved by incubating larvae in 120 mM EDTA in 0.1 M phosphate buffer pH 6.9. For tissues other than the ear, this step can be omitted or shortened. Embryos can also be permeabilized by cold ($-20°C$) acetone treatment for 7 min (embryos up to 30 hpf) or 15 min (embryos up to 5 dpf).
5. Wash 1x in PBSTr for 5 min.
6. Incubate in the BS for 1–2 h.
7. Incubate specimen overnight at 4°C in the BS containing appropriate concentration of the primary antibody (for anti-acetylated tubulin antibodies, use 1:500–1:1000 dilution).
8. Wash 4x in PBSTr, 2 min each and then 4x, 30 min each.
9. Incubate specimen in the dark for 2 h at RT or overnight at 4°C in the BS containing appropriate concentration of the secondary antibody. Optional: If F-actin counter-staining is needed, add phalloidin to the secondary antibody solution (dilute stock solution according to manufacturer's suggestions). Overnight secondary antibody incubation at 4°C usually produces a better signal.
10. Wash 4x in PBSTr, 2 min each and then 4x, 30 min each.

Embedding in agarose for imaging

1. Pour 1% agarose into a 35-mm plastic Petri dish.
2. Place the mold on top of the agarose so that it floats on the surface. Let the agarose harden completely.
3. Remove the mold from agarose. It will leave imprinted wells on the surface.
4. Depending on the organ to be examined, orient embryos appropriately. For example, to examine the cilia of the ear, position embryos on their sides in the imprinted wells.
5. Once embryos are positioned properly, immobilize them by overlaying with 1.5% low-melting point agarose. If necessary, quickly adjust embryo orientation with a probe before agarose hardens.
6. Flood embryos embedded in agarose with water and view using a water-dipping lens. Embryos can be stored for several days at 4°C before imaging, although signal quality will gradually deteriorate.

This embedding protocol can also be used for live imaging of embryos. In this case, add Tricaine before mounting, and embed embryos using agarose at

30–35°C. If embryos still move, add additional 1 mL of Tricaine stock solution to the working solution.

2.1.2 Method 1b: staining of whole larvae at 5–7 dpf

Antibody penetration poses difficulties when working with animals at 5 dpf and older. Staining protocol has to be modified for these stages as outlined below.

Additional Materials and Equipment (not listed in method 1a)
- Collagenase: 10-100 µg in PBSTD; Gibco-Life technologies, 17018-029
- PBSTD (1x PBS + 1% Triton X-100 + 1% DMSO)
- PBSTw (1x PBS + 0.1% Tween-20)
- TBSTr (150 mM NaCl + 50 mM Tris (pH 7.5) + 2% Triton X-100 in distilled water)

Protocol

Use of shaking and mixing as in Method 1a

1. Euthanize embryos with Tricaine solution and immediately transfer to freshly made cold fixative.
2. Fix fish overnight with 4% PFA at 4°C, without rocking; place the tube horizontally to allow the PFA to diffuse to all the fish.
3. Wash 2x in PBSTw, 10 min each
4. Wash 2x in PBSTr, 10 min each
5. Permeabilize larvae for 8 h or more at RT with 1% Triton X-100 in PBSTr. Optional: If the purpose of staining is to image the cilia of inner ear maculae, permeabilize fish by storing larvae in TBSTr solution for 3–4 days at 4°C. Change TBSTr daily.
6. Treat with collagenase (10–100 µg/mL in PBSTD) for 30 min at RT on the nutating mixer.
7. Wash 3x in PBSTr, 2 min each and then 4x, 30 min each.
8. Block for 2 h at RT in BS.
9. Remove the BS and incubate with primary antibody in the BS overnight at 4°C.
10. Wash 3x in PBSTr, 2 min each and then 4x, 30 min each.
11. Incubate the fish with secondary antibody diluted in the BS for 4 h at RT in the dark or overnight at 4°C.
12. Wash 3x in PBSTr, 2 min each and then 4x, 30 min each.
13. If phalloidin staining is required, add it to the secondary Ab as in Method 1a.
14. Embed as in Method 1a.

2.1.3 Method 1c: staining of dissected adult tissues (ear and retina)

While staining adult tissues, the challenge is to fix the specimen uniformly and dissect without causing damage. To ensure fixative penetration, fixation conditions are different compared to those for larvae. The protocols below are provided for the staining of the dissected ear and retina.

Dissection, Fixation, and Staining of the Adult Ear
Additional Materials (not listed in 1a)

- Microscope slides (inexpensive ones)
- Vectashield Antifade mounting medium H-1000, Vector Laboratories.

Protocol

Perform ear dissection as described (Einarsson et al., 2012).

1. Fix the dissected ear chambers (utricle, sacculus, and lagena) using 4% PFA for 1 h at 4°C.
2. Remove PFA and wash 3x in 1x PBS, 5 min each.
3. Incubate in PBSTr for 1 h.
4. Block in the BS for 45 min and incubate with the primary antibody (for anti-acetylated tubulin staining, use 1:500 dilution) overnight at 4°C.
5. Wash 3x in 1x PBS, 2 min each and then 4x, 20 min each.
6. Incubate in secondary antibody for 2 h at RT.
7. Wash 3x in 1x PBS, 2 min each and then 4x, 20 min each.
8. Add fluorophore-conjugated phalloidin and incubate for 30 min at RT in darkness.
9. Wash 5x in 1x PBS, 15 min each, in darkness.
10. To mount ear chambers, place tissue on a microscope glass slide in a drop of 1x PBS. Remove as much PBS as you can with a pipetman, add Vectashield mounting medium, and quickly add coverslip. For the utricle, make an incision in the anterior region, and flatten it before using the coverslip.

Dissection and fixation of the adult retina

Additional Materials (not listed in 1a)

- U-100 Insulin Syringes Terumo, BS N1H2913
- 30% Sucrose in PBSTw solution. Add 0.05% (w/v) of sodium azide and store at 4°C.
- Fine forceps, Dumont, no 5.

Protocol

1. Euthanize adult zebrafish with 10 mL of Tricaine stock solution and place in a Petri dish with dorsal side up.
2. Under a stereomicroscope, insert fine forceps inside the eye socket and cut the optic nerve, which connects the retina with the brain. Remove the eye with forceps.
3. Slowly inject 2 mL of 4% PFA into the eye between the retina and the pigmented epithelium to facilitate efficient fixation of the tissue. Most of the fixative will come out of the eye during the injection.
4. Place the dissected eyes in 4% PFA overnight at 4°C on an orbital shaker.
5. Remove PFA and inject 2 mL of 30% sucrose as described earlier for PFA. Incubate eyes in 30% sucrose overnight at 4°C on an orbital shaker. Embed for sectioning as below.

2.1.4 Method 1d: staining of cryosections (all stages)

This protocol is applicable to many organs. For studies of the retina, it is preferable to cut sections between 10 and 20 μm thick. To prepare thinner sections a mixture of embedding medium and sucrose can be used (Barthel & Raymond, 1990).

Additional Materials (not listed in 1a)
- Cryostat, such as Leica CM1860
- Embedding medium for frozen sections, such as OCT Compound; VWR 361603E
- Microscope Slides, Superfrost Plus Precleaned: Fisher, 12-550-15. The choice of slides is essential to avoid losing sections during washes.
- Nail polish, transparent
- Optional: Antigen retrieval solution: 100 mM sodium citrate in distilled water
- Pipettes, 3 mL dispensable polyethylene, eg, BD Falcon, 357524
- PTU, Sigma Aldrich, P7629 (0.003% Phenylthiourea in embryo medium)
- Razor blades, eg, Agar Scientific, T585
- Sucrose, 30% in PBSTw. Add 0.05% (w/v) of sodium azide and store at 4°C.
- Tip box, empty with perforated tip holding surface and a cover: eg, USA Scientific, 1111-1800.
- Vectashield Mounting Medium: Vector Labs, H-1000

Fixing, embedding, and sectioning:

1. To visualize photoreceptor outer segments, pigmentation needs to be eliminated. To block pigment formation, raise embryos in the presence of PTU starting at 24 hpf or earlier.
2. When embryos/larvae reach the desired stage, euthanize them in Tricaine stock solution and immediately transfer to fixative.
3. Fix embryos in 4% PFA for 2 h at RT or overnight at 4°C. Use the orbital shaker. Subsequent steps are at RT unless stated otherwise.
4. Remove PFA and wash 3x in PBSTw, 5 min each.
5. Place embryos in 30% sucrose solution overnight at 4°C or until specimen sink.
6. Replace sucrose with medium for frozen sectioning. This is a very viscous substance. Use Pasteur pipette.
7. Cut a polyethylene pipette perpendicular to its long axis into 6–10 mm wide rings. Place the rings flat on a glass slide and fill them halfway with frozen section medium.
8. Using Pasteur pipette, transfer specimen into the ring described above. Some sectioning medium will transfer too, which will aid in filling up the ring entirely.
9. Use a dissecting needle to orient specimen. To obtain transverse sections, specimen head must point straight down. Cool slides with specimen to −20°C in the cryostat chamber. Medium will solidify and turn white in about 5–10 min.
10. Prior to sectioning, remove polyethylene rings by cutting through their walls with a razor blade. This will leave you with a cylindrical block of the embedding medium containing embedded specimen.

11. Mount frozen blocks on a cryostat specimen holder using additional sectioning medium and collect sections on a prechilled slide inside the cryostat chamber set to −20°C. When brought to RT, sections will thaw and adhere to the slide.
12. Allow slides to dry at RT for 2 h. Then proceed with staining of cryosections.

    Staining of cryosections
13. Optional Step: Detection of some antigens may require special treatment such as placing slides into boiling-hot 10% sodium citrate solution in distilled water for 20 min or 5 min incubation in 1% SDS (these procedures are known as antigen retrieval) (Brzica et al., 2009). A useful heat treatment method has also been described (Inoue & Wittbrodt, 2011). Following antigen retrieval, thorough washing in PBS, 4x, 10 min each, is recommended.
14. (Optional) Using a PAP pen (available from several manufacturers), create a hydrophobic grease border around the edge of slides with sections. Allow the grease to dry for 2–3 min.
15. Briefly rehydrate slides by gently pipetting about 300 μL of PBSTw on the slide surface. This will also dissolve the frozen section medium.
16. In the meantime, prepare a humidified chamber. For this purpose, we use a tip box as described in Materials and Methods earlier. Fill the bottom compartment of the tip box with tap water. This will serve as a homemade humidifying chamber for antibody incubations.
17. Open the tip box cover. Place slides on the top of the perforated, tip holding surface.
18. Overlay each slide with 150 μL of the BS. The grease border will help to keep the BS on the slide but is not absolutely necessary. Incubate slides with the BS for 45 min at RT.
19. Dilute primary antibody(ies) to appropriate concentration(s) in BS.
20. Remove the BS by positioning slides vertically and collecting excess solution with a Kimwipe. Apply 100 μL of primary antibody solution as in Step 15. Incubate overnight at 4°C in a humidified chamber.
21. Next day wash slides 3x in PBSTw, 5 min each.
22. Incubate with secondary antibody(ies) at RT for 2 h. Optional: Add phalloidin to counterstain F-actin (follow the manufacturer's recommendation for dilutions).
23. Wash slides 5x in PBSTw, 5 min each.
24. Use a Kimwipe to remove excess PBSTw and apply mounting medium. Place a coverslip over the immunostained sections and seal with nail polish.
25. Image using confocal or conventional fluorescence microscopy.
26. This preparation may be stored at 4°C for at least a week, although the quality of the signal deteriorates over time.

Information regarding antibodies used to analyze photoreceptors and hair cells is provided in Section 3 below.

2.2 METHOD 2: *LIVE IMAGING OF CILIA AND BASAL BODIES USING TRANSGENIC LINES*

The use of transgenic lines expressing cilia-targeted polypeptides fused with fluorescent proteins, such as GFP, is an effective approach to image cilia in many organs, including the otic vesicle, the brain, somites, and the heart (Fig. 3). Some useful transgenic lines are listed in Table 1. The Arl13b-GFP fusion, which highlights the ciliary membrane, has been one of the most popular tools for cilia visualization in living zebrafish embryos thus far (Borovina et al., 2010; Duldulao et al., 2009). Similarly, the *centrin:GFP* fusion protein has been used to visualize and track the subcellular localization of ciliary basal bodies (Borovina et al., 2010; Zolessi, Poggi, Wilkinson, Chien, & Harris, 2006). Yet another option is Iguana-GFP fusion, which localizes to the cilia base (Tay et al., 2009). In the mouse, an SSTR3-GFP fusion can be used to visualize cilia (O'Connor et al., 2013; Paridaen, Wilsch-Bräuninger, &

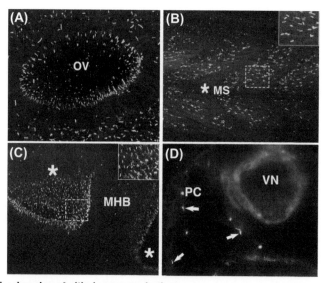

FIGURE 3 Live imaging of cilia in transgenic lines.

Confocal images of *Hsp70:Arl13b-GFP* transgenic embryos at 30 hpf. Embryos were heat shocked at 26 hpf for 30 min (A) Image of cilia in the otic vesicle (OV). (B) The same transgene labels cilia in muscle cells. The *asterisk* indicates the myoseptum (MS). Inset shows cilia in muscle fibers at higher magnification. (C) Brain ventricle cilia visualized using the same transgenic line. *Asterisks* indicate midbrain and hindbrain ventricles. The midbrain–hindbrain boundary (MHB) is also indicated. Inset shows midbrain ventricle cilia in detail. (D) Light sheet microscopy image of the heart in a 4 dpf larva expressing *kdrl-RFP* (red) and *Arl13-GFP* transgene (green). At this stage, cilia are present mostly in the pericardium (PC). The heart ventricle (VN) is indicated. *kdrl* labels endothelial cells while *Arl13-GFP* labels cilia. *Images courtesy of Malicki lab members, Xiaoming Fang and Eleni Leventea. (See color plate)*

Huttner, 2013), and Arl13b-mKate2 was applied to label ciliary axonemes in tissue culture cells (Diggle et al., 2014). mKate emits in the far red, which makes it suitable for dual-color imaging with green fluorophores (Shcherbo et al., 2009). The SSTR3-GFP and Arl13b-mKate2 fusions have not been tested in zebrafish so far but are likely to be useful in future applications. In some cases, it is useful to visualize not just cilia but ciliated cells. The *Tg(brn3c:GFP)* transgene, serves such a purpose for mechanosensory hair cells (Xiao, Roeser, Staub, & Baier, 2005).

The use of transgenic lines that constitutively express cilia-targeted fusion proteins, such as *Tg(Arl13b:GFP)*, is not without shortcomings. The analysis of *Tg(Arl13b:GFP)* transgenic fish revealed that both motile and primary cilia were significantly longer in a number of tissues, compared to those in the wild type (Lu et al., 2015). The magnitude of the cilia length increase is proportional to the strength of GFP signal. Similar artifacts are seen in cell culture when serotonin receptor ($5HT_6$) and fibrocystin cytoplasmic tail are used to target proteins to cilia (Su et al., 2013). Moreover, in our experience the *Tg(Arl13b:GFP)* transgene is silenced in some tissues and so cannot be used to consistently visualize all cilia. These observations reveal that transgenic lines that constitutively overexpress ciliary proteins should be used with caution.

The use of inducible transgenic lines to visualize cilia is likely to be less prone to artifacts. As an inducible transgene is activated only for a short time, cilia length and morphology artifacts are less likely to occur. We have used heat-shock promoter driven *Alr13b:GFP* transgenes to visualize cilia in the central nervous system, the otic vesicle, somites, skin, and many other tissues (Fig. 3). Heat shocking can be performed throughout development. We usually heat shock at 37–39°C for 15–60 min (Zhao & Malicki, 2011). Multiple heat shocks can be also applied at 30 min intervals. The duration of heat shock has to be empirically optimized for each transgenic line. For instance, the *iguana* transgenic strain *Tg(hs:gfp-igu)* was heat shocked at 37°C for 40 min at 12 hpf. This produced a uniform level of expression of the GFP-tagged Iguana protein in all cell types (Tay et al., 2009).

An alternative method to the use of transgenic lines is the injection of RNA transcript encoding *Arl13b:GFP* fusion into embryos at the 1–2 cell stage (Duldulao et al., 2009; Tay et al., 2009). This is, however, frequently less effective compared to the use of transgenic lines because the injected RNA is gradually degraded during embryonic development. Consequently, this method is usually limited to experiments performed during the first few days of development.

2.3 METHOD 3: *LIVE IMAGING OF CILIA MOVEMENT*

This protocol is provided for the pronephros but can also be used to visualize the movement of olfactory or spinal canal cilia. Imaging of cilia in the ear is described in Volume 134 of this series (Baxendale & Whitfield, 2016). We perform cilia imaging of the pronephric duct at 250 frames per second between 24 and 60 hpf. Slower cameras can also be used to visualize cilia movement, but the resulting recordings are less informative as they do not make it possible to analyze the

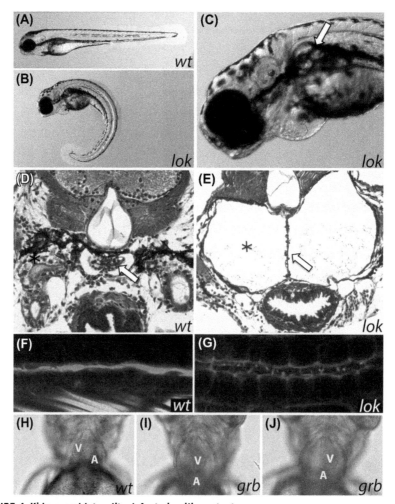

FIGURE 4 Kidney and laterality defects in cilia mutants.

Lateral views of wild-type (A) and *locke* mutants (*lok*) (B) at 3 dpf. (C) High magnification of the anterior region of a *lok* larva. *Arrow* indicates a kidney cyst. (D, E) JB4 plastic transverse sections through the glomerulus (*arrows*) and the pronephric tubule (*asterisks*) in wild-type (wt) (D) and lok (E) mutant embryos at 5 dpf. (F, G) Posterior pronephric duct cilia visualized by immunostaining for acetylated tubulin (red) and phalloidin (green) at 30 hpf in wt (F) and the *lok* (G) mutant embryos. (H–J) Ventral views of the heart at 72 hpf in wt (H) and *garbus* (*grb*) mutant embryos with absent (I) or inverted heart looping (J). (See color plate)

Images reprinted with permission from Zhao, C., & Malicki, J. (2007). Genetic defects of pronephric cilia in zebrafish. Mechanisms of Development, 124(7–8), 605–616.

frequency or the waveform of cilia movement in detail. The pronephric cilia move with the frequency of 20 ± 3 Hz (Kramer-Zucker et al., 2005).

Materials and Equipment
- BDM (2,3-butanedione monoxime, Sigma, B0753-25G), final concentration 20 μM.
- Camera, such as Photron FASTCAM-PCI500
- Microscope equipped with a 63x water immersion lens.
- Microscope slides: such as VWR 48311-950

Protocol
1. Liquefy low-melting point agarose in a microwave oven and keep at 37°C in a water bath.
2. To stop heartbeat and circulation, add BDM to the embryo medium for 1 min.
3. Place several drops of low-melting point agarose on a microscope slide, and then quickly place an embryo on the surface of the agarose.
4. Position embryo on its side as quickly as possible.
5. Transfer the slide to the stage of a microscope. Add a drop of water on top of the embryo before observing with water immersion lens.
6. Film using a high-speed digital video camera mounted on the microscope. Imaging is performed roughly in the middle of the pronephric duct. Images of moving cilia are acquired at 250 frames per second for the duration of 1 s and analyzed in slow motion (15 frames per second) to determnie the frequency of cilia movement.

2.4 METHOD 4: *LIVE IMAGING OF CILIA USING LIGHT SHEET MICROSCOPY*

Single plane illumination microscopy (SPIM), also known as light sheet microscopy, can be used to image cilia in whole organs of living specimen, including these in the spinal canal, the ear, the pericardium of the heart, and many others (Fig. 3). Data acquisition using SPIM is fast and results in low phototoxicity. SPIM can be used to generate long time-lapse recordings in 3D and to image fairly fast movements, such as the heartbeat. Current SPIM cameras record up to 80 frames per second (Carl Zeiss Microscopy GmbH). SPIM 3D reconstructions are also suitable to evaluate fine characteristics of cells, such as cilia position in the plane of epithelia (planar polarity). The protocol below is provided for fixed ear tissue but can also be applied to live specimens.

Materials and Equipment
- Agarose 1%, Bioline, BIO-41025
- Capillaries, Zeiss (reusable, provided with the microscope)
- Glass dish, for example, 1 5/8 inch square watch glass (Carolina Biological Supply)
- Glycerol, VWR, 444485B
- Microscope. We use a Zeiss Model Z.1. Other commercially available microscopes are Nikon Ti-diSPIM and Leica TCS SP8 DLS.

- PBS, 1x
- Phalloidin, Alexa Fluor 546 conjugate: Invitrogen, A22283
- Stereomicroscope, such as Zeiss Stemi 1000
- Transgenic zebrafish larvae or dissected tissues that express fluorescent proteins in cilia (see previous discussion). Some dissections, such as that of the adult ear, are technically demanding (Einarsson et al., 2012). Alternatively, antibody-stained specimen can be used.
- Tricaine stock solution (4 mg/mL), for final concentration add 1.5 mL into 20 mL of embryo medium.

Protocol

1. Dissect and stain ear tissues as described earlier and place them in a glass dish (Method 1c).
2. Using forceps and a stereomicroscope, remove as much of the nerve tissue as possible. Nerves connect to sensory epithelia and are darker than the epithelial walls of the ear. Be careful not to damage the ear tissue by pulling on nerves.
3. Incubate in 100% glycerol for 1 day at 4°C without rocking.
4. Remove excess glycerol from the glass dish using a pipette and replace with 1 mL warm 1% agarose.

   For live imaging of zebrafish larvae, select specimen with the strongest transgene expression and anaesthetize them using Tricaine.
5. Before agarose solidifies, aspirate the tissue specimen or fish embryo into the microscope capillary and load it into the microscope following manufacturer's instructions. Make sure that no air bubbles are present in the capillary.
6. Turn the capillary that contains the specimen around its axis to find the optimal viewing angle.
7. During imaging Z-stacks can be collected. In some cases it is preferable to collect Z-stacks from fewer focal planes to reduce background. The preferable Z-stack size should be empirically determined. Images can be opened and processed using the ImageJ software or the ZEN software provided by Zeiss.

Live imaging using this technique does not cause specimen damage. Animals can be recovered from agarose and used for further analysis.

2.5 METHOD 5: *ANALYSIS OF CILIARY TRANSPORT USING INDUCIBLE TRANSGENES*

Cilia-mediated signal transduction relies on a variety of receptors that localize to the ciliary membrane (Schou et al., 2015). Mechanisms that localize these proteins to cilia are of considerable interest because their defects lead to developmental and degenerative abnormalities both in humans and model animals. Opsin is a particularly abundant ciliary GPCR. Defects in opsin transport and the resulting opsin mislocalization to the cell body are a cause for severe photoreceptor degeneration in man (Bessant et al., 1999). Likewise, defects of the ciliary transition zone, which controls trafficking into and out of the ciliary compartment, are associated with opsin mislocalization and photoreceptor degeneration (Bachmann-Gagescu et al., 2011; Chang et al., 2006; Zhao & Malicki, 2011) (Fig. 5N).

In this section, we describe a quick method to evaluate opsin transport from the cytoplasm into photoreceptor cilia. The C-terminal sequence of rod opsin is sufficient to mediate ciliary transport in photoreceptor cells (Perkins, Kainz, O'Malley, & Dowling, 2002; Tam, Moritz, Hurd, & Papermaster, 2000). We add this sequence to the C-terminus of GFP and express the resulting fusion protein from a heat-shock promoter. This allows one to generate a short pulse of expression (Zhao & Malicki, 2011). Following heat shock, zebrafish larvae are collected at several time points to examine the localization of GFP signal in photoreceptors.

Materials and Equipment
- Cryosectioning reagents as in Method 1d.
- Embryos, wild-type, and/or mutant
- Fixative: 4% PFA (Sigma, P6148) in PBSTr
- Gateway BP Clonase II Enzyme Mix (Invitrogen, 11789020)
- Gateway constructs for zebrafish: p5E-*hsp70l*, pME-EGFP no stop, p3E-EGFPpA, pDestTol2pA2 (provided in Tol2 kit from the Chien Lab) (Kwan et al., 2007)
- Gateway entry vectors: pDONR221 and pDONR P2R-P3 (Invitrogen)
- Gateway LR Clonase II Plus enzyme (Invitrogen, 12538120)
- ImageJ (NIH) software
- Microinjection reagents as in Yuan and Sun (2009).
- Microscope, Olympus confocal FV1000, or similar.
- Phalloidin, Alexa Fluor 546 conjugate (Invitrogen, A22283)
- PTU, Sigma Aldrich, P7629 (0.003% Phenylthiourea in embryo medium)
- Transposase mRNA (prepared from pCS2FA or pCS-TP vectors using mMessage mMachine SP6 transcription Kit, Ambion AM1340) (Kwan et al., 2007)
- Tricaine stock solution (4 mg/mL)
- Water baths or Incubators (28°C and 37°C)

Protocol

Making GFP fusion constructs

We use Gateway cloning to construct expression vectors for GFP fusions. The Gateway-based kit assembled by Kwan et al. is very useful for this purpose (Kwan et al., 2007). The cloning strategy will depend on the type of fusion that you intend to generate. To fuse GFP to the C-terminus of the gene of interest, first clone the open reading frame (do not include stop codon) of the gene that you intend to target to cilia (such as opsin) into the pDONR221 entry vector using BP clonase, then make the final construct with one step MultiSite Gateway reaction according to published protocols (Kwan et al., 2007) and instructions provided by the manufacturer of Gateway reagents (currently Invitrogen/Thermo Fisher). The following constructs are used: p5E-*hsp70l* (provides promoter sequences), pDONR221-*gene of interest open reading frame*, p3E-EGFPpA (provides fluorescent tag, such as GFP and polyadenylation site), and pDestTol2pA2 (Tol2 transposon-based backbone vector that mediates integration in the zebrafish genome).

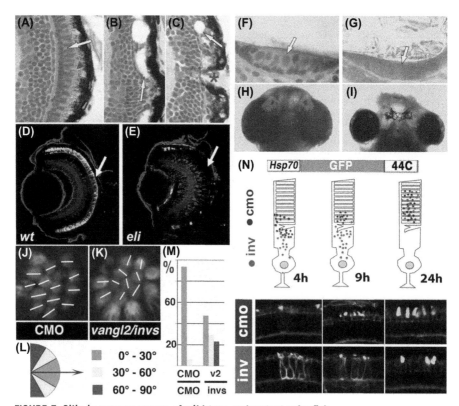

FIGURE 5 Cilia in sensory organs of wild-type and mutant zebrafish.

(A–C) Plastic (JB4) sections of wild-type and mutant retinae stained with Methylene Blue/Azure II. (A) The wild-type retina features elongated photoreceptor cell nuclei (*arrow*). The black tissue next to photoreceptor cells is the retinal-pigmented epithelium. Lightly stained features (*arrowhead*) are outer segments. (B, C) The photoreceptor cell layer of animals mutant for IFT genes, *flr* (B) and *ovl* (C), displays defective photoreceptor morphology and lacks outer segments. Cell death results in the appearance of empty spaces (*arrows* in B, C). *Asterisk* indicates the optic nerve. (D, E) Confocal images of transverse cryosections though the retina. Double cones are visualized with Zpr-1 antibody in wild-type (D) and eli^{tp49} mutant (E) retinae (green signal). Mueller glia are identified using anti-carbonic anhydrase antibody (red). (E) Mutations in the gene encoding an IFT particle component, *elipsa* (*eli*), lead to rapid photoreceptor degeneration. Older photoreceptors in the central retina are lost first. (F, G) Transverse plastic sections through the ear of a wild type (F) and its *oval/ift88* mutant sibling (G) at 5 dpf. Larvae were embedded in JB4, sectioned, and stained with Methylene Blue/Azure II. *Arrow* points to the position of hair cell nuclei, which are absent in *ovl* mutants. (H, I) DiI application (Method 10, red signal) to whole embryos stains olfactory pits of wild-type (H) but not *oval/ift88* mutant (I) animals. (J–M) Results of planar cell polarity assay in mechanosensory neuromast epithelia of fish treated with control morpholino (CMO) (J) and a mix of morpholinos directed to *vangl2* and *inversin* genes (K). Shown are confocal images of

To analyze opsin targeting to cilia, GFP has to be placed at the N-terminus of opsin cilia targeting sequence. To accomplish this, the DNA fragment encoding the last 44 amino acids of the *Xenopus rod* opsin gene was cloned into pDONR-P2R-P3 vector and then combined in a multisite reaction with three constructs from the Kwan et al. kit: p5E-*hsp70l*, pME-EGFP no stop, and pDestTol2pA2. A similar strategy was used to target peripherin to photoreceptor cilia (Zhao & Malicki, 2011).

Microinjection, heat shocking, cryosectioning, and data collection

1. Prepare the injection mix: transposase mRNA (100 ng/μL, final concentration), opsin-GFP constructs (40 ng/μL, final concentration) and, if necessary, morpholinos (concentration varies depending on the target locus).
2. At the one cell stage, inject ~2 nL volume into wild-type or mutant zebrafish embryos.
3. Raise embryos in 0.003% Phenylthiourea (PTU) in embryo medium at 28°C, change medium every day. PTU can be added at 24 hpf.
4. At 3 dpf, transfer larvae into 37°C prewarmed embryo medium and incubate at 37°C for about 30 min. The stage of embryos and heat-shock time can be

lateral line neuromasts in whole embryos at 4 dpf stained with anti-acetylated tubulin antibodies (green) and phalloidin (red). (L) The deviation of hair cell orientation from a common axis (*arrow*) is assigned to three categories and color coded as follows: blue, deviation of 0–30 degrees; yellow, 30–60 degrees; red, 60–90 degrees. (M) Graph showing frequencies of misoriented hair cells in control animals and morphants in each of the three categories outlined in panel (L). (N) Ciliary trafficking of opsin assay (Method 5). The upper portion of this panel schematically shows opsin trafficking in the wild-type control (*blue dots*) and morphant (*red dots*) photoreceptors. To perform this assay, 1–2 cell stage embryos are injected with the Hsp70-GFP-44C construct (above) and heat shocked at 3 dpf or later. The lower portion of this panel shows confocal images of transverse sections through retinae of these animals. To test the function of inversin in opsin trafficking, morpholinos against inversin (inv) and control morpholinos (CMO) were used as indicated. The subcellular localization of GFP-Rhodopsin is shown 4, 9, and 24 h after a heat shock (in green). Sections are counterstained with phalloidin (in red). Note that following inversin knockdown, opsin translocates slower into outer segments of photoreceptor cilia. (See color plate)

Panels A–E are reprinted with permission from Doerre, G., & Malicki, J. (2001). A mutation of early photoreceptor development, *mikre oko*, reveals cell–cell interactions involved in the survival and differentiation of zebrafish photoreceptors. Journal of Neuroscience, 21(17), 6745–6757, F–I from Tsujikawa, M., & Malicki, J. (2004b). Intraflagellar transport genes are essential for differentiation and survival of vertebrate sensory neurons. Neuron, 42(5), 703–716, and J–M, and N from Zhao, C., & Malicki, J. (2011) Nephrocystins and MKS proteins interact with IFT particle and facilitate transport of selected ciliary cargos. EMBO J., 30(13), 2532–2544,

adjusted according to experimental needs. We found that a 30-min heat shock is sufficient to produce robust expression.
5. Transfer larvae back to a 28°C incubator.
6. Collect larvae at 4, 9, and 24 h after heat shock or as needed.
7. Fix embryos with 4% PFA overnight at 4°C or 2–4 h at RT and proceed with sectioning as described in Method 1d.
8. In case of retina, counterstain sections with Alexa Fluor 546 Phalloidin in PBSTw for 1 h at RT. This helps to visualize layering of the retina.
9. Image cover-slipped sections on a confocal microscope using a 100x lens. Make sure to collect all images using the same settings (resolution, magnification, and laser power). Use low to medium laser power to avoid over oversaturating the signal. Proceed to data analysis.

Data analysis

Open confocal images with ImageJ software, in the Image option, select color → split channels. In the green channel, use the line tool to draw a transverse line at 25%, 50%, and 75% in the photoreceptor cell body length of each photoreceptor that has GFP expression. Then measure the maximum GFP intensity at each line. This value is used to evaluate opsin localization in the photoreceptor cell body. For each experimental group, we usually collect data from at least 6 embryos and 100 photoreceptors.

Understanding the dynamics of opsin transport and its genetic control will help to determine mechanisms of photoreceptor degeneration. The method outlined above is quick and can be used to evaluate opsin transport in various genetic conditions using morphants or mutant lines. We have also used this method to evaluate transport of other photoreceptor proteins, such as peripherin/rds (Zhao & Malicki, 2011). We anticipate that this approach is broadly applicable to study the transport dynamics and turnover of other ciliary receptors in other tissues. Lists of known ciliary targeting sequences that can be fused to fluorescent proteins to analyze cilia-directed transport are available in the literature (Malicki & Avidor-Reiss, 2014; Nachury, Seeley, & Jin, 2010).

3. PHENOTYPES OF CILIA MUTANTS IN ZEBRAFISH

Genetic screens in zebrafish have identified a broad assortment of cilia mutants (Drummond et al., 1998; Malicki, Neuhauss, et al., 1996; Sun et al., 2004). The cloning and phenotypic analysis of these mutant genes identified a number of important mechanisms that influence cilia formation and function (Duldulao et al., 2009; Kishimoto et al., 2008; Omori et al., 2008; Pathak, Obara, Mangos, Liu, & Drummond, 2007; Pooranachandran & Malicki, 2016; Zhao & Malicki, 2011). In addition, antisense morpholino gene knockdowns in zebrafish have been frequently used to evaluate the function of human ciliopathy genes (Khanna et al., 2009; Otto et al., 2003; Sayer et al., 2006; Zaghloul et al., 2010). Below we present several protocols that can be used to evaluate cilia-associated phenotypes in zebrafish embryos and larvae.

As a general concern, the maternal contribution is a complicating factor in the analysis of zebrafish mutants, especially during early stages of development, because it is known to obscure mutant phenotypes (Cao et al., 2010; Huang & Schier, 2009). Due to the presence of the maternal contribution, some mutants display normal cilia in the pronephric duct and sensory neurons early in development, but do not maintain cilia later on (Cao et al., 2010; Tsujikawa & Malicki, 2004b). Consequently, some cilia mutants do not display LR asymmetry defects. Antisense morpholinos targeted to the translation initiation site can be used to block expression of maternally deposited mRNA. This is not always effective, however, as morpholinos frequently cause partial loss of function, and they do not eliminate proteins deposited in the egg. A more reliable, but also substantially more laborious approach, is to make a maternal-zygotic mutant. Blastomere transplantation is used for this purpose in zebrafish (Ciruna et al., 2002; Huang & Schier, 2009).

3.1 METHOD 6: *EVALUATION OF HEART POSITION IN LIVE EMBRYOS*

Evaluation of heart position is a very simple method that can be applied at early developmental stages to identify laterality defects in cilia mutants. Proper controls need to be included to ensure the accuracy of the observations because some wild-type zebrafish strains have a higher incidence of laterality defects than others (see Section 1).

The simplest way to evaluate heart position is described in the protocol below. Another method includes in situ hybridization using markers for laterality. The most common marker used is *southpaw (spaw)*, which is first detectable at 4 to 6 somite stage (Long, Ahmad, & Rebagliati, 2003). Other markers include *lefty1, lefty2, nodal, and pitx2*, which display asymmetric expression during somitogenesis (Chang, Zwijsen, Vogel, Huylebroeck, & Matzuk, 2000; Liang et al., 2000; Long et al., 2003).

Materials
- Embryo medium
- Embryos at c. 30 hpf to evaluate jogging of the heart and 48–72 hpf to evaluate heart looping.
- Forceps, Dumont, no 5.
- Stereoscope, Zeiss Stemi 2000-C or similar
- Tricaine stock solution (4 mg/mL), for final concentration add 1.5 mL into 20 mL of embryo medium.

Protocol
1. Dechorionate embryos using forceps. The heartbeat of the zebrafish embryo is detectable using a stereomicroscope by 24 hpf.
2. Add Tricaine to stop embryo movements.
3. Turn the embryo ventral side up and observe the two heart chambers: the atrium and the ventricle. At this stage, the heart of wild-type embryos is positioned to the left of the body axis.

4. If possible, separate cilia mutants/morphants based on other phenotypes first (body curvature, for example) and then evaluate the position of the heart.
5. Compare the frequencies of heart position in mutant embryos with those of their wild-type siblings. Repeat the analysis in multiple crosses.

3.2 METHOD 7: *EVALUATION OF KIDNEY FUNCTION*

The method described below is used for quantitative evaluation of zebrafish kidney function using a fluorescent dye clearance assay.

Materials
- Embryo medium
- Embryos
- Glass capillaries, such as Harvard Apparatus 30-0038
- Methylcellulose, 3%, Sigma Aldrich, M0512, in embryo medium
- Microinjector, such as WPI PV800 Pneumatic PicoPump
- Molds to imprint agarose as in Method 1a.
- Plates for tissue culture, 24 well
- PTU, Sigma Aldrich, P7629 (0.003% Phenylthiourea in embryo medium)
- Rhodamine dextran (RD) 50 mg/mL, Life Technologies D1824
- Software, ImageJ (NIH)
- Stereoscope for fluorescence imaging equipped with a rhodamine filter.
- Tricaine stock solution (4 mg/mL), for final concentration add 1.5 mL into 20 mL of embryo medium.

Protocol
1. Prepare microinjection needles as described (Yuan & Sun, 2009). Load them with 8 μL of RD at the final concentration of 5 mg/mL.
2. At 8 hpf, transfer embryos into a Petri dish filled with embryo medium containing PTU to inhibit melanogenesis and incubate until 72 hpf.
3. At 72 hpf, add Tricaine.
4. Place embryos left side up on an agarose bed imprinted using a mold, and inject roughly 1 nL of RD into the heart by piercing the pericardium.
5. Transfer injected embryos to a 24-well plate containing 0.5–1.0 mL of fresh embryo medium with PTU in each well. Incubate at 28.5°C for 3 h.
6. Three hours after injection, anaesthetize embryos with Tricaine as in 3, and position them lateral side up in 3% methylcellulose. Image the entire body using a stereomicroscope equipped with a UV light source and rhodamine filter.
7. Allow embryos to recover and place them back in assigned wells.
8. At 24 h postinjection, acquire a second round of images as described above.
9. Quantify fluorescence intensity of each embryo at 3 and 24 h postinjection by specifying a 100 × 100 pixel region of interest (roi) in ImageJ. Position the heart in the center of the roi and measure fluorescent intensity.

3.3 METHOD 8: *ANALYSIS OF SENSORY CELL MORPHOLOGY*

Mutations in zebrafish ciliary genes lead to degeneration of sensory neurons in the visual, olfactory, and auditory systems (Doerre & Malicki, 2002; Gross et al., 2005; Krock & Perkins, 2008; Omori et al., 2008; Tsujikawa & Malicki, 2004b; Zhao et al., 2012). This correlates with defects seen in human ciliopathies such as Bardet−Biedl syndrome and nephronophthisis (reviewed in Oh & Katsanis, 2012). While studying phenotypes of cilia mutants, it is therefore valuable to examine morphology and survival of sensory cells. Below we describe reagents that can be used for this purpose.

3.3.1 Photoreceptor cells

Cilia defects frequently result in photoreceptor degeneration in animal models and man (Hartong, Berson, & Dryja, 2006; Oh & Katsanis, 2012; Tsujikawa & Malicki, 2004b). This can be monitored using several approaches. The simplest one is to evaluate the gross appearance of a mutant retina by analyzing plastic histology sections (Fig. 5A−C). Using a microscope equipped with Nomarski optics to view plastic sections, one can distinguish photoreceptor outer segments, a feature frequently abnormal in cilia mutants (Doerre & Malicki, 2001). In severely affected mutant retinae, photoreceptor loss results in the appearance of obvious holes in the outer retina (arrows in Fig. 5A−C).

Antibody staining is an informative way to follow histological studies. The Zpr-1 antibody stains the entire cell body of red−green double cone photoreceptor cells. This antibody can be used both on whole embryos and on frozen sections (Larison & Bremiller, 1990). Zpr-1-positive cells are found throughout the entire photoreceptor cell layer at 3 dpf and later (Fig. 5D)(Larison & Bremiller, 1990). In addition, staining of retinae with anti-opsin antibodies is very informative, as opsin transport is almost invariably affected in cilia mutants. Defects in IFT, for example, lead to opsin accumulation in the cell body (Doerre & Malicki, 2002; Tsujikawa & Malicki, 2004b). Several anti-opsin antibodies are available for zebrafish (Table 1, and accompanying chapter by Malicki et al., 2016). A more sophisticated transgene-based assay for opsin transport is provided in Method 5 (Zhao & Malicki, 2011).

Another very informative approach to the analysis of ciliary defects in zebrafish photoreceptors is electron microscopy. Photoreceptor outer segments differentiate a distinctive array of parallel membrane folds (Fig. 1J and Fig. 5N), which can be visualized on electron micrographs in exquisite detail. This membrane architecture is often disrupted or absent in cilia mutants (Doerre & Malicki, 2002; Krock & Perkins, 2008; Zhao et al., 2012). Electron microscopy also makes it possible to analyze details of microtubule arrangement in the outer segment, the transition fibers at the cilia base, and membrane folds of the periciliary ridge area (Insinna et al., 2008; Zhao et al., 2012; Pooranachandran & Malicki, 2016). Electron microscopy of photoreceptors is performed using standard approaches (Doerre & Malicki, 2001; Schmitt & Dowling, 1999).

In rare but interesting cases, the phenotypes of cilia-related mutations may vary in different photoreceptor types. For example, rods may degenerate while cones remain fairly intact (Zhao et al., 2012). To distinguish photoreceptors subtypes, one can use anti-opsin antibodies mentioned earlier. In situ hybridization with anti-opsin probes can also be used for the same purpose. Alternatively, a variety of transgenes that express fluorescent proteins in a subset of photoreceptors are available (see, for example, Fadool, 2003; Tsujimura, Masuda, Ashino, & Kawamura, 2015; Table 1, and accompanying chapter by Malicki et al., 2016).

Photoreceptor function can be also analyzed using behavioral tests, such as optomotor or optokinetic response assays and by performing electroretinography (Avanesov, Dahm, Sewell, & Malicki, 2005; Brockerhoff, Dowling, & Hurley, 1998; Brockerhoff et al., 1995; Morris, Schroeter, Bilotta, Wong, & Fadool, 2005; Muto et al., 2005). More information on photoreceptor cell markers, behavioral assays, and tests of electrical responses in the retina is provided in chapter "Analysis of the retina in the zebrafish model" by Malicki et al., (2016).

3.3.2 Auditory system hair cells

Similar to photoreceptors, hair cell survival can be evaluated on sections through maculae and cristae using conventional (Fig. 5F and G) or electron (Fig. 2J) microscopy (Jing & Malicki, 2009; Pooranachandran & Malicki, 2016; Tsujikawa & Malicki, 2004b). Anti-acetylated tubulin antibodies stain not only cilia but also cell bodies in hair cells (Figs. 1F and 2F). Because anti-acetylated tubulin staining is not specific to hair cells and does not visualize the entire cell body in these cells, a better option is the HCS-1 antibody, which specifically and uniformly marks zebrafish hair cell somata (Gale, Meyers, Periasamy, & Corwin, 2002; Goodyear et al., 2010; Schibler & Malicki, 2007). Staining with the aforementioned antibodies can be performed both in whole animals and on sections as described in Method 1. Hair cells can be also visualized using transgenic lines, such as the Tg(*brn3c:GFP*) line (Xiao et al., 2005), or Tg(*sqET4:gfp*) that marks hair cells in the lateral line (Loh et al., 2014).

Stereociliary bundles on the apical surface of hair cells are actin rich and can be visualized by staining with fluorophore-conjugated phalloidin. Phalloidin can be applied during or after staining with secondary antibodies. The relative positions of antibody-stained kinocilia and phalloidin-stained stereocilia are good indicators of planar cell polarity in mechanosensory epithelia. Planar cell polarity is compromised in some cilia mutants (Kishimoto et al., 2008; Marshall & Kintner, 2008; Zhao & Malicki, 2011). It can be evaluated on photographs of sensory epithelia by drawing a line connecting the kinocilium to the center of the stereociliary bundle in each cell. In polarized tissue, lines drawn on neighboring cells are parallel to each other, and kinocilia localize to the same side of cells.

Finally, a complication of hair cell analysis in the ear is the presence of otoliths, which form above the surface of maculae. Zebrafish otoliths have been described in detail (Platt, 1993; Stooke-Vaughan, Huang, Hammond, Schier, & Whitfield, 2012). They are highly mineralized opaque structures, which have to be removed to visualize hair cells. They also cause mechanical damage to ear tissue preparations. Otoliths can be dissolved by incubating embryos in 1% Triton or 120 mM EDTA (Method 1a). For better visualization of hair cells positioned deep in the inner ear, a 2% Triton treatment is recommended over 3—4 days (Method 1b).

3.3.3 Lateral line hair cells
Neuromasts differentiate on the surface of the skin and thus can be easily stained in whole animals. Neuromast hair cells can be visualized with some of the same reagents used to stain hair cells in the ear. In live animals they stain with DASPEI dye (Murakami et al., 2003; Whitfield et al., 1996). We describe a DASPEI staining method below.

3.3.4 Olfactory sensory neurons
Whole-mount antibody staining of nasal cilia is particularly straightforward as these cilia are present on the body surface and thus are easily accessible to staining reagents (Fig. 1B). Following staining, nasal cilia can be easily visualized using confocal microscopy (Omori et al., 2008; Zhao et al., 2012). Phalloidin and DAPI can be used as counterstains in such experiments to highlight the apical surface of cells and their nuclei, respectively. Similar to photoreceptors and mechanosensory hair cells, olfactory neurons degenerate in some cilia mutants (Tsujikawa & Malicki, 2004b). Survival of ciliated olfactory neurons can be evaluated by in situ hybridization with probes to the *olfactory marker protein* gene (*omp*), or via DiI labeling as described in Method 10 below (Dynes & Ngai, 1998; Hansen & Zeiske, 1993; Tsujikawa & Malicki, 2004b). Olfactory neurons can also be evaluated using behavioral tests as outlined in Method 11 below or a transgenic *omp* line (DeMaria et al., 2013).

3.4 METHOD 9: *STAINING OF NEUROMAST HAIR CELLS IN LIVE SPECIMEN*

Neuromasts in live zebrafish larvae can be visualized using a simple procedure (outlined below).

Materials
- DASPEI (2-(4-(dimethylamino)styryl)-N-Ethylpyridinium Iodide), Invitrogen, D426
- Embryo medium
- Methylcellulose, 3%, Sigma Aldrich, M0512, in embryo medium
- Tricaine stock solution (4 mg/mL), add 1.5 mL into 20 mL of embryo medium.

Protocol
1. Transfer larvae to 1 mM DASPEI solution in embryo medium.
2. Incubate for 20 min.
3. Rinse thoroughly in embryo medium.
4. Anaesthetize in Tricaine, and mount in methylcellulose or agarose as described in Method 1a.

5. View using fluorescence microscopy using DASPEI filter cube (excitation 450–490 nm and barrier 515 nm).

3.5 METHOD 10: *LABELING OF OLFACTORY NEURONS BY DiI INCORPORATION*

The larval olfactory pit is open to external environment. As their cilia protrude into the olfactory pit, olfactory neurons can be labeled by exposing zebrafish larvae to dye droplets suspended in water (Tsujikawa & Malicki, 2004b) (Fig. 5H, I). This assay is analogous to the so-called dye-filling assay used to evaluate cilia in *C. elegans* (Lambacher et al., 2015; Williams, Winkelbauer, Schafer, Michaud, & Yoder, 2008).

Materials
- Agarose, low melting point
- DiI Neuro Trace Tissue Labeling Paste: Invitrogen, N22880
- DMSO: Sigma, D5879
- Embryo medium
- Ethanol: Sigma, 279741-1L
- Incubator set at 30°C
- Pipette, disposable glass Pasteur pipette, VWR 612-1701
- Tricaine stock solution (4 mg/mL), add 1.5 mL into 20 mL of embryo medium.

Protocol
1. Prepare DiI/DMSO stock by mixing c. 60 mg of DiI paste (one scoop with a 200-μL tip) in 500–1000 μL of DMSO.
2. Prepare embryo medium containing 7% DiI/DMSO stock and 1% ethanol. Warm up to 30°C.
3. Incubate live larvae in the solution from Step 2 at 30°C for 10 min.
4. Wash larvae in embryo medium at 28°C. They will not look healthy, but will remain alive.
5. Add Tricaine to immobilize larvae.
6. To visualize olfactory pits, larvae should be positioned vertically, heads up, in agarose wells. In this case, wells can be made by punching holes in an agarose bed with a glass Pasteur pipette. Mounted embryos can be imaged by confocal or conventional fluorescence microscopy using a water dipping lens.

3.6 METHOD 11: *TESTS OF OLFACTION*

Anosmia, or loss of the sense of smell, is a common ciliopathy phenotype in man (Kulaga et al., 2004). It is caused by cilia abnormalities in the olfactory epithelium. The zebrafish olfactory system is an attractive model to study olfactory cilia. Bile acids are stimulants for zebrafish larvae, which respond with a characteristic swimming behavior which involves 90-degree turns and extensive fast paced swimming

(Braubach, Wood, Gadbois, Fine, & Croll, 2009; DeMaria et al., 2013; Michel & Derbidge, 1997; Yaksi, 2013). The first olfactory responses appear by 3 dpf (Li, Field, & Raisman, 2005). Using the ZebraBox system manufactured by ViewPoint, one can evaluate swimming behavior of both larvae and adults in response to olfactory stimuli.

Materials
- Bile Acids (Sigma, B8756), 10^{-4} M final concentration
- Embryo medium
- Plates for tissue culture, 24 well
- ZebraBox, ViewPoint, Lyon, France (www.viewpoint.fr)
- Zebrafish larvae at 4–5 dpf

Protocol
1. Add 1 ml of embryo medium to wells of a 24-well plate. Place a single larva in each well.
2. Transfer the plate in the ViewPoint ZebraBox chamber and let the larvae acclimatize for 45 min in darkness (turn lights off using ViewPoint control panel).
3. Start recording using the tracking option of the ViewPoint Software for 10 min. The system records movement in infrared light.
4. Open the ZebraBox chamber and add olfactory stimulant (for example, bile acids) to each well as quickly as possible using pipetman.
5. Record response for 2 min. The motility of wild-type larvae should be increased.
6. Data can be exported in Excel format, which provides information about the distance traveled, the duration of movement, and the speed of each larva in the plate.

Responses of larvae are quite variable in this assay. A large sample size is required to produce meaningful data.

4. FUTURE DIRECTIONS

The zebrafish has proven to be an excellent model to study many aspects of cilia function. The ease of generating zebrafish mutants in ciliary genes using forward and reverse genetic approaches has led to a number of important findings (Kishimoto et al., 2008; Omori et al., 2008; Pathak, Austin, & Drummond, 2011; Zhao & Malicki, 2011). Although many aspects of ciliogenesis have become much better understood in recent years, numerous questions are yet to be answered. Animal models, including the zebrafish in particular, will be indispensable in this regard. Cilia are well characterized in a number of organs, but the understanding of what they do varies greatly depending on the context. Photoreceptor cilia are among the best understood in terms of function and structure. In contrast to that, very little is known about the role cilia in the brain or the bone. The understanding of what cilia do in these organs will benefit from live imaging of intact animals. Such imaging experiments are the strength of the zebrafish model. Advances in imaging, such as light sheet microscopy and the use of ever more sophisticated combinations of mutant genotypes and transgenic tools to monitor cell behavior in live animals have created a fertile ground for the zebrafish model to continue generating insights into the mechanisms of ciliogenesis.

ACKNOWLEDGMENTS

We thank Julian Swatler for help with preparing figures. This work was supported by NEI (RO1EY018176) and MRC (MR/N000714/1) research grants.

REFERENCES

Afzelius, B. A. (2004). Cilia-related diseases. *Journal of Pathology, 204*(4), 470−477.

Aizawa, H., Goto, M., Sato, T., & Okamoto, H. (2007). Temporally regulated asymmetric neurogenesis causes left-right difference in the zebrafish habenular structures. *Developmental Cell, 12*(1), 87−98.

Avanesov, A., Dahm, R., Sewell, W. F., & Malicki, J. J. (2005). Mutations that affect the survival of selected amacrine cell subpopulations define a new class of genetic defects in the vertebrate retina. *Developmental Biology, 285*(1), 138−155.

Bachmann-Gagescu, R., Phelps, I. G., Stearns, G., Link, B. A., Brockerhoff, S. E., Moens, C. B., & Doherty, D. (2011). The ciliopathy gene cc2d2a controls zebrafish photoreceptor outer segment development through a role in Rab8-dependent vesicle trafficking. *Human Molecular Genetics, 20*(20), 4041−4055.

Bang, P. I., Sewell, W. F., & Malicki, J. J. (2001). Morphology and cell type heterogeneities of the inner ear epithelia in adult and juvenile zebrafish (*Danio rerio*). *Journal of Comparative Neurology, 438*(2), 173−190.

Barthel, L. K., & Raymond, P. A. (1990). Improved method for obtaining 3-microns cryosections for immunocytochemistry. *Journal of Histochemistry and Cytochemistry, 38*(9), 1383−1388.

Basu, B., & Brueckner, M. (2008). Cilia multifunctional organelles at the center of vertebrate left-right asymmetry. *Current Topics in Developmental Biology, 85*, 151−174.

Baxendale, S., & Whitfield, T. (2016). Methods to study the development, anatomy and function of the zebrafish inner ear. In . In H. W. Detrich, III, M. Westerfield, & L. Zon (Eds.), *The Zebrafish: Cellular and Developmental Biology, Part B Cellular Biology* (Vol. 134, pp. 165−210).

Becker-Heck, A., Zohn, I. E., Okabe, N., Pollock, A., Lenhart, K. B., Sullivan-Brown, J.,...Burdine, R. D. (2010). The coiled-coil domain containing protein CCDC40 is essential for motile cilia function and left-right axis formation. *Nature Genetics, 43*(1), 79−84.

Bessant, D. A., Khaliq, S., Hameed, A., Anwar, K., Payne, A. M., Mehdi, S. Q., & Bhattacharya, S. S. (1999). Severe autosomal dominant retinitis pigmentosa caused by a novel rhodopsin mutation (Ter349Glu). Mutations in brief no. 208. Online. *Human Mutation, 13*(1), 83.

Borovina, A., Superina, S., Voskas, D., & Ciruna, B. (2010). Vangl2 directs the posterior tilting and asymmetric localization of motile primary cilia. *Nature Cell Biology, 12*(4), 407–412.

Branchek, T., & Bremiller, R. (1984). The development of photoreceptors in the zebrafish, *Brachydanio rerio*. I. Structure. *Journal of Comparative Neurology, 224*, 107–115.

Brand, M., Heisenberg, C. P., Jiang, Y. J., Beuchle, D., Lun, K., Furutani-Seiki, M., ... Nüsslein-Volhard, C. (1996). Mutations in zebrafish genes affecting the formation of the boundary between midbrain and hindbrain. *Development, 123*, 179–190.

Braubach, O. R., Wood, H. D., Gadbois, S., Fine, A., & Croll, R. P. (2009). Olfactory conditioning in the zebrafish (*Danio rerio*). *Behavioural Brain Research, 198*(1), 190–198.

Brockerhoff, S. E., Dowling, J. E., & Hurley, J. B. (1998). Zebrafish retinal mutants. *Vision Research, 38*(10), 1335–1339.

Brockerhoff, S. E., Hurley, J. B., Janssen-Bienhold, U., Neuhauss, S. C., Driever, W., & Dowling, J. E. (1995). A behavioral screen for isolating zebrafish mutants with visual system defects. *Proceedings of the National Academy of Sciences of the United States of America, 92*(23), 10545–10549.

Brockerhoff, S. E., Rieke, F., Matthews, H. R., Taylor, M. R., Kennedy, B., Ankoudinova, I., ... Hurley, J. B. (2003). Light stimulates a transducin-independent increase of cytoplasmic Ca^{2+} and suppression of current in cones from the zebrafish mutant nof. *Journal of Neuroscience, 23*(2), 470–480.

Brzica, H., Breljak, D., Ljubojevic, M., Balen, D., Micek, V., Anzai, N., & Sabolić, I. (2009). Optimal methods of antigen retrieval for organic anion transporters in cryosections of the rat kidney. *Arhiv Za Higijenu Rada I Toksikologiju, 60*(1), 7–17.

Callen, A. M., Adoutte, A., Andrew, J. M., Baroin-Tourancheau, A., Bre, M. H., Ruiz, P. C., ... Levilliers, N. (1994). Isolation and characterization of libraries of monoclonal antibodies directed against various forms of tubulin in *Paramecium*. *Biologie Cellulaire, 81*(2), 95–119.

Cao, Y., Park, A., & Sun, Z. (2010). Intraflagellar transport proteins are essential for cilia formation and for planar cell polarity. *Journal of the American Society of Nephrology, 21*(8), 1326–1333.

Cardenas-Rodriguez, M., & Badano, J. L. (2009). Ciliary biology: understanding the cellular and genetic basis of human ciliopathies. *American Journal of Medical Genetics. Part C, Seminars in Medical Genetics, 151C*(4), 263–280.

Carrillo, S. A., Anguita-Salinas, C., Pena, O. A., Morales, R. A., Munoz-Sanchez, S., Munoz-Montecinos, C., ... Allende, M. L. (2016). Macrophage recruitment contributes to regeneration of mechanosensory hair cells in the zebrafish lateral line. *Journal of Cellular Biochemistry*. http://dx.doi.org/10.1002/jcb.25487.

Chang, H., Zwijsen, A., Vogel, H., Huylebroeck, D., & Matzuk, M. M. (2000). Smad5 is essential for left-right asymmetry in mice. *Developmental Biology, 219*(1), 71–78.

Chang, B., Khanna, H., Hawes, N., Jimeno, D., He, S., Lillo, C., ... Swaroop, A. (2006). In-frame deletion in a novel centrosomal/ciliary protein CEP290/NPHP6 perturbs its interaction with RPGR and results in early-onset retinal degeneration in the rd16 mouse. *Human Molecular Genetics, 15*(11), 1847–1857.

Chen, J. N., van Eeden, F. J., Warren, K. S., Chin, A., Nusslein-Volhard, C., Haffter, P., & Fishman, M. C. (1997). Left-right pattern of cardiac BMP4 may drive asymmetry of the heart in zebrafish. *Development, 124*(21), 4373–4382.

Chen, J. N., van Bebber, F., Goldstein, A. M., Serluca, F. C., Jackson, D., Childs, S., … Fishman, M. C. (2001). Genetic steps to organ laterality in zebrafish. *Comparative and Functional Genomics, 2*(2), 60–68.

Chuang, J. Z., Hsu, Y. C., & Sung, C. H. (2015). Ultrastructural visualization of trans-ciliary rhodopsin cargoes in mammalian rods. *Cilia, 4*, 4.

Ciruna, B., Weidinger, G., Knaut, H., Thisse, B., Thisse, C., Raz, E., & Schier, A. F. (2002). Production of maternal-zygotic mutant zebrafish by germ-line replacement. *Proceedings of the National Academy of Sciences of the United States of America, 99*(23), 14919–14924.

Dambly-Chaudiere, C., Sapede, D., Soubiran, F., Decorde, K., Gompel, N., & Ghysen, A. (2003). The lateral line of zebrafish: a model system for the analysis of morphogenesis and neural development in vertebrates. *Biologie Cellulaire, 95*(9), 579–587.

DeMaria, S., Berke, A. P., Van Name, E., Heravian, A., Ferreira, T., & Ngai, J. (2013). Role of a ubiquitously expressed receptor in the vertebrate olfactory system. *Journal of Neuroscience, 33*(38), 15235–15247.

Diggle, C. P., Moore, D. J., Mali, G., zur Lage, P., Ait-Lounis, A., Schmidts, M., … Mill, P. (2014). HEATR2 plays a conserved role in assembly of the ciliary motile apparatus. *PLoS Genetics, 10*(9), e1004577.

Ding, J. D., Salinas, R. Y., & Arshavsky, V. Y. (2015). Discs of mammalian rod photoreceptors form through the membrane evagination mechanism. *Journal of Cell Biology, 211*(3), 495–502.

Doerre, G., & Malicki, J. (2001). A mutation of early photoreceptor development, *mikre oko*, reveals cell–cell interactions involved in the survival and differentiation of zebrafish photoreceptors. *Journal of Neuroscience, 21*(17), 6745–6757.

Doerre, G., & Malicki, J. (2002). Genetic analysis of photoreceptor cell development in the zebrafish retina. *Mechanisms of Development, 110*(1–2), 125–138.

Doll, C. A., Burkart, J. T., Hope, K. D., Halpern, M. E., & Gamse, J. T. (2011). Subnuclear development of the zebrafish habenular nuclei requires ER translocon function. *Developmental Biology, 360*(1), 44–57.

Drummond, I. A., Majumdar, A., Hentschel, H., Elger, M., Solnica-Krezel, L., Schier, A. F., … Fishman, M. C. (1998). Early development of the zebrafish pronephros and analysis of mutations affecting pronephric function. *Development, 125*(23), 4655–4667.

Duldulao, N. A., Lee, S., & Sun, Z. (2009). Cilia localization is essential for in vivo functions of the Joubert syndrome protein Arl13b/Scorpion. *Development, 136*(23), 4033–4042.

Duval, M. G., Chung, H., Lehmann, O. J., & Allison, W. T. (2013). Longitudinal fluorescent observation of retinal degeneration and regeneration in zebrafish using fundus lens imaging. *Molecular Vision, 19*, 1082–1095.

Dynes, J. L., & Ngai, J. (1998). Pathfinding of olfactory neuron axons to stereotyped glomerular targets revealed by dynamic imaging in living zebrafish embryos. *Neuron, 20*(6), 1081–1091.

Einarsson, R., Haden, M., DiCiolli, G., Lim, A., Mah-Ginn, K., Aguilar, K., … Yazejian, B. (2012). Patch clamp recordings in inner ear hair cells isolated from zebrafish. *Journal of Visualized Experiments, 68*.

Eisen, J. S., & Smith, J. C. (2008). Controlling morpholino experiments: don't stop making antisense. *Development, 135*(10), 1735–1743.

Essner, J. J., Amack, J. D., Nyholm, M. K., Harris, E. B., & Yost, H. J. (2005). Kupffer's vesicle is a ciliated organ of asymmetry in the zebrafish embryo that initiates left-right development of the brain, heart and gut. *Development, 132*(6), 1247–1260.

Fadool, J. M. (2003). Development of a rod photoreceptor mosaic revealed in transgenic zebrafish. *Developmental Biology, 258*(2), 277–290.

Falk, N., Lösl, M., Schröder, N., & Giessl, A. (2015). Specialized cilia in mammalian sensory systems. *Cells, 4*(3), 500–519.

Fisch, C., & Dupuis-Williams, P. (2011). Ultrastructure of cilia and flagella – back to the future! *Biologie Cellulaire, 103*(6), 249–270.

Fotiadis, D., Liang, Y., Filipek, S., Saperstein, D. A., Engel, A., & Palczewski, K. (2003). Atomic-force microscopy: rhodopsin dimers in native disc membranes. *Nature, 421*(6919), 127–128.

Gale, J. E., Meyers, J. R., & Corwin, J. T. (2000). Solitary hair cells are distributed throughout the extramacular epithelium in the bullfrog's saccule. *Journal of the Association for Research in Otolaryngology, 1*(2), 172–182.

Gale, J. E., Meyers, J. R., Periasamy, A., & Corwin, J. T. (2002). Survival of bundleless hair cells and subsequent bundle replacement in the bullfrog's saccule. *Journal of Neurobiology, 50*(2), 81–92.

Garnaas, M. K., Cutting, C. C., Meyers, A., Kelsey, P. B., Jr., Harris, J. M., North, T. E., & Goessling, W. (2012). Rargb regulates organ laterality in a zebrafish model of right atrial isomerism. *Developmental Biology, 372*(2), 178–189.

Gayoso, J., Castro, A., Anadon, R., & Manso, M. J. (2012). Crypt cells of the zebrafish *Danio rerio* mainly project to the dorsomedial glomerular field of the olfactory bulb. *Chemical Senses, 37*(4), 357–369.

Goetz, S. C., & Anderson, K. V. (2010). The primary cilium: a signalling centre during vertebrate development. *Nature Reviews Genetics, 11*(5), 331–344.

Goodyear, R. J., Legan, P. K., Christiansen, J. R., Xia, B., Korchagina, J., Gale, J. E., … Richardson, G. P. (2010). Identification of the hair cell soma-1 antigen, HCS-1, as otoferlin. *Journal of the Association for Research in Otolaryngology, 11*(4), 573–586.

Green, J., & Mykytyn, K. (2010). Neuronal ciliary signaling in homeostasis and disease. *Cellular and Molecular Life Sciences, 67*.

Gross, J. M., Perkins, B. D., Amsterdam, A., Egana, A., Darland, T., Matsui, J. I., … Dowling, J. E. (2005). Identification of zebrafish insertional mutants with defects in visual system development and function. *Genetics, 170*(1), 245–261.

Guemez-Gamboa, A., Coufal, N. G., & Gleeson, J. G. (2014). Primary cilia in the developing and mature brain. *Neuron, 82*(3), 511–521.

Haddon, C., & Lewis, J. (1996). Early ear development in the embryo of the zebrafish, *Danio rerio*. *Journal of Comparative Neurology, 365*(1), 113–128.

Hansen, A., & Zeiske, E. (1993). Development of the olfactory organ in the zebrafish, *Brachydanio rerio*. *Journal of Comparative Neurology, 333*(2), 289–300.

Hansen, A., & Zeiske, E. (1998). The peripheral olfactory organ of the zebrafish, *Danio rerio*: an ultrastructural study. *Chemical Senses, 23*(1), 39–48.

Hartong, D. T., Berson, E. L., & Dryja, T. P. (2006). Retinitis pigmentosa. *Lancet, 368*(9549), 1795–1809.

Haycraft, C. J., Banizs, B., Aydin-Son, Y., Zhang, Q., Michaud, E. J., & Yoder, B. K. (2005). Gli2 and Gli3 localize to cilia and require the intraflagellar transport protein polaris for processing and function. *PLoS Genetics, 1*(4), e53.

Huang, P., & Schier, A. F. (2009). Dampened Hedgehog signaling but normal Wnt signaling in zebrafish without cilia. *Development, 136*(18), 3089–3098.

Hudspeth, A. J. (1989). How the ear's works work. *Nature, 341*(6241), 397–404.

Inoue, D., & Wittbrodt, J. (2011). One for all—a highly efficient and versatile method for fluorescent immunostaining in fish embryos. *PLoS One, 6*(5), e19713.

Insinna, C., Pathak, N., Perkins, B., Drummond, I., & Besharse, J. C. (2008). The homodimeric kinesin, Kif17, is essential for vertebrate photoreceptor sensory outer segment development. *Developmental Biology, 316*(1), 160–170.

Jaffe, K. M., Grimes, D. T., Schottenfeld-Roames, J., Werner, M. E., Ku, T. S., Kim, S. K., ... Burdine, R. D. (2016). c21orf59/kurly controls both cilia motility and polarization. *Cell Reports [electronic Resource], 14*(8), 1841–1849.

Jenkins, P. M., McEwen, D. P., & Martens, J. R. (2009). Olfactory cilia: linking sensory cilia function and human disease. *Chemical Senses, 34*(5), 451–464.

Jin, D., Ni, T. T., Sun, J., Wan, H., Amack, J. D., Yu, G., ... Zhong, T. P. (2014). Prostaglandin signalling regulates ciliogenesis by modulating intraflagellar transport [Article] *Nature Cell Biology, 16*(9), 841–851.

Jing, X., & Malicki, J. (2009). Zebrafish ale oko, an essential determinant of sensory neuron survival and the polarity of retinal radial glia, encodes the p50 subunit of dynactin. *Development, 136*(17), 2955–2964.

Jones, C., Roper, V. C., Foucher, I., Qian, D., Banizs, B., Petit, C., ... Chen, P. (2008). Ciliary proteins link basal body polarization to planar cell polarity regulation. *Nature Genetics, 40*(1), 69–77.

Kennedy, B., & Malicki, J. (2009). What drives cell morphogenesis: a look inside the vertebrate photoreceptor. *Developmental Dynamics, 238*(9), 2115–2138.

Kermen, F., Franco, L. M., Wyatt, C., & Yaksi, E. (2013). Neural circuits mediating olfactory-driven behavior in fish. *Frontiers in Neural Circuits, 7*, 62.

Khanna, H., Davis, E. E., Murga-Zamalloa, C. A., Estrada-Cuzcano, A., Lopez, I., den Hollander, A. I., ... Katsanis, N. (2009). A common allele in RPGRIP1L is a modifier of retinal degeneration in ciliopathies. *Nature Genetics, 41*(6), 739–745.

Khodiyar, V. K., Howe, D., Talmud, P. J., Breckenridge, R., & Lovering, R. C. (2013). From zebrafish heart jogging genes to mouse and human orthologs: using Gene Ontology to investigate mammalian heart development. *F1000Res, 2*, 242.

Kikuchi, K., & Hilding, D. (1965). The development of the organ of Corti in the mouse. *Acta Oto-laryngologica, 60*(3), 207–222.

Kimmel, C. B., Ballard, W. W., Kimmel, S. R., Ullmann, B., & Schilling, T. F. (July 1995). Stages of embryonic development of the zebrafish. *Developmental Dynamics, 203*(3), 253–310.

Kimura, R. S. (1969). Distribution, structure, and function of dark cells in the vestibular labyrinth. *Annals of Otology, Rhinology, and Laryngology, 78*(3), 542–561.

Kindt, K. S., Finch, G., & Nicolson, T. (2012). Kinocilia mediate mechanosensitivity in developing zebrafish hair cells. *Developmental Cell, 23*(2), 329–341.

Kishi, S., Slack, B. E., Uchiyama, J., & Zhdanova, I. V. (2009). Zebrafish as a genetic model in biological and behavioral gerontology: where development meets aging in vertebrates—a mini-review. *Gerontology, 55*(4), 430–441.

Kishimoto, N., Cao, Y., Park, A., & Sun, Z. (2008). Cystic kidney gene seahorse regulates cilia-mediated processes and Wnt pathways. *Developmental Cell, 14*(6), 954–961.

Kobayashi, D., & Takeda, H. (2012). Ciliary motility the components and cytoplasmic preassembly mechanisms of the axonemal dyneins. *Differentiation; Research in Biological Diversity, 83*(2), S23–S29.

Kramer-Zucker, A. G., Olale, F., Haycraft, C. J., Yoder, B. K., Schier, A. F., & Drummond, I. A. (2005). Cilia-driven fluid flow in the zebrafish pronephros, brain and Kupffer's vesicle is required for normal organogenesis. *Development, 132*(8), 1907−1921.

Kreiling, J. A., Prabhat, Williams, G., & Creton, R. (2007). Analysis of Kupffer's vesicle in zebrafish embryos using a cave automated virtual environment. *Developmental Dynamics, 236*(7), 1963−1969.

Krock, B. L., & Perkins, B. D. (2008). The intraflagellar transport protein IFT57 is required for cilia maintenance and regulates IFT-particle-kinesin-II dissociation in vertebrate photoreceptors. *Journal of Cell Science, 121*(Pt 11), 1907−1915.

Kulaga, H. M., Leitch, C. C., Eichers, E. R., Badano, J. L., Lesemann, A., Hoskins, B. E., … Katsanis, N. (2004). Loss of BBS proteins causes anosmia in humans and defects in olfactory cilia structure and function in the mouse. *Nature Genetics, 36*(9), 994−998.

Kwan, K. M., Fujimoto, E., Grabher, C., Mangum, B. D., Hardy, M. E., Campbell, D. S., … Chien, C. B. (2007). The Tol2kit: a multisite gateway-based construction kit for Tol2 transposon transgenesis constructs. *Developmental Dynamics, 236*(11), 3088−3099.

Lagman, D., Callado-Pérez, A., Franzén, I. E., Larhammar, D., & Abalo, X. M. (2015). Transducin duplicates in the zebrafish retina and pineal complex: differential specialisation after the teleost tetraploidisation. *PLoS One*, 1−23.

Lambacher, N. J., Bruel, A.-L., van Dam, T. J. P., Szymanska, K., Slaats, G. G., Kuhns, S., … Blacque, O. E. (2015). TMEM107 recruits ciliopathy proteins to subdomains of the ciliary transition zone and causes Joubert syndrome. *Nature Cell Biology, 18*(1), 122−131.

Larison, K., & Bremiller, R. (1990). Early onset of phenotype and cell patterning in the embryonic zebrafish retina. *Development, 109*, 567−576.

Ledent, V. (2002). Postembryonic development of the posterior lateral line in zebrafish. *Development, 129*(3), 597−604.

Leightner, A. C., Hommerding, C. J., Peng, Y., Salisbury, J. L., Gainullin, V. G., Czarnecki, P. G., … Harris, P. C. (2013). The Meckel syndrome protein meckelin (TMEM67) is a key regulator of cilia function but is not required for tissue planar polarity. *Human Molecular Genetics, 22*(10), 2024−2040.

Li, Y., Field, P. M., & Raisman, G. (2005). Olfactory ensheathing cells and olfactory nerve fibroblasts maintain continuous open channels for regrowth of olfactory nerve fibres. *Glia, 52*(3), 245−251.

Liang, J. O., Etheridge, A., Hantsoo, L., Rubinstein, A. L., Nowak, S. J., Izpisua Belmonte, J. C., & Halpern, M. E. (2000). Asymmetric nodal signaling in the zebrafish diencephalon positions the pineal organ. *Development, 127*(23), 5101−5112.

Lidow, M. S., & Menco, B. P. (1984). Observations on axonemes and membranes of olfactory and respiratory cilia in frogs and rats using tannic acid-supplemented fixation and photographic rotation. *Journal of Ultrastructure Research, 86*(1), 18−30.

Lipschitz, D. L., & Michel, W. C. (2002). Amino acid odorants stimulate microvillar sensory neurons. *Chemical Senses, 27*(3), 277−286.

Liu, Y., Pathak, N., Kramer-Zucker, A., & Drummond, I. A. (2007). Notch signaling controls the differentiation of transporting epithelia and multiciliated cells in the zebrafish pronephros. *Development, 134*, 1111−1122.

Loh, S. L., Teh, C., Muller, J., Guccione, E., Hong, W., & Korzh, V. (2014). Zebrafish yap1 plays a role in differentiation of hair cells in posterior lateral line. *Scientific Reports, 4*, 4289.

Long, S., Ahmad, N., & Rebagliati, M. (2003). The zebrafish nodal-related gene southpaw is required for visceral and diencephalic left-right asymmetry. *Development, 130*(11), 2303−2316.

Lopez-Schier, H., & Hudspeth, A. J. (2006). A two-step mechanism underlies the planar polarization of regenerating sensory hair cells. *Proceedings of the National Academy of Sciences of the United States of America, 103*(49), 18615−18620.

Lopez-Schier, H., Starr, C. J., Kappler, J. A., Kollmar, R., & Hudspeth, A. J. (2004). Directional cell migration establishes the axes of planar polarity in the posterior lateral-line organ of the zebrafish. *Developmental Cell, 7*(3), 401−412.

Louvi, A., & Grove, E. A. (2011). Cilia in the CNS: the quiet organelle claims center stage. *Neuron, 69*(6), 1046−1060.

Lu, H., Toh, M. T., Narasimhan, V., Thamilselvam, S. K., Choksi, S. P., & Roy, S. (2015). A function for the Joubert syndrome protein Arl13b in ciliary membrane extension and ciliary length regulation. *Developmental Biology, 397*(2), 225−236.

Lush, M. E., & Piotrowski, T. (2014). Sensory hair cell regeneration in the zebrafish lateral line. *Developmental Dynamics, 243*(10), 1187−1202.

Ma, M., & Jiang, Y. J. (2007). Jagged2a-notch signaling mediates cell fate choice in the zebrafish pronephric duct. *PLoS Genetics, 3*(1), e18.

Malicki, J., & Avidor-Reiss, T. (2014). From the cytoplasm into the cilium: bon voyage. *Organogenesis, 10*(1), 138−157.

Malicki, J., Neuhauss, S. C., Schier, A. F., Solnica-Krezel, L., Stemple, D. L., Stainier, D. Y., ... Driver, W. (1996). Mutations affecting development of the zebrafish retina. *Development, 123*, 263−273.

Malicki, J., Pooranachandran, N., Nikolaev, A., Fang, X., & Avanesov, A. (2016). Analysis of the retina in the zebrafish model. In H. W. Detrich, III, M. Westerfield, & L. Zon (Eds.). *The Zebrafish: Cellular and Developmental Biology, Part A Cellular Biology* (Vol. 134).

Malicki, J., Schier, A. F., Solnica-Krezel, L., Stemple, D. L., Neuhauss, S. C., Stainier, D. Y., ... Driver, W. (1996). Mutations affecting development of the zebrafish ear. *Development, 123*, 275−283.

Marshall, W. F., & Kintner, C. (2008). Cilia orientation and the fluid mechanics of development. *Current Opinion in Cell Biology, 20*(1), 48−52.

Matsui, T., Ishikawa, H., & Bessho, Y. (2015). Cell collectivity regulation within migrating cell cluster during Kupffer's vesicle formation in zebrafish. *Frontiers in Cell and Developmental Biology, 3*, 27.

May-Simera, H. L., Kai, M., Hernandez, V., Osborn, D. P., Tada, M., & Beales, P. L. (2010). Bbs8, together with the planar cell polarity protein Vangl2, is required to establish left-right asymmetry in zebrafish. *Developmental Biology, 345*(2), 215−225.

May-Simera, H., & Kelley, M. W. (2012). Planar cell polarity in the inner ear. *Current Topics in Developmental Biology, 101*, 111−140.

McDermott, B. M., Jr., Asai, Y., Baucom, J. M., Jani, S. D., Castellanos, Y., Gomez, G., ... Hudspeth, A. J. (2010). Transgenic labeling of hair cells in the zebrafish acousticolateralis system. *Gene Expression Patterns, 10*(2−3), 113−118.

McIntyre, J. C., Davis, E. E., Joiner, A., Williams, C. L., Tsai, I. C., Jenkins, P. M., ... Martens, J. R. (2012). Gene therapy rescues cilia defects and restores olfactory function in a mammalian ciliopathy model. *Nature Medicine, 18*(9), 1423−1428.

Metcalfe, W. K. (1985). Sensory neuron growth cones comigrate with posterior lateral line primordial cells in zebrafish. *Journal of Comparative Neurology, 238*(2), 218−224.

Michel, W. C., & Derbidge, D. S. (1997). Evidence of distinct amino acid and bile salt receptors in the olfactory system of the zebrafish, *Danio rerio*. *Brain Research*, 1−9.

Mitchison, T., Wuhr, M., Nguyen, P., Ishihara, K., Groen, A., & Field, C. M. (2012). Growth, interaction, and positioning of microtubule asters in extremely large vertebrate embryo cells. *Cytoskeleton, 69*(10), 738−750.

Montgomery, J. C., Macdonald, F., Baker, C. F., & Carton, A. G. (2002). Hydrodynamic contributions to multimodal guidance of prey capture behavior in fish. *Brain, Behavior and Evolution, 59*(4), 190−198.

Morris, A. C., Schroeter, E. H., Bilotta, J., Wong, R. O., & Fadool, J. M. (2005). Cone survival despite rod degeneration in XOPS-mCFP transgenic zebrafish. *Investigative Ophthalmology and Visual Science, 46*(12), 4762−4771.

Murakami, S. L., Cunningham, L. L., Werner, L. A., Bauer, E., Pujol, R., Raible, D. W., & Rubel, E. W. (2003). Developmental differences in susceptibility to neomycin-induced hair cell death in the lateral line neuromasts of zebrafish (*Danio rerio*). *Hearing Research, 186*(1−2), 47−56.

Muto, A., Orger, M. B., Wehman, A. M., Smear, M. C., Kay, J. N., Page-McCaw, P. S., ... Baier, H. (2005). Forward genetic analysis of visual behavior in zebrafish. *PLoS Genetics, 1*(5), e66.

Nachury, M. V., Seeley, E. S., & Jin, H. (2010). Trafficking to the ciliary membrane: how to get across the periciliary diffusion barrier? *Annual Review of Cell and Developmental Biology, 26*(1), 59−87.

Nakayama, K., & Loomis, J. M. (1974). Optical velocity patterns, velocity-sensitive neurons, and space perception: a hypothesis. *Perception, 3*(1), 63−80.

Nasevicius, A., & Ekker, S. C. (2000). Effective targeted gene 'knockdown' in zebrafish. *Nature Genetics, 26*(2), 216−220.

Nicolson, T., Rusch, A., Friedrich, R. W., Granato, M., Ruppersberg, J. P., & Nusslein-Volhard, C. (1998). Genetic analysis of vertebrate sensory hair cell mechanosensation: the zebrafish circler mutants. *Neuron, 20*(2), 271−283.

Nuñez, V. A., Sarrazin, A. F., Cubedo, N., Allende, M. L., Dambly-Chaudière, C., & Ghysen, A. (2009). Postembryonic development of the posterior lateral line in the zebrafish. *Evolution & Development, 11*(4), 391−404.

O'Connor, A. K., Malarkey, E. B., Berbari, N. F., Croyle, M. J., Haycraft, C. J., Bell, P. D., ... Yoder, B. K. (2013). An inducible CiliaGFP mouse model for in vivo visualization and analysis of cilia in live tissue. *Cilia, 2*(1), 8.

Oh, E. C., & Katsanis, N. (2012). Cilia in vertebrate development and disease. *Development, 139*(3), 443−448.

Okabe, N., Xu, B., & Burdine, R. D. (2008). Fluid dynamics in zebrafish Kupffer's vesicle. *Developmental Dynamics, 237*(12), 3602−3612.

Omori, Y., & Malicki, J. (2006). Oko meduzy and related crumbs genes are determinants of apical cell features in the vertebrate embryo. *Current Biology, 16*(10), 945−957.

Omori, Y., Zhao, C., Saras, A., Mukhopadhyay, S., Kim, W., Furukawa, T., ... Malicki, J. (2008). Elipsa is an early determinant of ciliogenesis that links the IFT particle to membrane-associated small GTPase Rab8. *Nature Cell Biology, 10*(4), 437−444.

Oteiza, P., Koppen, M., Krieg, M., Pulgar, E., Farias, C., Melo, C., ... Concha, M. L. (2010). Planar cell polarity signalling regulates cell adhesion properties in progenitors of the zebrafish laterality organ. *Development, 137*(20), 3459−3468.

Otto, E. A., Schermer, B., Obara, T., O'Toole, J. F., Hiller, K. S., Mueller, A. M., ... Hildebrandt, F. (2003). Mutations in INVS encoding inversin cause nephronophthisis type 2, linking renal cystic disease to the function of primary cilia and left-right axis determination. *Nature Genetics, 34*(4), 413—420.

Panizzi, J. R., Becker-Heck, A., Castleman, V. H., Al-Mutairi, D. A., Liu, Y., Loges, N. T., … Drummond, I. A. (2012). CCDC103 mutations cause primary ciliary dyskinesia by disrupting assembly of ciliary dynein arms. *Nature Genetics, 44*(6), 714—719.

Paridaen, J. T. M. L., Wilsch-Bräuninger, M., & Huttner, W. B. (2013). Asymmetric inheritance of centrosome-associated primary cilium membrane directs ciliogenesis after cell division. *Cell, 155*(2), 333—344.

Pathak, N., Austin, C. A., & Drummond, I. A. (2011). Tubulin tyrosine ligase-like genes ttll3 and ttll6 maintain zebrafish cilia structure and motility. *Journal of Biological Chemistry, 286*(13), 11685—11695.

Pathak, N., Austin-Tse, C. A., Liu, Y., Vasilyev, A., & Drummond, I. A. (2014). Cytoplasmic carboxypeptidase 5 regulates tubulin glutamylation and zebrafish cilia formation and function. *Molecular Biology of the Cell, 25*(12), 1836—1844.

Pathak, N., Obara, T., Mangos, S., Liu, Y., & Drummond, I. A. (2007). The zebrafish fleer gene encodes an essential regulator of cilia tubulin polyglutamylation. *Molecular Biology of the Cell, 18*(11), 4353—4364.

Pazour, G. J., Baker, S. A., Deane, J. A., Cole, D. G., Dickert, B. L., Rosenbaum, J. L., … Besharse, J. C. (2002). The intraflagellar transport protein, IFT88, is essential for vertebrate photoreceptor assembly and maintenance. *Journal of Cell Biology, 157*(1), 103—113.

Perkins, B. D., Kainz, P. M., O'Malley, D. M., & Dowling, J. E. (2002). Transgenic expression of a GFP-rhodopsin COOH-terminal fusion protein in zebrafish rod photoreceptors. *Visual Neuroscience, 19*(3), 257—264.

Platt, C. (1993). Zebrafish inner ear sensory surfaces are similar to those in goldfish. *Hearing Research, 65*, 133—140.

Pooranachandran, N., & Malicki, J. (2016). Unexpected roles for ciliary kinesins and intraflagellar transport proteins. *Genetics, 115*.

Pugh, E., & Lamb, T. (2000). *Phototransduction in vertebrate rods and cones: Handbook of biological physics* (Vol. 3, pp. 183—255). Elsevier Science B.V..

Quinlan, R. J., Tobin, J. L., & Beales, P. L. (2008). Modeling ciliopathies: primary cilia in development and disease. *Current Topics in Developmental Biology, 84*, 249—310.

Raible, D. W., & Kruse, G. J. (2000). Organization of the lateral line system in embryonic zebrafish. *Journal of Comparative Neurology, 421*(2), 189—198.

Ramsey, M., & Perkins, B. D. (2013). Basal bodies exhibit polarized positioning in zebrafish cone photoreceptors. *Journal of Comparative Neurology, 521*(8), 1803—1816.

Randlett, O., Poggi, L., Zolessi, F. R., & Harris, W. A. (2011). The oriented emergence of axons from retinal ganglion cells is directed by laminin contact in vivo. *Neuron, 70*(2), 266—280.

Raymond, P., Barthel, L., Rounsifer, M., Sullivan, S., & Knight, J. (1993). Expression of rod and cone visual pigments in godfish and zebrafish: a rhodopsin-like gene is expressed in cones. *Neuron, 10*, 1161—1174.

Reese, T. S. (1965). Olfactory cilia in the frog. *Journal of Cell Biology, 25*(2), 209—230.

Riley, B. B., Zhu, C., Janetopoulos, C., & Aufderheide, K. J. (1997). A critical period of ear development controlled by distinct populations of ciliated cells in the zebrafish. *Developmental Biology, 191*(2), 191—201.

Saade, C. J., Alvarez-Delfin, K., & Fadool, J. M. (2013). Rod photoreceptors protect from cone degeneration-induced retinal remodeling and restore visual responses in zebrafish. *Journal of Neuroscience, 33*(5), 1804—1814.

Sayer, J. A., Otto, E. A., O'Toole, J. F., Nurnberg, G., Kennedy, M. A., Becker, C., ... Hildebrandt, F. (2006). The centrosomal protein nephrocystin-6 is mutated in Joubert syndrome and activates transcription factor ATF4. *Nature Genetics, 38*(6), 674−681.

Schibler, A., & Malicki, J. (2007). A screen for genetic defects of the zebrafish ear. *Mechanisms of Development, 124*(7−8), 592−604.

Schmitt, E. A., & Dowling, J. E. (1996). Comparison of topographical patterns of ganglion and photoreceptor cell differentiation in the retina of the zebrafish, *Danio rerio*. *Journal of Comparative Neurology, 371*(2), 222−234.

Schmitt, E. A., & Dowling, J. E. (1999). Early retinal development in the zebrafish, *Danio rerio*: light and electron microscopic analyses. *Journal of Comparative Neurology, 404*(4), 515−536.

Schottenfeld, J., Sullivan-Brown, J., & Burdine, R. D. (2007). Zebrafish curly up encodes a Pkd2 ortholog that restricts left-side-specific expression of southpaw. *Development, 134*(8), 1605−1615.

Schou, K. B., Pedersen, L. B., & Christensen, S. T. (2015). Ins and outs of GPCR signaling in primary cilia. *EMBO Reports, 16*.

Shcherbo, D., Murphy, C. S., Ermakova, G. V., Solovieva, E. A., Chepurnykh, T. V., Shcheglov, A. S., ... Chudakov, D. M. (2009). Far-red fluorescent tags for protein imaging in living tissues. *Biochemical Journal, 418*(3), 567.

Smith, D. J., Montenegro-Johnson, T. D., & Lopes, S. S. (2014). Organized chaos in Kupffer's vesicle: how a heterogeneous structure achieves consistent left-right patterning. *Bioarchitecture, 4*(3), 119−125.

Smyth, V. A., Di Lorenzo, D., & Kennedy, B. N. (2008). A novel, evolutionarily conserved enhancer of cone photoreceptor-specific expression. *Journal of Biological Chemistry, 283*(16), 10881−10891.

Stooke-Vaughan, G. A., Huang, P., Hammond, K. L., Schier, A. F., & Whitfield, T. T. (2012). The role of hair cells, cilia and ciliary motility in otolith formation in the zebrafish otic vesicle. *Development, 139*(10), 1777−1787.

Stooke-Vaughan, G. A., Obholzer, N. D., Baxendale, S., Megason, S. G., & Whitfield, T. T. (2015). Otolith tethering in the zebrafish otic vesicle requires Otogelin and α-Tectorin. *Development, 142*(6), 1137−1145.

Su, S., Phua, S. C., Derose, R., Chiba, S., Narita, K., Kalugin, P. N., ... Inoue, T. (2013). Genetically encoded calcium indicator illuminates calcium dynamics in primary cilia. *Nature Methods, 10*(11), 1105−1107.

Sun, Z., Amsterdam, A., Pazour, G. J., Cole, D. G., Miller, M. S., & Hopkins, N. (2004). A genetic screen in zebrafish identifies cilia genes as a principal cause of cystic kidney. *Development, 131*(16), 4085−4093.

Sweet, E. M., Vemaraju, S., & Riley, B. B. (2011). Sox2 and Fgf interact with Atoh1 to promote sensory competence throughout the zebrafish inner ear. *Developmental Biology, 358*(1), 113−121.

Szymanska, K., & Johnson, C. A. (2012). The transition zone: an essential functional compartment of cilia. *Cilia, 1*(1), 10.

Tam, B. M., Moritz, O. L., Hurd, L. B., & Papermaster, D. S. (2000). Identification of an outer segment targeting signal in the COOH terminus of rhodopsin using transgenic *Xenopus laevis*. *Journal of Cell Biology, 151*(7), 1369−1380.

Tanimoto, M., Ota, Y., Inoue, M., & Oda, Y. (2011). Origin of inner ear hair cells: morphological and functional differentiation from ciliary cells into hair cells in zebrafish inner ear. *Journal of Neuroscience, 31*(10), 3784−3794.

Tay, S. Y., Yu, X., Wong, K. N., Panse, P., Ng, C. P., & Roy, S. (2009). The iguana/DZIP1 protein is a novel component of the ciliogenic pathway essential for axonemal biogenesis. *Developmental Dynamics, 239*(2), 527—534.

Tissir, F., Qu, Y., Montcouquiol, M., Zhou, L., Komatsu, K., Shi, D., ... Goffinet, A. M. (2010). Lack of cadherins Celsr2 and Celsr3 impairs ependymal ciliogenesis, leading to fatal hydrocephalus. *Nature Neuroscience, 13*(6), 700—707.

Tsujikawa, M., & Malicki, J. (2004a). Genetics of photoreceptor development and function in zebrafish. *International Journal of Developmental Biology, 48*(8—9), 925—934.

Tsujikawa, M., & Malicki, J. (2004b). Intraflagellar transport genes are essential for differentiation and survival of vertebrate sensory neurons. *Neuron, 42*(5), 703—716.

Tsujimura, T., Masuda, R., Ashino, R., & Kawamura, S. (2015). Spatially differentiated expression of quadruplicated green-sensitive RH2 opsin genes in zebrafish is determined by proximal regulatory regions and gene order to the locus control region. *BMC Genetics, 16*, 130.

Vihtelic, T. S., Doro, C. J., & Hyde, D. R. (1999). Cloning and characterization of six zebrafish photoreceptor opsin cDNAs and immunolocalization of their corresponding proteins. *Vis Neurosci, 16*(3), 571—585.

Walczak-Sztulpa, J., Eggenschwiler, J., Osborn, D., Brown, D. A., Emma, F., Klingenberg, C., ... Kuss, A. W. (2010). Cranioectodermal dysplasia, Sensenbrenner syndrome, is a ciliopathy caused by mutations in the IFT122 gene. *American Journal of Human Genetics, 86*(6), 949—956.

Wen, G. Y., Soifer, D., & Wisniewski, H. M. (1982). The doublet microtubules of rods of the rabbit retina. *Anatomy and Embryology (Berlin), 165*(3), 315—328.

Westerfield, M. (2007). *The zebrafish book. A guide for the laboratory use of zebrafish (Danio rerio)* (4th ed.) (Eugene).

Whitfield, T. T., Granato, M., van Eeden, F. J., Schach, U., Brand, M., Furutani-Seiki, M., ... Nüsslein-Volhard, C. (1996). Mutations affecting development of the zebrafish inner ear and lateral line. *Development, 123*, 241—254.

Williams, C. L., Winkelbauer, M. E., Schafer, J. C., Michaud, E. J., & Yoder, B. K. (2008). Functional redundancy of the B9 proteins and nephrocystins in *Caenorhabditis elegans* ciliogenesis. *Molecular Biology of the Cell, 19*(5), 2154—2168.

Wloga, D., Dave, D., Meagley, J., Rogowski, K., Jerka-Dziadosz, M., & Gaertig, J. (2010). Hyperglutamylation of tubulin can either stabilize or destabilize microtubules in the same cell. *Eukaryotic Cell, 9*(1), 184—193.

Wloga, D., Webster, D. M., Rogowski, K., Bre, M. H., Levilliers, N., Jerka-Dziadosz, M., ... Gaertig, J. (2009). TTLL3 is a tubulin glycine ligase that regulates the assembly of cilia. *Developmetal Cell, 16*(6), 867—876.

Wolff, A., de Nechaud, B., Chillet, D., Mazarguil, H., Desbruyeres, E., Audebert, S., ... Denoulet, P. (1992). Distribution of glutamylated alpha and beta-tubulin in mouse tissues using a specific monoclonal antibody, GT335. *European Journal of Cell Biology, 59*(2), 425—432.

Worthington, W. C., Jr., & Cathcart, R. S., 3rd (1963). Ependymal cilia: distribution and activity in the adult human brain. *Science, 139*(3551), 221—222.

Xiao, T., Roeser, T., Staub, W., & Baier, H. (2005). A GFP-based genetic screen reveals mutations that disrupt the architecture of the zebrafish retinotectal projection. *Development, 132*(13), 2955—2967.

Yaksi, E. (2013). Neural circuits mediating olfactory-driven behavior in fish. *Frontiers in Neural Circuits*, 1—9.

Yin, J., Brocher, J., Linder, B., Hirmer, A., Sundaramurthi, H., Fischer, U., & Winkler, C. (2012). The 1D4 antibody labels outer segments of long double cone but not rod photoreceptors in zebrafish. *Investigative Ophthalmology and Visual Science, 53*(8), 4943−4951.

Yoder, B. K., Hou, X., & Guay-Woodford, L. M. (2002). The polycystic kidney disease proteins, polycystin-1, polycystin-2, polaris, and cystin, are co-localized in renal cilia. *Journal of the American Society of Nephrology, 13*(10), 2508−2516.

Yuan, S., & Sun, Z. (2009). Microinjection of mRNA and morpholino antisense oligonucleotides in zebrafish embryos. *Journal of Visualized Experiments, 27*.

Yuan, S., Li, J., Diener, D. R., Choma, M. A., Rosenbaum, J. L., & Sun, Z. (2012). Target-of-rapamycin complex 1 (Torc1) signaling modulates cilia size and function through protein synthesis regulation. *Proceedings of the National Academy of Sciences of the United States of America, 109*(6), 2021−2026.

Yuan, S., Zhao, L., Brueckner, M., & Sun, Z. (2015). Intraciliary calcium oscillations initiate vertebrate left-right asymmetry. *Current Biology*, 1−13.

Zaghloul, N. A., Liu, Y., Gerdes, J. M., Gascue, C., Oh, E. C., Leitch, C. C., … Katsanis, N. (2010). Functional analyses of variants reveal a significant role for dominant negative and common alleles in oligogenic Bardet-Biedl syndrome. *Proceedings of the National Academy of Sciences of the United States of America, 107*(23), 10602−10607.

Zhai, G., Gu, Q., He, J., Lou, Q., Chen, X., Jin, X., … Yin, Z. (2014). Sept6 is required for ciliogenesis in Kupffer's vesicle, the pronephros, and the neural tube during early embryonic development. *Molecular and Cellular Biology, 34*(7), 1310−1321.

Zhao, C., & Malicki, J. (2007). Genetic defects of pronephric cilia in zebrafish. *Mechanisms of Development, 124*(7−8), 605−616.

Zhao, C., & Malicki, J. (2011). Nephrocystins and MKS proteins interact with IFT particle and facilitate transport of selected ciliary cargos. *EMBO Journal, 30*(13), 2532−2544.

Zhao, C., Omori, Y., Brodowska, K., Kovach, P., & Malicki, J. (2012). Kinesin-2 family in vertebrate ciliogenesis. *Proceedings of the National Academy of Sciences of the United States of America, 109*(7), 2388−2393.

Zimmerman, K., & Yoder, B. K. (2015). SnapShot: sensing and signaling by cilia. *Cell, 161*(3), 692−2.e1.

Zolessi, F. R., Poggi, L., Wilkinson, C. J., Chien, C. B., & Harris, W. A. (2006). Polarization and orientation of retinal ganglion cells in vivo. *Neural Development, 1*, 2.

CHAPTER 10

Functional calcium imaging in zebrafish lateral-line hair cells

Q.X. Zhang, X.J. He, H.C. Wong, K.S. Kindt[1]

National Institute on Deafness and Other Communication Disorders,
NIH, Bethesda, MD, United States
[1]*Corresponding author: E-mail: katie.kindt@nih.gov*

CHAPTER OUTLINE

Introduction 230
1. Calcium Indicator Selection and Comparison 231
   1.1 Selecting a Calcium Indicator 231
   1.2 Comparison of Calcium Indicators 232
   1.3 Validating a Relevant Calcium Signal 236
2. Imaging Systems and Optimal Parameters 237
   2.1 General Microscope and Equipment Requirements 237
   2.2 Choosing a Specific Imaging System 239
   2.3 Determining Optimal Imaging Parameters 240
   2.4 Synchronizing a Stimulus With Image Acquisition 242
3. Image Processing 242
   3.1 Image Registration Using ImageJ 243
   3.2 Signal Detection and Representation in ImageJ 244
   3.3 Spatial Detection and Visualization Using MATLAB 244
Summary 248
Discussion 248
Acknowledgments 249
References 249

Abstract

Sensory hair-cell development, function, and regeneration are fundamental processes that are challenging to study in mammalian systems. Zebrafish are an excellent alternative model to study hair cells because they have an external auxiliary organ called the lateral line. The hair cells of the lateral line are easily accessible, which makes them suitable for live, function-based fluorescence imaging. In this chapter, we describe methods to perform functional calcium imaging in zebrafish lateral-line hair cells. We compare genetically encoded calcium indicators that have been used previously to measure calcium in lateral-line hair cells. We also outline equipment required for calcium imaging and compare different imaging systems. Lastly, we discuss how to set up optimal imaging parameters and how to process and visualize calcium signals. Overall, using these methods, in vivo calcium imaging is a powerful tool to examine sensory hair-cell function in an intact organism.

INTRODUCTION

The zebrafish model system has many attributes that make it an ideal system for live, physiology-based fluorescence imaging. Zebrafish larvae are easy to genetically manipulate and are optically transparent (Patton & Zon, 2001). These features have allowed researchers to use the zebrafish system to express genetically encoded calcium indicators (GECIs) to visualize functional activity in vivo (Amsterdam, Lin, & Hopkins, 1995; Higashijima, Masino, Mandel, & Fetcho, 2003). While these advantages are now being rapidly applied to explore whole brain imaging in zebrafish, they are also applicable to studying hair cells of the auditory and vestibular organs in zebrafish (Ahrens, Orger, Robson, Li, & Keller, 2013).

Zebrafish are particularly well suited for studying hair cells because they have an external vestibular organ, the lateral line, which is easily accessible and readily amenable to live imaging. The lateral line is composed of groups of sensory hair cells arranged in rosette structures known as neuromasts. Neuromasts have hair bundles—apical projections from hair cells—that protrude from the surface of the body and detect water movements which is important for schooling, mating, and predator avoidance behaviors (Freeman, 1928). The accessibility of neuromast hair cells in vivo offers a significant advantage over mammalian auditory and vestibular systems, where the sensory organs are encased in a bony labyrinth and must be excised for study.

As a result of its easy accessibility, researchers have used the lateral line to study a diversity of topics, including the genetics of hearing loss, hair-cell development, function, ototoxicity, and regeneration (Kindt, Finch, & Nicolson, 2012; Lush & Piotrowski, 2014; Owens et al., 2008; Sarrazin et al., 2010). However, relatively few studies have comprehensively examined hair-cell function. Many studies have relied on styryl dyes such as DASPEI and FM 1-43 to indicate whether hair cells have mechanotransduction channels and are viable (Lush & Piotrowski, 2014; Owens et al., 2008). FM-dye labeling is a simple and straightforward method, but it lacks detailed and quantifiable information regarding hair-cell function. Alternatively, extracellular microphonic measurements or whole-cell recording of hair-cell currents have been used to examine hair-cell function (Corey et al., 2004; Olt, Johnson, & Marcotti, 2014; Ricci et al., 2013; Trapani & Nicolson, 2010). Although extremely powerful, these techniques are technically challenging and require highly specialized equipment and expertise. In the case of microphonics, measurements represent the summed response of all hair cells in a neuromast, and in the case of whole-cell recordings, measurements are from a single hair cell. On the other hand, techniques that use GECIs provide better spatial resolution. Specifically, calcium imaging methods enable measurements of mechanically evoked activity in all of the individual hair cells in a neuromast simultaneously in a reproducible and quantifiable manner. In addition, an imaging-based approach can be a more straightforward way to examine hair-cell activity for researchers more familiar with microscopy than electrophysiology.

In this chapter, we describe methods used to image calcium activity in hair cells of the zebrafish lateral line. We also discuss how to select a GECI and compare several indicators that have been used previously for calcium imaging in the lateral line. We then outline several suitable optical imaging setups and how to optimize settings for capturing calcium signals. Finally, we address how to process the captured images. We hope our description of these methods will extend the possibilities for hair-cell research in zebrafish.

1. CALCIUM INDICATOR SELECTION AND COMPARISON
1.1 SELECTING A CALCIUM INDICATOR

GECIs are a powerful tool for simultaneous measurement of mechanically evoked calcium responses in multiple hair cells. To appreciate the implications of these signals in the context of a hair cell, it is important to consider what a mechanically evoked calcium response represents. In hair cells, a mechanically evoked calcium response is the sum of many different calcium sources, mainly sites of mechanotransduction and synaptic transmission (Beurg et al., 2010, 2010; Gillespie & Müller, 2009; Jaalouk & Lammerding, 2009; LeMasurier & Gillespie, 2005; Vollrath, Kwan, & Corey, 2007). In response to mechanical hair-bundle deflection, calcium and other cations enter apically through mechanotransduction channels located on the sensory hair bundle. This apical influx of cations changes the receptor potential and triggers activation of voltage-gated calcium channels that then allow the calcium influx required for synaptic transmission at the basal end of the cell. Thus, calcium measurements can offer a good representation of the significant, activity-dependent signals in hair cells.

Currently there are two main types of GECIs. One type includes the fluorescence resonance energy transfer (FRET)-based indicators that rely on two spectrally distinct fluorescent proteins such as yellow fluorescent protein (YFP) and cyan fluorescent protein (CFP). The second type of GECIs is non-FRET-based and relies on a single fluorescent protein. Both types of indicators include a calmodulin–calcium binding domain that alters the fluorescence of the indicator in response to calcium concentration changes (Tian, Hires, & Looger, 2011). For FRET-based calcium indicators, such as cameleons, an increase in the YFP/CFP ratio indicates an increase in calcium (Miyawaki et al., 1997). For non-FRET-based indicators, such as GCaMPs, calcium binding directly alters fluorescence of the single fluorescent protein in a dose-dependent manner. FRET indicators require additional equipment beyond non-FRET-based indicators so that the two fluorescent proteins (eg, YFP and CFP) can be recorded simultaneously. These devices include a "beam splitter" or a dual-chip camera to capture the two emission spectra simultaneously, or alternatively, a high-speed filter turret for sequential image capture. Despite these equipment requirements, FRET-based indicators can be very powerful when measuring fluorescent signals in a live organism. The main advantage is that FRET-based indicators make it easy to identify and discard image acquisitions with movement artifacts. With FRET-based indicators symmetrical changes of both fluorescent signals represent movement, while reciprocal changes in the fluorescent signals indicate true calcium responses.

When choosing a calcium indicator, it is important to consider the calcium concentration that is being measured, including calcium levels at rest and peak calcium levels upon stimulation. Previous work has estimated baseline calcium levels to be approximately 100 nM within the hair-cell cytosol (Ikeda, Sunose, & Takasaka, 1993). Upon stimulation, calcium levels can transiently reach the micromolar range; the calcium concentration varies dramatically according to subcellular location, such as in the hair bundle and the ribbon synapse where peak calcium levels are estimated to be much higher than the cytosol (Bortolozzi, Lelli, & Mammano, 2008; Lumpkin & Hudspeth, 1998). Various reviews have discussed in detail how to appropriately select a GECI (Pérez Koldenkova & Nagai, 2013; Tian et al., 2011). In general, it is important to choose an indicator with a dynamic range appropriate to the calcium signals under investigation. We recommend trying more than one indicator as it is often unclear from in vitro work how a particular indicator will translate in vivo. In this chapter, we focus specifically on GECIs suitable for measuring cytosolic calcium responses in lateral-line hair cells.

For GECI selection in general, in addition to understanding expected changes in calcium concentration upon stimulation, it is also important to understand the calcium dynamics of a biologically relevant stimulus. Recent advances in calcium indicators have improved the speed and sensitivity of the indicators to allow detection of single action potentials (Chen et al., 2013). Although mature hair cells do not fire action potentials, indicator speed and sensitivity could be informative in detecting receptor potential oscillations, which can resonate at high frequency speeds similar to those of action potentials. Lateral-line hair cells are required to sense local water movement along the fish and function as velocity sensors important for rheotactic behaviors (Montgomery, Carton, Voigt, Baker, & Diebel, 2000). Some of these stimuli are likely to be relatively long and slow (seconds), similar to a stimulus in mammalian vestibular organs where hair cells are activated by fluid flow through the semicircular canals. In addition, the lateral line has been shown to encode frequencies up to 100–200 Hz, which is beyond the range current GECIs can reliably encode (Trump & McHenry, 2008; Weeg & Bass, 2002). Therefore, GECIs are suitable for some but not all relevant lateral-line stimuli. Overall, GECIs enable simultaneous activity measurements from populations of hair cells with reasonably high spatial and temporal efficiency.

1.2 COMPARISON OF CALCIUM INDICATORS

To date, four different GECIs have been expressed transgenically in zebrafish to measure calcium activity in hair cells: the FRET-based indicator cameleon D3 and the non-FRET-based indicators GCaMP3, RGECO (a red-shifted variant based on GCaMP3), and GCaMP7a (newer GCaMP variant) (Esterberg, Hailey, Coffin, Raible, & Rubel, 2013; Kindt et al., 2012; Maeda et al., 2014; Pujol-Martí et al., 2014). In these studies, the *myosinVIb* or *pou4f3 (brn3c)* promoters were used to express the calcium indicators in hair cells. In the case of the *brn3c* promoter, a *pou4f3:GAL4* driver line was used in combination with a *UAS:GCaMP7*a transgenic (Pujol-Martí et al., 2014).

Here we provide a side-by-side comparison of three (D3, RGECO, GCaMP3) of these indicators to assist researchers in GECI selection (Fig. 1). Each of these three indicators has a similar dissociation constant (Kd) (approximately 600, 480, and 540 nM respectively), and dynamic range (ratio of calcium-saturated fluorescence to calcium-free fluorescence: 10, 12, and 10, respectively) (Palmer et al., 2006; Tian et al., 2009; Zhao et al., 2011). To examine the activity dependence of each of these different indicator lines, we mount, immobilize, and mechanically stimulate fish identical to the methods described in a later chapter "Physiological recordings from the zebrafish ear and lateral line" in this volume as well as in a previously published methods chapter detailing electrophysiological recordings in lateral-line hair cells (Trapani & Nicolson, 2010). For our preparation, we anesthetize and pin larvae to a Sylgard recording chamber. Larvae are then paralyzed by cardiac injection of alpha-bungarotoxin. Proper pinning and paralysis are essential for high-quality image acquisitions. To ensure reproducible and quantifiable mechanically evoked calcium responses, it is important to prepare and calibrate the fluid jet for each sample. We set our stimulus intensity to produce a near-saturating stimulus. To do this, we position the fluid jet pipette 100 μm away from the edge of the neuromast and limit the pipette tip opening to 30—40 μm. Then, using *differential interference contrast* optics we locate the tips of the kinocilia and apply a short (200 ms) fluid stimulus and measure the displacement of the kinocilia tips using a camera and micrometer. For these measurements, on our imaging systems we utilize a built in scale bar option within the software to measure the distance deflected. Alternatively, we calibrate our imaging field with a stage micrometer slide to measure deflection distances. The fluid jet should displace the tips of the kinocilia ~2—3 μm. This deflection distance is sufficient to produce a near-saturating stimulus with respect to hair-bundle deflection. Distance and intensity (see Section 2.4) of the fluid jet can be adjusted slightly to achieve the proper tip-deflection distance. We caution against using stronger stimuli which are potentially damaging to hair cells and can cause motion artifacts that lead to false fluorescence intensity changes. For our comparison of these indicators, we chose three different stimulus durations: 0.1-, 2-, and 4-s step stimuli (static hair-bundle deflection for the stimulus duration). Hair cells in neuromasts of the primary posterior lateral line are oriented to respond to either an anterior- or a posterior-directed stimulus. For simplicity, in Fig. 1 and Fig. 4 we have only presented data for an anterior stimulus (stimulus towards the anterior of the fish). The range of stimuli was selected to show the low (0.1 s) and high (4 s) end of the dynamic range of the indicators. In general it is good practice to choose stimuli that fall within the dynamic range of the indicator to ensure that the

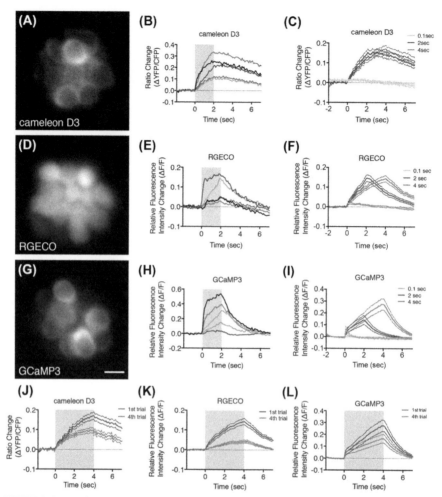

FIGURE 1 Comparison of cameleon D3, RGECO, and GCaMP3 calcium signals in zebrafish hair cells.

(A, D, G) Widefield images of neuromasts expressing cameleon D3, RGECO, and GCaMP3, respectively. (B, E, H) Traces of five representative, individual hair cells for each indicator. (C, F, I) Average calcium response for each indicator in response to a 0.1, 2, and 4 s stimulus, $n = 10-15$ hair cells. (J, K, L) Calcium signals in response to a 4-s stimulus using each indicator after the first and fourth trial. Larvae were imaged at day 4. *Gray boxes* denote stimulus period. Scale bar = 5 μm. (See color plate)

majority of responses are detected and the indicator is not saturated. We also assayed the photostability of each indicator after repeated image acquisitions.

For our indicator comparison, images were acquired on an upright fixed-stage Nikon widefield microscope with settings and equipment similar to those described in Section 2 and Table 1. A Hamamatsu ORCA-D2 dual-chip camera was used to simultaneously image CFP and YFP for the D3 cameleon indicator. One commonality between all indicators is that the calcium signals detected are extremely heterogenous (Fig. 1B, E, and H) (Kindt et al., 2012; Olt et al., 2014). This heterogeneity is likely due to a combination of the developmental stage and the functional properties of each individual hair cell, as well as the exact imaging plane examined. Using these indicators on average, we found that GCaMP3 was the most sensitive and offered the best dynamic range (Fig. 1C, F, and I). GCaMP3 was able to detect the 0.1 s stimulus reliably and did not saturate even after 4 s of stimulation. Lastly we found that GCaMP3 and D3 are more suitable for repeated imaging than RGECO (Fig. 1J–L). RGECO suffers the same poor photostability as the red fluorescent proteins upon which it is based and is more susceptible to photobleaching than the GCaMP series (Akerboom et al., 2013; Drobizhev, Makarov, Tillo, Hughes, & Rebane, 2011). However, RGECO remains a very useful calcium indicator because it is red-shifted and therefore can be used in combination with other GFP lines to provide additional subcellular information (Maeda et al., 2014). From this comparison we conclude that all of the previously described indicators are suitable for measuring mechanically evoked calcium responses in lateral-line hair cells.

Table 1 Parameters for Three Zebrafish Hair-Cell Calcium Imaging Systems

	Nikon Widefield	Bruker Swept-field Confocal	Nikon C2 Line-Scanning Confocal
Detector	Camera — Hamamatsu ORCA-R2 Gain 50/100	Camera — Rolera EM-C^2 EM gain 3700/4000	PMT — Nikon Nikon C2 Gain 75/255
Bin	2 × 2	2 × 2	—
ROI	50 × 50 µm	50 × 50 µm	50 × 50 µm
X-Y resolution	0.43 µm/pixel	0.53 µm/pixel	0.33 µm/pixel
Slit size	—	35 µm slit	—
Light source power	Intensilight (mercury) 25% with cy3 filter	Solid state 561 nm laser 5%	Solid state 561 nm laser 5%
Speed	10 fps	10 fps	7.5 fps
Optical section	Largest, worst	Variable depending on pinhole or slit size	Thinnest, best
Max speed	100 fps	100 fps	10 fps
Main advantages	Inexpensive	Fast, high optical resolution	Highest optical resolution
Main disadvantages	Low optical resolution	Cost	Slow speed, cost, photobleaching

Of the three, GCaMP3 is the most sensitive, has the best dynamic range, and is extremely photostable.

1.3 VALIDATING A RELEVANT CALCIUM SIGNAL

A primary consideration when measuring calcium signals in a live organism is to ensure that changes in fluorescence intensity are not due to movement artifacts but rather represent genuine activity-dependent changes in calcium (Greenberg & Kerr, 2009). This represents one of the major sources of error in measuring calcium levels in all cells. As mentioned above, the use of FRET-based indicators can help solve this problem, but if using a FRET-based indicator is not an option, there are several ways to determine whether fluorescence changes in an image acquisition sequence represent an activity-dependent calcium signal or movement artifacts.

Many variables can cause movement artifacts. In most cases, movement occurs because larvae are not sufficiently immobilized, leading to unwanted drift in X, Y, or Z directions. In other cases, debris from a clogged fluid jet blown over the neuromast during stimulus and image acquisition can cause false changes in fluorescence. Regardless of how well the sample is prepared, some movement will be unavoidable. Therefore, it is important to carefully examine each image acquisition sequence. Acquisitions with excessive Z-movements cannot be corrected by postprocessing and must be eliminated. Acquisitions with minor X-Y movement can be corrected (see Section 3). When evaluating image acquisitions, it is also important to observe the shape and kinetics of the apparent response. If fluorescent signals are the result of movement artifacts, the rise and fall of the fluorescent signal will be abrupt and plateau sharply (this can also occur if the indicator is saturated) (Fig. 2). By contrast, in the case of a true mechanically evoked calcium response for measurements made in the cytosol, the fluorescent signal will rise for the duration of the stimulus and decline slowly after the stimulus has ended (Fig. 2).

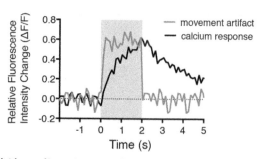

FIGURE 2 Distinguishing artifacts from functional calcium signals.

Example of traces resembling a movement artifact is shown in gray, while a true, mechanically evoked hair-cell calcium response is shown in black. Note the slow rise and decay in the black traces versus the abrupt rise and fall in signal in the gray trace. *Gray box* denotes stimulus period.

In addition to careful sample preparation and acquisition evaluation, it is also possible to verify the legitimacy of an observed signal by repeating the stimulus while the fish is bathed in a mechanotransduction channel blocker such as amiloride or dihydrostreptomycin, or agents that disrupt the tip links that are required for mechanosensitive function (Assad, Shepherd, & Corey, 1991; Ricci, 2002). A 10-min incubation with 5 mM BAPTA tetrasodium salt, and subsequent washout is sufficient to break all tip links and eliminate true mechanically evoked calcium responses (Kindt et al., 2012). It is important to note that 30 mins after BAPTA washout, mechanosensitive responses will start to return as the tip links reform. Other agents, such as amiloride (1 mM) or dihydrostreptomycin (1 mM), may be used to block mechanotransduction but must be added to the solution in the fluid jet and maintained in the bath during image acquisition. Fluorescence changes resulting from movement artifacts will be unaffected by treatment with mechanotransduction channel blockers, and associated image acquisitions can be discarded. Overall, diligent sample preparation, careful examination of image acquisitions, and drug controls will help ensure that any fluorescent signals measured are true, activity-dependent calcium changes in lateral-line hair cells.

2. IMAGING SYSTEMS AND OPTIMAL PARAMETERS

It is ideal but not necessary to have a microscope dedicated exclusively to imaging fluorescence activity of zebrafish hair cells. The options for microscopes and associated accessories are numerous. In this section, we outline three microscope systems suitable for imaging-evoked calcium responses in zebrafish hair cells: widefield, swept-field confocal (SFC), and point-scanning confocal systems. In addition we describe the pros and cons of each imaging system, as well as recommended parameters. Despite the differences in the microscope setups listed above, there are many components in common and the general concepts underlying parameter optimization are the same. Below in Section 2.1 and in Fig. 3, we describe, in general terms, equipment components that are required for all imaging systems. The equipment listed below highlights only a subset of the many options available and the provided parameters are designed to offer a starting point for optimizing imaging settings.

2.1 GENERAL MICROSCOPE AND EQUIPMENT REQUIREMENTS

1. Upright rather than inverted microscope body
    Olympus BX51W1, Nikon Eclipse FN1, Zeiss Axioexaminer, Leica DM6FS
2. 60x water immersion objective with a high NA (0.9–1.1) and suitable working distance of ~2 mm
3. Fixed stage
    Sutter MT-1000 or MT1078, Burleigh Gibraltar Platforms, Prior Z-Deck

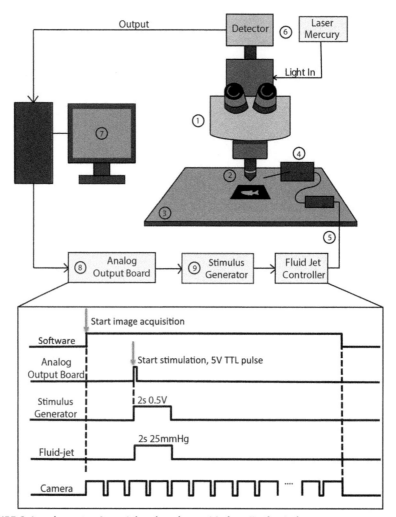

FIGURE 3 Imaging setup for calcium imaging and hair-cell stimulation.
Schematic of the imaging system requirements outlined in Section 2.1. (1) upright microscope; (2) water immersion objective; (3) fixed stage; (4) manipulator to position stimulation pipette; (5) fluid jet (and controller) to stimulate hair cells; (6) light source and detector; (7) imaging software; (8) analog output board; (9) stimulus generator. The bottom panel outlines the signals used to coordinate imaging with hair-cell stimulation. The software starts the image acquisition. This triggers the camera to capture at a given frame rate. During the time series to start hair-cell stimulation, a 5 V TTL pulse is sent from the software at a defined frame rate to the stimulus generator using an analog output board. The stimulus generator converts the TTL pulse into an analog voltage signal with a defined duration and magnitude that is then sent to the fluid jet controller. The fluid jet controller converts the voltage signal into fluid flow by applying positive or negative pressure on the fluid within the stimulating pipette. This, in turn, deflects hair bundles and activates hair cells.

4. Manipulator to position fluid jet or piezo pipette for stimulation
   Sutter MP-225, Scientific Patchstar, Sensapex triple axis, Burleigh PCS 6200
5. Fluid jet or piezo to stimulate hair cells
   Fluid jet: ALA HSPC-1 and HSPV-P
   Piezo: PI instruments PICMA P-882.11-888.11 series plus controller
6. Light source and detector (see Table 1 for detector examples)
   Widefield: LED or mercury light (Nikon Intensilight, Excelitas X-Cite XLED1 and X-cite120Q) and camera for detection
   Confocal: Lasers and photomultiplier tubes (PMTs) or camera for detection
7. Imaging software
   Molecular Devices Metamorph, Micro-Manager, Zeiss ZEN, Nikon Elements
8. Digital to analog output board to send a transistor−transistor logic (TTL) trigger from the imaging software to a stimulus generator
   Consult your microscope software representative
   Example: National Instruments PCI 6700 series
9. Stimulus generator to trigger fluid jet or piezo
   Tektronix AFG1000-3000 series, BK Precision 4000 series
10. Vibration isolation table

2.2 CHOOSING A SPECIFIC IMAGING SYSTEM

In addition to these essential components that are required for all imaging systems, a specific imaging system must be chosen. This chapter highlights three different imaging systems: widefield, SFC, and point- or line-scanning confocal systems (Table 1). A widefield imaging system is the simplest and least-expensive system, but the main drawback is the low optical or axial resolution. Point- or line-scanning confocal systems are the most expensive and offer the highest optical resolution, but at the expense of speed. SFC imaging systems, on the other hand, can scan samples using pinholes or slits to balance optical resolution and speed respectively (Castellano-Muñoz, Peng, Salles, & Ricci, 2012). This is similar in some aspects to spinning-disk systems, except the pinhole and therefore optical resolution and light collection are fixed on a spinning-disk system while the pinhole is adjustable on an SFC system.

Widefield and SFC systems both require a camera to detect signals, and the overall speed of image acquisition is limited mainly by camera choice. Current camera-based systems are faster than point-scanning confocal systems that use PMTs to detect signals. Camera choices are ever expanding and many options are available. Commonly used CMOS, EM-CCD, and CCD cameras are all adequate for the imaging described in this chapter. It is best practice to request a demonstration of a specific camera from the manufacturer or vendor to determine if it is suitable for your imaging requirements. When selecting a camera, a relatively high frame rate (10−100 fps) and low noise is desirable, similar to other calcium imaging applications. In the case of zebrafish hair cell imaging, the pixel size is another important consideration. Zebrafish hair cells are quite small: lateral-line hair cells are approximately 5 μm across × 10 μm apex-to-base (magnified 60× = 300 × 600 μm). Many cameras are designed for whole brain, full-field imaging and have a relatively large pixel size ~16 μm. Using a camera with a 16 μm pixel, each hair cell would be represented by (300/16 × 600/16) 18.74 × 37.5 pixels. Cameras with 6–8 μm pixel size offer a more suitable with (300/6 × 600/6) 50 × 100 pixels per hair cell. This is especially true for calcium imaging, as binning is almost always a required imaging parameter and will result in even fewer pixels represented per hair cell. If necessary, large pixel size can be overcome using a magnifying coupling lens to decrease pixel size and increase resolution.

In addition to camera-based imaging systems, an alternative is a point- or line-scanning confocal system. Although these systems offer the best optical sectioning, they use PMTs to detect signals, which have a slower frame rate than cameras. Although faster and more sensitive PMTs are being rapidly developed, camera-based imaging currently offers the fastest acquisition speeds. Additionally, hair-cell calcium imaging is possible with 2-photon systems although due to the superficial location of lateral-line hair cells, the depth advantages gained using a 2-photon system are not utilized and these systems are not discussed further here. With rapid advances in optics and image processing, we expect an ever-greater variety of imaging components, beyond the systems we have discussed, that will satisfy the resolution and speed requirements for hair-cell calcium imaging.

2.3 DETERMINING OPTIMAL IMAGING PARAMETERS

Any imaging setup, including the three systems described here, is useless without determining proper parameters. For a given system, there are optimal settings that balance the amount of light (laser, mercury, or LED) reaching the sample, resolution at the detector (camera, PMT), and speed of acquisition. In all systems, light intensity and detector settings must be identified that are able to detect changes in fluorescent signals at a frame rate relevant for hair-cell activity while minimizing photobleaching. To aid in this process, we have compared the three imaging systems listed in Table 1 using *Tg[myo6b:RGECO]* larvae. For this comparison, we fixed the exposure time at 100–150 ms. For spontaneous and evoked calcium responses in zebrafish hair cells, a frame rate of at least 10 Hz (100 ms exposure time) is required to detect calcium-dependent activity. Light intensity should be sufficiently high to have adequate signal above noise yet low enough to prevent pixel saturation. For this comparison, an 8-s streaming time series was acquired on each system. A 2-s step stimulus (0.5 V, 25 mmHg fluid jet, 2 μm kinocilia tip deflection) was triggered 2 s into the acquisition. Traces from all three systems are shown in Fig. 4.

Comparing the three imaging systems, it is apparent that patterns of calcium signals are similar: all of the same hair cells respond and the overall pattern of heterogeneity is conserved among systems. For example, on all three systems cell 5 has the largest calcium response, while cell 3 has one of the smallest calcium responses. Despite these similarities, there are differences in the overall magnitude of the calcium signals, the signal-to-noise ratio (SNR), and the kinetics of the calcium signals.

2. Imaging systems and optimal parameters

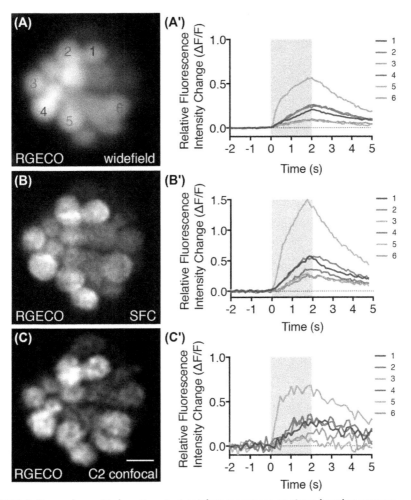

FIGURE 4 Comparison of hair-cell evoked calcium responses on three imaging systems.

(A, B, C) Tg[myo6b:RGECO] images from the same neuromast acquired on a Nikon widefield microscope, Bruker SFC confocal microscope, and Nikon C2 confocal microscope, respectively. (A′, B′, C′) Evoked calcium responses in response to a 2-s step deflection measured on each system using the parameters outlined in Table 1. Larvae were examined at day 5, scale bar = 5 μm. (See color plate)

All these differences stem mainly from differences in optical resolution between the systems, which increases from widefield to SFC to point-scanning systems (Fig. 4A′–C′). With higher optical resolution, it is more likely that the calcium signals observed are in the imaging plane, rather than due to out-of-focus light from other planes. Increased optical resolution comes at a cost because it decreases the amount of calcium indicator imaged per pixel.

With regards to overall signal magnitude, calcium imaging using the SFC system revealed the largest signals. Both the SFC and widefield systems have a high SNR while the SNR is the lowest on the point-scanning systems when the optical section is smallest (Fig. 4C′). The SNR is high on the widefield system because there is more indicator per optical section compared to the point-scanning confocal system. A large optical section boosts the signal detection but also collects a lot of out-of-focus light which can negatively impact signal magnitude. The increase in signal and high SNR gained by a thicker optical section in widefield (Fig. 4A′) and SFC (Fig. 4B′) systems is at a cost. It is less likely that the signals detected are restricted to the imaging plane. This is the main advantage of point-scanning confocal systems with high optical resolution. Although thinner sectioning and higher optical resolution can impair signal detection, it is advantageous for detecting calcium signals restricted to a particular imaging plane. This trade-off is most apparent in Fig. 4, cell 5. It is very likely that the large calcium signal in cell 5 is because the optical section contains synaptic ribbons, a site of rapid and sizable calcium concentration changes (Bortolozzi et al., 2008). As the optical section tightens from Fig. 4A′ to B′ to C′, the onset time of the calcium signal becomes faster and more representative of a synaptic plane where calcium signals are large and occur rapidly. Overall, of the three systems, SFC offers an optimal compromise between optical resolution and signal detection.

2.4 SYNCHRONIZING A STIMULUS WITH IMAGE ACQUISITION

One major challenge to examining activity while imaging is coordinating image acquisition with stimulus delivery. Major imaging companies are now aware of this gap, especially with the upsurge of neuroscience research and are working to address this issue. The main limitation lies in constraints built into imaging software. We outline a method for image acquisition and stimulus coordination in Fig. 3. In this example the imaging software is programmed to send a 5 V TTL output during the time series image acquisition. This generally requires a digital to analog output board in order for the imaging software to send the 5 V TTL output. The analog output board relays this 5 V TTL output to an external device, such as a stimulus generator. The stimulus generator accepts the 5 V signal as an input or trigger and generates a new output or stimulus waveform. For example, the stimulus generator can be configured to create a variety of stimuli such as steps and ramps, and to control the intensity and duration of the stimulus. The stimulus configured in the stimulus generator is then output as an analog voltage signal which is sent to the fluid jet or piezo controller where the voltage levels and durations are converted into fluid flow or piezo movement and hair-cell stimulation. This pathway and timing of communication are outlined in Fig. 3.

3. IMAGE PROCESSING

There are numerous methods for processing and analyzing raw calcium imaging data. In this section, we discuss the importance of image registration and outline an approach to do this using the free, open-source image processing software ImageJ (Section 3.1, Schneider, Rasband, & Eliceiri, 2012). Then we describe two approaches for visualizing calcium responses. The first is a simple and straightforward method to detect calcium signals using ImageJ (Section 3.2; Schneider et al., 2012). The second method describes algorithms to visualize the spatial dynamics of calcium signals that can be accomplished using MATLAB R2014b software (Section 3.3; The MathWorks, Natick, MA, USA).

3.1 IMAGE REGISTRATION USING IMAGEJ

During image acquisition in a live organism, some movement is unavoidable (example in Fig. 5A). While it is possible to correct for movement in the lateral (X-Y) direction, it is difficult to correct for large changes in the Z direction. When evaluating the quality of the acquired images, it is important to discard acquisitions with excessive Z-drift and perform proper controls (see Section 1) to ensure the signals measured are true calcium signals. Lateral movements (X-Y) can also cause artifacts that misrepresent calcium signals, but can be corrected through image registration. Because measurements of fluorescence changes are made throughout the image acquisition, at the same hair cell within a region of interest (ROI), it is

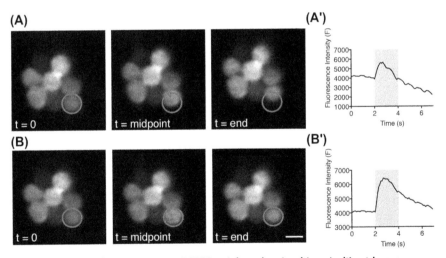

FIGURE 5 Live imaging of neuromast RGECO calcium signals with and without image registration.

(A) Representative images captured at the onset, midpoint, and end of image acquisition period before image registration. (A′) Temporal curve of the fluorescence intensity before image registration. (B) Representative images captured at the onset, midpoint, and end of image acquisition period after ImageJ StackReg image registration. (B′) Temporal curve of the fluorescence intensity after StackReg image registration. *Red circles* (gray in print versions) indicate the same ROI used for calculating the fluorescence intensity. *Gray bars* in (A′) and (B′) indicate the 2-s fluid jet stimulus. Scale bar = 5 μm.

essential to perform image registration to remove lateral movements and keep the hair cell within the specified ROI (see example in Fig. 5). One simple method to perform image registration is to utilize the *StackReg* plugin in ImageJ (Thévenaz, Ruttimann, & Unser, 1998). *StackReg* functions by performing recursive alignment of the acquired image sequence. The *StackReg* plugin requires that a second plugin, named *TurboReg*, is installed, which can be downloaded from the website, http://bigwww.epfl.ch/thevenaz/turboreg/. The underlying principle is to use each slice within a time series as the template to align the next slice, so that the alignment proceeds iteratively. An example of this registration is shown in Fig. 5, where the three images indicate the onset, midpoint, and end of an image acquisition. After image registration, the upward movement (Y-direction) is corrected and the ROI marked by the red circle maintains its original position over time. Fig. 5B′ demonstrates the temporal curve of the fluorescence intensity after registration. As indicated in Fig. 5, without image registration, the signal will be greatly degraded (compare Fig. 5A′ vs. Fig. 5B′) or provide inaccurate information regarding the calcium signals.

3.2 SIGNAL DETECTION AND REPRESENTATION IN IMAGEJ

After image registration, to obtain a quick readout of the temporal changes of the calcium signals in ImageJ, select an ROI and go to *Image > Stacks > plot z-axis Profile*. A new window will appear, showing the mean of the fluorescence intensity of the selected ROI over the time series (Fig. 6). If the *Live* (marked by the arrow) option is selected, the displayed temporal curve will change as the ROI is moved. The values can be copied (select *copy* option) and saved for further processing.

The temporal curve of the fluorescence intensity shows how the signals change during the time series. To have a better comparison among different ROIs or fluorescence intensities obtained with different imaging systems, the relative fluorescence intensity change ($\Delta F/F_0$) is more commonly used. However, when ROIs have different baseline fluorescence intensity (F_0), that reflects resting calcium availability rather than indicator concentration, $\Delta F/F_0$ may not be an appropriate measure. As shown in Fig. 7, three different ROIs with varied baseline intensity have a similar fluorescence intensity change ΔF (Fig. 7B″), but their relative values ($\Delta F/F_0$) vary considerably. Therefore, we suggest taking both of these measurements ($\Delta F/F_0$ and ΔF) into consideration for signal magnitude comparison.

3.3 SPATIAL DETECTION AND VISUALIZATION USING MATLAB

A major advantage of optical imaging of calcium signals is high spatial resolution. In the case of genetically encoded indicators expressed in lateral-line hair cells, it is possible to examine the responses of all hair cells within a neuromast at once. To better visualize the signal distribution patterns of calcium within the whole

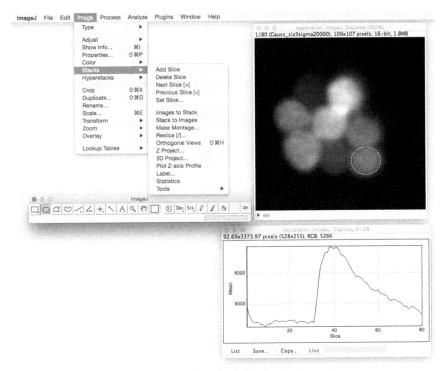

FIGURE 6 Plotting the temporal curve of a selected ROI using ImageJ.
To view the intensity changes over time within a ROI, draw a ROI, then under Image-Stacks, select Plot z-axis profile. Select Live to refresh the graph while moving the ROI between cells. Select save to save the intensity values within an ROI.

neuromast upon fluid jet stimulus, we have outlined the following image processing algorithms that can be used in MATLAB (Yao & Zhao, 2008):

1. Create the baseline image

The images obtained before applying fluid jet stimulus are averaged pixel by pixel over the whole prestimulus period. If there are m frames of prestimulus images, then in the baseline image the fluorescence intensity F_0 at the pixel (x,y) is calculated as:

$$F_0(x,y) = \frac{1}{m}\sum_{t=1}^{t=m} F_t(x,y) \qquad [1]$$

2. Temporally bin the entire image sequence

To increase the SNR, the whole image sequence is binned temporally. For example, if there are n frames of images in the whole sequence, including the

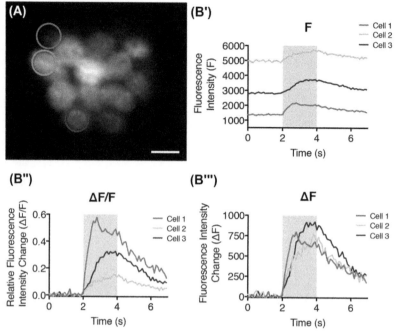

FIGURE 7 Temporal representation of RGECO calcium signal at three selected ROIs.
(A) A representative RGECO neuromast image with three ROIs (hair cells) indicated by *red, green,* and *blue circles* respectively. (B′) Temporal curve of the fluorescence intensity. (B″) Temporal curve of the relative fluorescence intensity change. (B‴) Temporal curve of the fluorescence intensity change. Scale bar = 5 μm. (See color plate)

prestimulus, stimulus, and poststimulus periods. We bin every k images into one to get n/k images, as shown in Fig. 8A and as follows:

$$F_i(x,y) = \frac{1}{k}\sum_{t=(i-1)*k+1}^{t=i*k} F_t(x,y) \quad (i = 1, 2, 3 \ldots n/k) \quad [2]$$

3. Generate images of fluorescence intensity changes

The baseline image is subtracted from each frame of binned images to obtain images that represent the change in fluorescent signal from baseline. In this way, the frames prior to the stimulus are defined to have no signal, as in:

$$\Delta F_i(x,y) = F_i(x,y) - F_0(x,y) \quad (i = 1, 2, 3 \ldots n/k) \quad [3]$$

To better visualize the fluorescence intensity changes, the changes can be scaled and encoded as a lookup table with blue indicating decreased fluorescence intensity

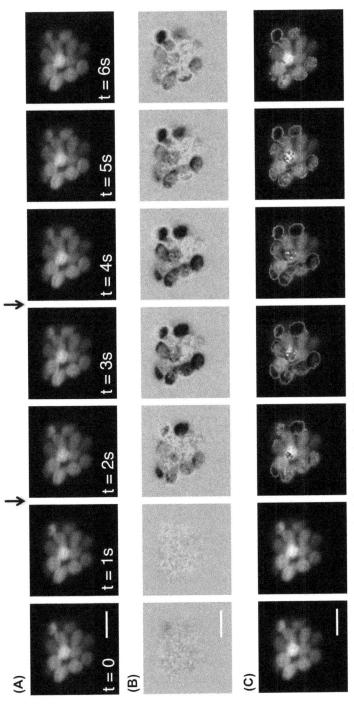

FIGURE 8 Spatiotemporal patterns of calcium signal using RGECO.

(A) Temporal binning of the whole image sequence; 7-s time series, acquired at 10 Hz, binned every five frames. A 2-s fluid jet stimulus was applied 2 s into the time series. Half of the binned frames are displayed. (B) Baseline-subtracted, colorized fluorescent-signal images. (C) Combined images created by overlaying the colorized fluorescent-signal images in B onto the baseline grayscale image. The calcium signal distributions can be visualized over the entire neuromast and in individual hair cells. The two *arrows* indicate the fluid jet stimulation duration. The scale bar = 10 μm. Images were processed in MATLAB. (See color plate) and red indicating an increase in intensity, as shown in Fig. 8B. By overlaying the baseline grayscale image with colorized signal images (Fig. 8C) it is possible to determine the fluorescence intensity changes within a neuromast and each individual hair cell.

SUMMARY

In this chapter, we have described methods for measuring mechanically evoked calcium responses in zebrafish lateral-line hair cells. We have compared the three published transgenic zebrafish lines that have been previously used to examine calcium responses in zebrafish hair cells. In addition, we have identified components required to set up a suitable imaging system for these calcium measurements. Then we performed a side-by-side comparison of three suitable imaging systems and provided suggested parameters for each system. Lastly, we have outlined two methods (one simple and one more complex) to visualize and measure calcium signals obtained from lateral-line hair-cell recordings. Overall this information should benefit any researcher looking to examine calcium responses in hair cells of the zebrafish.

DISCUSSION

The live-imaging methods described in this chapter represent an extremely valuable tool in assessing hair-cell function in zebrafish. Advances in imaging and development of faster and more sensitive calcium indicators, along with new transgenic zebrafish lines continue to push these techniques forward (Chen et al., 2013).

In addition to calcium indicators with increased sensitivity, recent work has begun to take advantage of the fact that GECIs can be targeted subcellularly. Here, activity can be examined in distinct subcellular domains, as has been accomplished in zebrafish to study the role of the mitochondria in hair-cell ototoxicity (Esterberg, Hailey, Rubel, & Raible, 2014). It is possible that localized calcium indicators will provide critical information regarding function of the apical hair bundle and the ribbon synapse, two subcellular compartments where local calcium dynamics are thought to play a critical role in hair-cell function (Ceriani & Mammano, 2012).

Recent work has created calcium indicators that function at different wavelengths, such as the blue-shifted BGECO and the red-shifted RGECO and RCaMP (Akerboom et al., 2013; Zhao et al., 2011). Using these new indicators, it is possible to produce double transgenic larvae that express two different colored calcium indicators in different structures, as has been done to examine concurrent calcium changes in the hair-cell cytosol and in the mitochondria in response to ototoxic insults (Esterberg et al., 2014). For faster stimulus-dependent hair-cell activity, this two-color imaging can be accomplished using the systems described in this chapter using a beam splitter (Photometrics dual-view) or a dual-chip camera (Hamamastu

ORCA-D2) which allows imaging of two channels simultaneously during a stimulus (Akerboom et al., 2013). Two-color imaging will also be useful in future work where activity measurements will likely extend past the sensory hair cells to analyses of the hair-cell afferent and efferent neurons. For example, a double transgenic zebrafish that expresses a red indicator in hair cells and green indicator in the afferent fiber could be used to image pre- and postsynaptic calcium simultaneously. Such an approach could provide critical information about how sensory hair cells encode information within the context of a sensory circuit and will further our knowledge of auditory and vestibular systems.

In addition to 2-color imaging and circuit analysis, future work will also benefit from rapid imaging in the Z-axis; in this case, data can be acquired with high spatial information in 3D and also over time, resulting in 4D image acquisitions. This can be accomplished using a piezoelectric motor attached to an objective to allow rapid imaging in multiple Z-planes so that differences in activity across the Z-axis can be recorded (Göbel, Kampa, & Helmchen, 2007). In addition to piezoelectric motors, light-sheet microscopy now allows large volumes to be imaged at relatively fast speeds at single-cell resolution (Ahrens et al., 2013). Light-sheet microscopy is particularly attractive because samples are protected from phototoxicity by low laser intensities; this feature will provide a powerful way to study functional activity repeatedly over large developmental windows.

Finally, in addition to the collection of calcium indicators currently available, there are a wealth of indicators now available to examine other functional hair-cell properties; for example, indicators to detect glutamate, vesicle release, and voltage signals (Marvin et al., 2013; Odermatt, Nikolaev, & Lagnado, 2012; St-Pierre et al., 2014). Using the general methods described in this chapter, these valuable optical tools could be used to provide critical information regarding how hair cells develop, function, succumb to ototoxic insults, and regenerate, all in an intact organism.

ACKNOWLEDGMENTS

We would like to acknowledge Alisha Beirl, Elyssa Monzack, Alain Silk, Josef Trapani, and Michael Waltman for their comments on the manuscript. This work was supported by the Intramural Research Program of the NIH, NIDCD 1ZIADC000085-01.

REFERENCES

Ahrens, M. B., Orger, M. B., Robson, D. N., Li, J. M., & Keller, P. J. (2013). Whole-brain functional imaging at cellular resolution using light-sheet microscopy. *Nature Methods, 10*, 413–420.

Akerboom, J., Carreras Calderón, N., Tian, L., Wabnig, S., Prigge, M., Tolö, J. ... Looger, L. L. (2013). Genetically encoded calcium indicators for multi-color neural activity imaging and combination with optogenetics. *Frontiers in Molecular Neuroscience, 6*.

Amsterdam, A., Lin, S., & Hopkins, N. (1995). The Aequorea victoria Green fluorescent protein can be used as a reporter in live zebrafish embryos. *Developmental Biology, 171*, 123–129.

Assad, J. A., Shepherd, G. M. G., & Corey, D. P. (1991). Tip-link integrity and mechanical transduction in vertebrate hair cells. *Neuron, 7*, 985–994.

Beurg, M., Michalski, N., Safieddine, S., Bouleau, Y., Schneggenburger, R., Chapman, E. R. … Dulon, D. (2010). Control of exocytosis by synaptotagmins and otoferlin in auditory hair cells. *The Journal of Neuroscience, 30*, 13281–13290.

Bortolozzi, M., Lelli, A., & Mammano, F. (2008). Calcium microdomains at presynaptic active zones of vertebrate hair cells unmasked by stochastic deconvolution. *Cell Calcium, 44*, 158–168.

Castellano-Muñoz, M., Peng, A. W., Salles, F. T., & Ricci, A. J. (2012). Swept field laser confocal microscopy for enhanced spatial and temporal resolution in live-cell imaging. *Microscopy and Microanalysis, 18*, 753–760.

Ceriani, F., & Mammano, F. (2012). Calcium signaling in the cochlea — molecular mechanisms and physiopathological implications. *Cell Communication and Signaling, 10*, 20.

Chen, T.-W., Wardill, T. J., Sun, Y., Pulver, S. R., Renninger, S. L., Baohan, A. … Kim, D. S. (2013). Ultra-sensitive fluorescent proteins for imaging neuronal activity. *Nature, 499*, 295–300.

Corey, D. P., García-Añoveros, J., Holt, J. R., Kwan, K. Y., Lin, S.-Y., Vollrath, M. A. … Zhang, D. S. (2004). TRPA1 is a candidate for the mechanosensitive transduction channel of vertebrate hair cells. *Nature, 432*, 723–730.

Drobizhev, M., Makarov, N. S., Tillo, S. E., Hughes, T. E., & Rebane, A. (2011). Two-photon absorption properties of fluorescent proteins. *Nature Methods, 8*, 393–399.

Esterberg, R., Hailey, D. W., Coffin, A. B., Raible, D. W., & Rubel, E. W. (2013). Disruption of intracellular calcium regulation is integral to aminoglycoside-induced hair cell death. *The Journal of Neuroscience, 33*, 7513–7525.

Esterberg, R., Hailey, D. W., Rubel, E. W., & Raible, D. W. (2014). ER–mitochondrial calcium flow underlies vulnerability of mechanosensory hair cells to damage. *The Journal of Neuroscience, 34*, 9703–9719.

Freeman, W. (1928). The function of the lateral line organs. *Science, 68*, 205.

Gillespie, P. G., & Müller, U. (2009). Mechanotransduction by hair cells: models, molecules, and mechanisms. *Cell, 139*, 33–44.

Göbel, W., Kampa, B. M., & Helmchen, F. (2007). Imaging cellular network dynamics in three dimensions using fast 3D laser scanning. *Nature Methods, 4*, 73–79.

Greenberg, D. S., & Kerr, J. N. D. (2009). Automated correction of fast motion artifacts for two-photon imaging of awake animals. *Journal of Neuroscience Methods, 176*, 1–15.

Higashijima, S., Masino, M. A., Mandel, G., & Fetcho, J. R. (2003). Imaging neuronal activity during zebrafish behavior with a genetically encoded calcium indicator. *Journal of Neurophysiology, 90*, 3986–3997.

Ikeda, K., Sunose, H., & Takasaka, T. (1993). Effects of free radicals on the intracellular calcium concentration in the isolated outer hair cell of the Guinea pig cochlea. *Acta Otolaryngologica (Stockholm), 113*, 137–141.

Jaalouk, D. E., & Lammerding, J. (2009). Mechanotransduction gone awry. *Nature Reviews. Molecular Cell Biology, 10*, 63–73.

Kindt, K. S., Finch, G., & Nicolson, T. (2012). Kinocilia mediate mechanosensitivity in developing zebrafish hair cells. *Developmental Cell, 23*, 329–341.

LeMasurier, M., & Gillespie, P. G. (2005). Hair-cell mechanotransduction and cochlear amplification. *Neuron, 48*, 403−415.

Lumpkin, E. A., & Hudspeth, A. J. (1998). Regulation of free Ca^{2+} concentration in hair-cell stereocilia. *The Journal of Neuroscience, 18*, 6300−6318.

Lush, M. E., & Piotrowski, T. (2014). Sensory hair cell regeneration in the zebrafish lateral line. *Developmental Dynamics, 243*, 1187−1202.

Maeda, R., Kindt, K. S., Mo, W., Morgan, C. P., Erickson, T., Zhao, H. ... Nicolson, T. (2014). Tip-link protein protocadherin 15 interacts with transmembrane channel-like proteins TMC1 and TMC2. *Proceedings of the National Academy of Sciences of the United States of America, 111*, 12907−12912.

Marvin, J. S., Borghuis, B. G., Tian, L., Cichon, J., Harnett, M. T., Akerboom, J. ... Looger, L. L. (2013). An optimized fluorescent probe for visualizing glutamate neurotransmission. *Nature Methods, 10*, 162−170.

Miyawaki, A., Llopis, J., Heim, R., McCaffery, J. M., Adams, J. A., Ikura, M., & Tsien, R. Y. (1997). Fluorescent indicators for Ca^{2+} based on green fluorescent proteins and calmodulin. *Nature, 388*, 882−887.

Montgomery, J., Carton, G., Voigt, R., Baker, C., & Diebel, C. (2000). Sensory processing of water currents by fishes. *Philosophical Transactions of the Royal Society of London B Biological Sciences, 355*, 1325−1327.

Odermatt, B., Nikolaev, A., & Lagnado, L. (2012). Encoding of luminance and contrast by linear and nonlinear synapses in the retina. *Neuron, 73*, 758−773.

Olt, J., Johnson, S. L., & Marcotti, W. (2014). In vivo and in vitro biophysical properties of hair cells from the lateral line and inner ear of developing and adult zebrafish. *The Journal of Physiology, 592*, 2041−2058.

Owens, K. N., Santos, F., Roberts, B., Linbo, T., Coffin, A. B., Knisely, A. J. ... Raible, D. W. (2008). Identification of genetic and chemical modulators of zebrafish mechanosensory hair cell death. *PLoS Genetics, 4*, e1000020.

Palmer, A. E., Giacomello, M., Kortemme, T., Hires, S. A., Lev-Ram, V., Baker, D., & Tsien, R. Y. (2006). Ca^{2+} indicators based on computationally redesigned calmodulin-peptide pairs. *Chemistry & Biology, 13*, 521−530.

Patton, E. E., & Zon, L. I. (2001). The art and design of genetic screens: zebrafish. *Nature Reviews Genetics, 2*, 956−966.

Pérez Koldenkova, V., & Nagai, T. (2013). Genetically encoded Ca^{2+} indicators: properties and evaluation. *Biochimica et Biophysica Acta BBA − Molecular Cell Research, 1833*, 1787−1797.

Pujol-Martí, J., Faucherre, A., Aziz-Bose, R., Asgharsharghi, A., Colombelli, J., Trapani, J. G., & López-Schier, H. (2014). Converging axons collectively initiate and maintain synaptic selectivity in a constantly remodeling sensory organ. *Current Biology, 24*, 2968−2974.

Ricci, A. (2002). Differences in mechano-transducer channel kinetics underlie tonotopic distribution of fast adaptation in auditory hair cells. *Journal of Neurophysiology, 87*, 1738−1748.

Ricci, A. J., Bai, J.-P., Song, L., Lv, C., Zenisek, D., & Santos-Sacchi, J. (2013). Patch-clamp recordings from lateral line neuromast hair cells of the living zebrafish. *The Journal of Neuroscience, 33*, 3131−3134.

Sarrazin, A. F., Nuñez, V. A., Sapède, D., Tassin, V., Dambly-Chaudière, C., & Ghysen, A. (2010). Origin and early development of the posterior lateral line system of zebrafish. *The Journal of Neuroscience, 30*, 8234−8244.

Schneider, C. A., Rasband, W. S., & Eliceiri, K. W. (2012). NIH Image to ImageJ: 25 years of image analysis. *Nature Methods, 9*, 671–675.

St-Pierre, F., Marshall, J. D., Yang, Y., Gong, Y., Schnitzer, M. J., & Lin, M. Z. (2014). High-fidelity optical reporting of neuronal electrical activity with an ultrafast fluorescent voltage sensor. *Nature Neuroscience, 17*, 884–889.

Thévenaz, P., Ruttimann, U. E., & Unser, M. (1998). A pyramid approach to subpixel registration based on intensity. *IEEE Transactions on Image Processing, 7*, 27–41.

Tian, L., Hires, S. A., & Looger, L. L. (2011). Imaging neuronal activity with genetically encoded calcium indicators. *Cold Spring Harbor Protocols*. http://dx.doi.org/10.1101/pdb.top069609.

Tian, L., Hires, S. A., Mao, T., Huber, D., Chiappe, M. E., Chalasani, S. H. ... Looger, L. L. (2009). Imaging neural activity in worms, flies and mice with improved GCaMP calcium indicators. *Nature Methods, 6*, 875–881.

Trapani, J. G., & Nicolson, T. (2010). Physiological recordings from zebrafish lateral-line hair cells and afferent neurons. *Methods in Cell Biology, 100*, 219–231.

Trump, W. J. V., & McHenry, M. J. (2008). The morphology and mechanical sensitivity of lateral line receptors in zebrafish larvae (*Danio rerio*). *The Journal of Experimental Biology, 211*, 2105–2115.

Vollrath, M. A., Kwan, K. Y., & Corey, D. P. (2007). The micromachinery of mechanotransduction in hair cells. *Annual Review of Neuroscience, 30*, 339–365.

Weeg, M. S., & Bass, A. H. (2002). Frequency response properties of lateral line superficial neuromasts in a vocal fish, with evidence for acoustic sensitivity. *Journal of Neurophysiology, 88*, 1252–1262.

Yao, X. C., & Zhao, Y. B. (2008). Optical dissection of stimulus-evoked retinal activation. *Opt Express, 16*, 12446–12459.

Zhao, Y., Araki, S., Wu, J., Teramoto, T., Chang, Y.-F., Nakano, M. ... Campbell, R. E. (2011). An expanded palette of genetically encoded Ca^{2+} indicators. *Science, 333*, 1888–1891.

CHAPTER 11

Physiological recordings from the zebrafish lateral line

J. Olt[*], A.J. Ordoobadi[§], W. Marcotti[*], J.G. Trapani[§,1]

[*]*University of Sheffield, Sheffield, United Kingdom*
[§]*Amherst College, Amherst, MA, United States*
[1]*Corresponding author: E-mail: jtrapani@amherst.edu*

CHAPTER OUTLINE

Introduction	254
1. Common Methods for Lateral Line Electrophysiology	255
1.1 Ethics Statement	255
1.2 Larval Tissue Preparation	256
1.3 Larval Immobilization With Bungarotoxin	256
1.4 Recording Chamber and Larval Mounting	257
1.5 Physiological Solutions	258
2. Stimulation of Neuromast Hair Cells	258
2.1 Mechanical Stimulation	259
2.2 Optical Stimulation	260
3. Recording Microphonic Potentials	261
3.1 Microphonics Equipment and Setup	261
3.2 Positioning the Microphonic Recording Electrode	262
3.3 Recording Microphonic Potentials	262
3.4 Analysis of Microphonic Potentials	264
4. In Vivo Hair Cell Physiology	265
4.1 Identification and Access	266
4.2 Electrophysiology Electrodes and Placement	266
4.3 Establishing Whole-Cell Recordings	268
4.4 Analysis of Whole-Cell Recording	270
5. Afferent Neuron Action Currents	270
5.1 Action Current Electrodes and Placement	271
5.2 Establishing an Action Current Recording	271
5.3 Analysis of Afferent Fiber Spiking	273

6. Summary	275
Discussion	275
Acknowledgments	276
References	276

Abstract

During sensory transduction, external physical stimuli are translated into an internal biological signal. In vertebrates, hair cells are specialized mechanosensory receptors that transduce sound, gravitational forces, and head movements into electrical signals that are transmitted with remarkable precision and efficiency to afferent neurons. Hair cells have a conserved structure between species and are also found in the lateral line system of fish, including zebrafish, which serve as an ideal animal model to study sensory transmission in vivo. In this chapter, we describe the methods required to investigate the biophysical properties underlying mechanosensation in the lateral line of the zebrafish in vivo from microphonic potentials and single hair cell patch-clamp recordings to single afferent neuron recordings. These techniques provide real-time measurements of hair-cell transduction and transmission following delivery of controlled and defined stimuli and their combined use on the intact zebrafish provides a powerful platform to investigate sensory encoding in vivo.

INTRODUCTION

Our auditory system has developed the capacity to process and transmit information with remarkable precision and fidelity. Most of our current knowledge about sensory transduction in hair cells comes from in vitro experiments using altricial rodents and other vertebrates including the bullfrog and turtle (for reviews see Fettiplace and Kim (2014), Howard, Roberts, and Hudspeth (1988), Pan and Holt (2015)). Currently, mice are still the preferred choice for investigating the development and function of the auditory and vestibular systems due to in part the availability of models linked to deafness and vestibular dysfunction in humans (Lenz & Avraham, 2011). However, while in vivo afferent neuron recordings have been performed in mammals, they have relied on tone delivery to stimulate hair cells in the ear (Buran et al., 2010; Taberner & Liberman, 2005) as direct access to innervated hair cells from an intact animal is technically very challenging. Recently, zebrafish (*Danio rerio*) have emerged as a powerful tool to study sensory hair-cell transduction in vivo (Nicolson et al., 1998; Olt, Johnson, & Marcotti, 2014; Ricci et al., 2013; Trapani & Nicolson, 2010) overcoming many of the technical challenges present in rodent models.

The lateral line is required in zebrafish to detect water movement and vibration up to about 300 Hz (Van Trump & McHenry, 2008), and the animal uses this information for schooling behavior, detection of prey, mating, and avoiding predators (Ghysen & Dambly-Chaudière, 2004). The hair cells in the lateral line are clustered in rosettelike structures termed neuromasts, which are deposited along the length of the fish during the first few days post fertilization (dpf) (Ghysen & Dambly-Chaudière, 2004). Each neuromast contains about 8 hair cells at around 4 dpf and increases to about 20—30 cells by 30 dpf (Kindt, Finch, & Nicolson, 2012; Olt et al., 2014; Sheets, Kindt, & Nicolson, 2012).

On the apical surface of each hair cell is the mechanosensory hair bundle. Each hair bundle is composed of microvilli-like structures called stereocilia that are organized in a staircased fashion together with a single, taller kinocilium located adjacent to the tallest stereocilium in fish, amphibians, and mammalian vestibular hair cells. The stereocilia are arranged in rows and are connected by extracellular links of several types, with the topmost link, known as the tip link, being crucial for the opening of the mechanoelectrical transducer (MET) channels (Assad, Shepherd, & Corey, 1991; Maeda et al., 2014; Pickles, Comis, & Osborne, 1984). While in mammalian systems, hair cells and associated hair bundles are encased in bone and are difficult to access, the hair cells within zebrafish neuromasts are located on the body surface. The accessibility of the hair cells and afferent neurons of the lateral-line system represents an ideal experimental model to investigate the molecular mechanisms underlying the development and function of this sophisticated sensory cell and how signals are encoded along the afferent neuronal pathway.

Here we describe the methods used to perform electrophysiological recordings from hair cells and afferent neurons in the intact larval zebrafish, which allow for functional studies of sensory transduction from the lateral line. **Microphonic potentials** are used as an assay for the functioning of the MET current at the top of the hair cell. We also show that this technique is sensitive to detect other transepithelial ionic currents, such as those through the light-gated ion channel, Channelrhodopsin-2. **Single cell patch-clamp** is used to investigate the biophysical properties of hair cells including basolateral membrane ion channels and vesicle fusion at ribbon synapses. **Extracellular, loose-patch recordings** of afferent neuron activity at rest and in response to stimulation allow for a better understanding of how hair cells accomplish the task of encoding sensory information into defined trains of action potentials.

1. COMMON METHODS FOR LATERAL LINE ELECTROPHYSIOLOGY

In this section, we describe methods that are common to the three techniques used to examine hair-cell transduction and sensory encoding in the lateral line of larval zebrafish.

1.1 ETHICS STATEMENT

All zebrafish studies should be approved by the appropriate sources at a researcher's home institution. Work performed in the UK was licensed by the Home Office under the Animals (Scientific Procedures) Act 1986 and was approved by the University of Sheffield Ethical Review Committee. In the United States, animal protocols were approved by the Institutional Animal Care and Use Committee (IAUC) at Amherst College under assurance number 3925-1 with the Office of Laboratory Animal Welfare.

1.2 LARVAL TISSUE PREPARATION

Larval zebrafish (3.0–5.2 dpf) should be anesthetized until they cease to respond to touch stimuli in a petri dish containing 0.016% or 0.04% tricaine methanesulfonate (MS-222; Henry Schein, Inc., Dumfries, UK; Sigma–Aldrich, St. Louis, MO, the United States) in normal extracellular solution A or E3 embryo medium (Westerfield, 2000) and then transferred to the recording chamber.

The larval zebrafish can be mounted using two similar approaches, both of which are suitable methods for all three physiological techniques. For hair cell patch-clamp experiments, a microscope coverslip (22 mm diameter) coated with a layer of ~0.5 mm thick silicone elastomer (Sylgard 1–84: Dow Corning, Midland, MI, the United States) was positioned at the bottom of a recording chamber filled with normal extracellular solution (see point II.E below). The coated coverslips are prepared in advance and only used once. For microphonic and afferent neuron recordings, larvae are transferred to a ~1-mm thick Sylgard-lined recording dish (PC-R; Siskiyou, Grants Pass, OR, the United States) within ~1 mL of E3 with MS-222.

The larva is then secured on its side to the recording chamber by using fine wire to pin it down onto the Sylgard using a stereomicroscope. Larvae can be pinned by using precut, fine tungsten wire (0.015 mm, Advent Research Materials Ltd, Oxford, UK), or pins can be fashioned by electrolytically sharpening the tips of short pieces of 0.050-mm diameter tungsten rod (A-M Systems, Sequim, WA, the United States) that is then bent to the shape of an "L." In the latter scenario, the final pin should have its sharpened end on the long side of the "L" while the short end of the "L" is used as a handle, allowing the pin to be grasped and manipulated with fine forceps.

One tungsten pin is inserted into the anterior end of the fish, either between the ear and the eye or at the region between the heart, inner ear, and mouth (anterior pin), the second pin is then inserted into the notochord at the tail close to the anal exit (posterior pin; Fig. 1). The viability of the larva should be monitored throughout the experiment by observing blood flow and heart rate, which can be verified by eye due to the transparent nature of the zebrafish larva.

1.3 LARVAL IMMOBILIZATION WITH BUNGAROTOXIN

To ensure that the larva remains immobilized during recordings, the pinned larva is paralyzed by microinjecting 125 μM α-bungarotoxin (Abcam, Cambridge, England or Tocris Bioscience, Bristol, UK) into the heart using a pressure injector (MPPI-3,

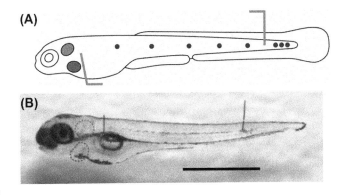

FIGURE 1

The larval zebrafish is pinned to the recording chamber. (A) Schematic drawing showing the approximate locations of tungsten pins (green) used to fix the larva to a Sylgard-lined recording chamber. One pin is inserted close to the head just posterior to the inner ear (*blue circle*) and above the heart (*orange circle*) and the second pin is inserted into the notochord toward the distal end of the tail. (B) Image of 5-day postfertilization zebrafish with pins in place and inner ear (*blue dashed circle*) and heart (*orange dashed circle*) indicated. Scale bar: 1 mm. (See color plate)

Drawing (top) modified from Olt, J., Johnson, S.L., & Marcotti, W. (2014). In vivo and in vitro biophysical properties of hair cells from the lateral line and inner ear of developing and adult zebrafish. The Journal of Physiology, 592, 2041–2058.

Applied Scientific Instrumentation, Eugene, OR, the United States). Microinjection is performed with a tapered micropipette fabricated using a pipette puller so that the tip diameter is 1–3 μm. The micropipette is then filled with α-bungarotoxin and directed through the skin and into the heart with a micromanipulator. Successful injection is confirmed by observing the heart and surrounding tissue swell slightly in size with the heart beat pausing for a brief moment. A low concentration of phenol red can be added to the α-bungarotoxin to aid in visualization (see Trapani and Nicolson (2010)). For electrophysiological recordings, the paralytic MS-222 should be carefully rinsed away and replaced with extracellular recording solution. If not using perfusion, we recommend several generous rinses to ensure that all MS-222 is removed. After rinsing, the larva is left within a ∼1 mL drop of extracellular solution.

1.4 RECORDING CHAMBER AND LARVAL MOUNTING

After the larva is pinned and paralyzed, the recording chamber is then mounted on a rotating platform (Märzhäuser Wetzlar GmbH, Germany or PC-A, Siskiyou, the United States) on a fixed-stage, upright microscope (Olympus BX51WI, Olympus) with a 40× or 60× long-working distance water immersion objective (LUMPlanFL

N, Olympus) and Nomarski optics. The rotating platform is ideal, as it allows the larvae to be rotated to align properly with the recording electrode and source of stimulation (see Section 2). Because the hair cells in the posterior lateral line have their activation polarity oriented along the anteroposterior axis, larvae are rotated so that the hair cell stereociliary bundles will receive mechanical stimulation from the fluid jet along this axis.

For accurate measurement of the biophysical properties of hair cells, it is important to maintain physiological temperatures, which for larval and adult zebrafish is ~28.5°C. This can be accomplished by modifying the center of the rotating platform in order to have small resistors mounted to the underside, so that they can be heated by passing an electrical current (Marcotti, 2012). It is also important to continuously perfuse the recording chamber containing the live zebrafish with extracellular solution using a peristaltic pump (Masterflex L/S, Cole Palmer, London, UK).

1.5 PHYSIOLOGICAL SOLUTIONS

The extracellular solution A used for hair cell patch-clamp recordings in zebrafish contains (in mM): 135 NaCl, 5.8 KCl, 1.3 $CaCl_2$, 0.9 $MgCl_2$, 0.7 NaH_2PO_4, 5.6 D-glucose and 10 HEPES-NaOH, with sodium pyruvate (2 mM), MEM amino acids solution (50×, without L-glutamine), and MEM vitamins solution (100×) added from stock solutions (Fisher Scientific UK Ltd, Loughborough, UK); final pH is brought to 7.5. For microphonic and afferent neuron recordings, extracellular solution B contains (in mM): 140 NaCl, 2 KCl, 2 $CaCl_2$, and 1 $MgCl_2$, and 10 HEPES-NaOH, with final pH = 7.8. The patch pipette intracellular solution used for current and voltage recordings contains (in mM): 131 KCl, 3 $MgCl_2$, 1 EGTA-KOH, 5 Na_2ATP, 5 HEPES-KOH, and 10 sodium phosphocreatine, with final pH = 7.3. For Ca^{2+} current recordings and capacitance measurements, the intracellular solution contains (in mM) 85 Cs-glutamate, 20 CsCl, 3 $MgCl_2$, 1 EGTA-CsOH, 5 Na_2ATP, 5 Hepes-CsOH, 10 Na_2-phosphocreatine; 0.3 Na_2GTP, 15 4-aminopyridine (4-AP); 20 TEA, with final pH = 7.3. Solutions can be filtered (0.2 μm) before storing at 4°C.

2. STIMULATION OF NEUROMAST HAIR CELLS

To investigate properties of hair cell mechanoelectrical transduction and hair cell–driven spiking in afferent neurons, the sensory receptor must be stimulated. Hair cells within lateral line neuromasts are traditionally stimulated with precisely controlled delivery of extracellular solution via a picospritzer, high-speed pressure clamp, or other fluid delivery system. Piezoelectric devices have also been used to deflect the neuromast cupula (Haehnel-Taguchi, Akanyeti, & Liao, 2014; Nicolson et al., 1998). In all cases, the movement of the cupula causes the kinocilia of the neuromast hair cells to deflect, which in turn causes the displacement of the stereocilia and the opening of MET channels. In addition to mechanical stimulation, optogenetic stimulation can be achieved by expression of light-activated proteins (such as ion channels) in hair cells, which provides remote, optical stimulation of hair cells (Monesson-Olson, Browning-Kamins, Aziz-Bose, Kreines, & Trapani, 2014; Monesson-Olson, Troconis, & Trapani, 2014).

2.1 MECHANICAL STIMULATION

An ideal form of mechanical stimulation of neuromast hair cells is through a fluid jet system, which consists of a fluid-filled micropipette connected to a device that can apply positive and negative output pressure. One commercial product available for this is a high-speed pressure clamp (HSPC-1, ALA Scientific Instruments, Farmingdale, NY, the United States), which connects to a pressure vacuum pump (PV-Pump, ALA Scientific Instruments, Farmingdale, NY, the United States) in order to supply pressure to the headstage of the pressure clamp. The headstage of the HSPC-1 is connected via thick-walled silicone tubing to a micropipette holder. The attached micropipette is then positioned to direct the fluid jet at the cupula of the neuromast. The micropipette is made from nonfilamented borosilicate glass using a single, hard pull delivered with a micropipette puller (eg, P-1000, Sutter Instrument, Novato, CA, the United States). The pulled micropipette will have a long, wispy tip that must be trimmed by rubbing it against either a ceramic tile (Sutter Instrument, Novato, CA, the United States) or another pulled pipette in order to cleanly break the glass and generate a tip with an outer diameter of approximately 40 μm. A clean break that creates a round and smooth tip is essential to ensure accurate, even fluid flow out of the micropipette. The micropipette is then filled with extracellular recording solution and fixed to the holder of a micromanipulator (eg, MPC-385, Sutter Instrument, Novato, CA, the United States) for its final positioning near the neuromast. For fluid jet stimulation of primary neuromasts of the posterior lateral line, it is crucial that the fluid jet pipette is aligned with the antereoposterior axis of the fish. The tip of the fluid jet pipette should be positioned at a distance of approximately 50–100 μm from the neuromast, and the bottom edge of the pipette should be aligned with the top of the kinocilia (Fig. 2). Sufficient pressure for the fluid jet will deflect the kinocilia ∼3 μm and constructing a dose–response relationship of fluid jet pressure and physiological output will allow for optimizing the final output pressure.

One method for controlling the pressure delivered to a fluid jet is with a high-speed pressure clamp, as mentioned above. For the HSPC-1, the output pressure of the device is determined by the input command voltage with a voltage-to-pressure conversion of 20 mV/mmHg. Therefore, a 200-mV voltage step would result in a pressure output of 10 mmHg, which can be subsequently measured and recorded along with the electrophysiological recording using the feedback voltage output provided by the HSPC-1 headstage. The length of the voltage pulse controls pressure duration, and trains of stimuli can be delivered via the waveform of the input command voltage.

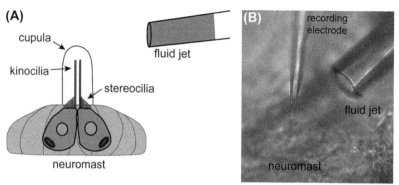

FIGURE 2

Placement of fluid jet for mechanical stimulation of hair cells in a lateral line neuromast. (A) In the vertical axis, the fluid jet is aligned with the tips of the kinocilia. In the horizontal axis, the fluid jet should be approximately 50–100 μm from the cupula (side view). Note that only two opposing hair cells are shown, but typically a 5-day postfertilization (dpf) larva has ~8–14 hair cells per neuromast. (B) DIC image of a recording electrode and fluid jet arranged for a microphonics experiment at the L1 neuromast in a 5 dpf zebrafish larva (top-down view).

2.2 OPTICAL STIMULATION

Optical stimulation of hair cells can be accomplished with transgenic zebrafish larvae that express a light-gated ion channel in hair cells. One possible zebrafish transgenic line, *Tg(myo6b:ChR2-EYFP)*, expresses the light-gated ion channel, Channelrhodopsin-2 (ChR2) in inner ear and lateral line hair cells (Monesson-Olson, Browning-Kamins, et al., 2014), but other lines could be created that utilize other promoters and/or optical activators including Chronos or Chrimson (Klapoetke et al., 2014; Smedemark-Margulies & Trapani, 2013). Previous work has shown that expression of ChR2 in neurons does not alter the health of the cell, membrane resistance, or dynamic properties of the membrane in the absence of light stimulation (Boyden, Zhang, Bamberg, Nagel, & Deisseroth, 2005). Consistent with this finding, transgenic zebrafish expressing ChR2 in hair cells develop normally, mate and propagate with normal fecundity, and live as adults with no observable difference in behaviors.

Light for optical stimulation of a neuromast can be delivered through the 40× or a 60× objective used to visualize the neuromast if using an epifluorescence microscope with adjustment of the fluorescence aperture. One ideal form of output light is an LED source such as a light engine (eg, Lumencor, Beaverton, OR, the United States) or collimated LED (Zhu et al., 2009) that can be turned on and off rapidly using a transistor–transistor logic (TTL) pulse via an analog output from the electrophysiology equipment. Alternatively, a standard fluorescence lamp can be used together with a shutter device capable of fast triggering (Lambda Shutter, Sutter

Instruments, Novato, CA, the United States). Because light-activated channels have an ideal spectrum for activation, an appropriate range of wavelengths from the output light can be achieved by using an excitation filter or by using an LED or laser light source of desired wavelength. The intensity of light delivered can also be set and varied at the light source as well as by using neutral density filters placed within the light path. A light power meter (eg, PMD100, Thorlabs, Newton, NJ, the United States) placed at the level of the sample can measure the light intensity delivered to the tissue.

3. RECORDING MICROPHONIC POTENTIALS

Microphonic potentials represent the collective, transepithelial flow of ions that occurs when hair cells are stimulated (Corey & Hudspeth, 1983). Following hair bundle deflection, the microphonic potential results from the current flowing through open MET channels and voltage-gated ion channels expressed in the basolateral membrane of hair cells, which in the zebrafish lateral line include both calcium and potassium channels (Olt et al., 2014). We have also observed microphonic potentials in transgenic larvae with hair-cell expression of ChR2 upon light activation (see Fig. 3). Overall, microphonic potentials provide information about the amplitude and adaptation of MET-channel currents and can be used to assay mechanotransduction in neuromast hair cells (Nicolson et al., 1998; Obholzer et al., 2008).

3.1 MICROPHONICS EQUIPMENT AND SETUP

The microphonic recording electrode is fabricated from nonfilamented, borosilicate glass (B150-86-10HP, Sutter Instruments, Novato, CA, the United States) using a micropipette puller with heat set to ramp temperature. Nonfilamented glass is chosen as the filament promotes fluid flow out of the tip of the micropipette, which can cause mechanical artifacts and obscure the currents generated during MET channel activation. The recording electrode is filled with extracellular solution B and should have a final tip resistance of 3–5 MΩ in extracellular solution.

Because microphonic potentials recorded from neuromasts have very small amplitudes ($\sim$2–10 µV), they must be amplified (eg, 50,000$\times$) through a combination of amplifiers. In our setup, microphonics are recorded using both an Axopatch 200B (Molecular Devices, Sunnyvale, CA, the United States) patch-clamp amplifier and an additional amplifier (Model 440, Brownlee Precision, Palo Alto, CA, the United States). The Axopatch 200B is set to current-clamp mode (I = 0) at 500x gain and the output signal is then further amplified (100$\times$) by the Brownlee amplifier. The output signal from the Brownlee is also high-pass filtered at 0.1 Hz to remove baseline drift and low-pass filtered at 100 Hz to further reduce noise. We do not filter 60 Hz noise as the microphonic signal may fall within this frequency range depending on the stimulus waveform. Instead,

60-cycle noise and other higher frequency electrical interference should be reduced through careful grounding and elimination of sources of alternating current within the recording room (Sherman-Gold, 1993). Mechanical sources of noise are reduced by performing recordings on a vibration-isolation air table (TMC, Peabody, MA the United States). Voltage signals are digitized with an ITC-16 DAQ device and recorded with data acquisition software (eg, Patchmaster, HEKA Elektronik). Because of a low signal-to-noise ratio due to the exceptionally small microphonic signal, 200 repeated sweeps are typically recorded and averaged to achieve a final microphonic trace for a given experiment. In some cases, high intensity stimuli coupled with a low-noise electrophysiology rig will result in microphonic signals that can be observed during individual sweeps.

3.2 POSITIONING THE MICROPHONIC RECORDING ELECTRODE

Microphonics can be recorded from any neuromast along the anterior or posterior lateral line (ALL or PLL) as well as from hair cells in the ear (for example, see Gleason et al. (2009)). After identifying an appropriate neuromast, the recording electrode should be placed at the base of the cupula so that it is as close as possible to the stereocilia, which are short and only protrude slightly above the apical surface of the neuromast hair cells (Fig. 3). To ensure that the recording electrode does not interfere with deflection of the cupula, it is important to position the electrode at a 45–90 degree angle to the fluid jet (see Fig. 2B). Also, to prevent the tip of the electrode from clogging during positioning, 20–40 mmHg of positive pressure can be applied using a three-way stopcock connected to a syringe or a pressure transducer (DPM1B, Fluke Biomedical, Everett, WA, the United States), but no pressure should be applied to the electrode during recordings.

3.3 RECORDING MICROPHONIC POTENTIALS

Although microphonics are traditionally obtained during mechanical stimulation of hair cells, they can also be recorded from transgenic larvae during optical stimulation. For mechanical stimuli, microphonic potentials are often recorded from neuromasts during sinusoidal mechanical stimulation and are characterized by a $2f$ response to a $1f$ stimulus. This $2f$ response occurs because neuromasts contain two groups of hair cells with opposing activation polarities (López-Schier, Starr, Kappler, Kollmar, & Hudspeth, 2004). When the neuromast is stimulated with a waveform of frequency f, the resulting microphonic potentials occur at twice ($2f$) that frequency, representing the alternate activation of the two groups of hair cells of opposing polarity (Nicolson et al., 1998). A $2f$ response will not be observed with step deflections that only activate only one population of hair cells or with optical stimulation that activates both populations of hair cells at the same time (Fig. 3).

To ensure that the recorded signals are not due to a mechanical artifact but represent the hair cell's physiological response to the stimulus, a permeant blocker

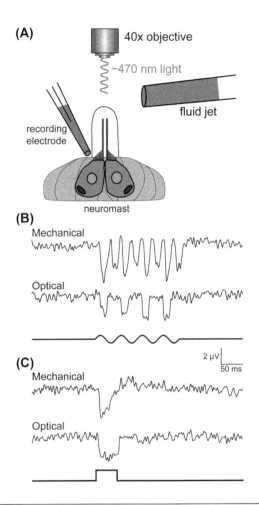

FIGURE 3

Microphonic recordings during both mechanical and optical neuromast stimulation. (A) Cartoon of the recording arrangement for either mechanical stimulation with a fluid jet or optical stimulation via ~470 nm light directed through a 40× objective centered over a single neuromast. (B) Mechanical stimulation with a 20-Hz sine wave produces a microphonic potential with a characteristic $2f$ response. Using a sine wave voltage protocol for optical stimulation only produces light when positive voltage reaches threshold for the light source, and the resulting microphonic is $1f$ as optical stimulation activates both populations of hair cells at once. (C) Stimulation with a 50-ms square wave pulse reveals kinetics of microphonic activation, inactivation, and deactivation. Recordings were performed using a *Tg(myo6b:ChR2-EYFP)* transgenic larva.

of the MET channel, such as dihydrostreptomycin (Kroese, Das, & Hudspeth, 1989; Marcotti, van Netten, & Kros, 2005) can be applied or hair cells can be killed with neomycin (see Fig. 4A; (Harris et al., 2003)). Both of these procedures will abolish or largely reduce the microphonic signal depending on the concentration used. However, aminoglycoside-induced hair cell death does not occur on a timescale appropriate for acute application. Alternatively, the preparation can be rotated 90 degrees so that the neuromast is stimulated perpendicular to its excitatory axis, which should significantly reduce the microphonic signal (Nicolson et al., 1998). Microphonics will also be diminished or absent when hair bundles are damaged, which can easily happen due to mechanical stress during the stimulation procedure or if the petri dishes that house the larvae are not kept clean from coleps and other parasites through daily media changes (William Detrich, Zon, & Westerfield, 2009).

The waveform of the microphonic potential varies depending on the position of the recording electrode within the neuromast. When the recording electrode is placed to one end of the neuromast, the $2f$ potentials will alternate in amplitude. Since the two populations of hair cells within a neuromast roughly segregate to opposite sides, the higher amplitude peaks correspond to the population of hair cells closest to the recording electrode (Fig. 4B). The alternating, smaller amplitude peaks then result from the other population of hair cells that are located toward the other end of the neuromast and have opposite activation polarities. If the recording electrode is placed in the center of the neuromast, its location is approximately centered between the two populations of hair cells and the peaks will have similar amplitudes (Fig. 4C).

3.4 ANALYSIS OF MICROPHONIC POTENTIALS

Quantification of recorded microphonics can provide information about the amplitude and kinetics of hair-cell transducer currents. Typically, to confirm the presence of the characteristic $2f$ response, a fast Fourier transform can be performed to generate a power spectrum across an appropriate range of frequencies (Nicolson et al., 1998). For microphonics recorded with a bidirectional stimulus waveform, the spectrum should have peaks at the stimulus frequency (f) and at twice the frequency ($2f$). Methods to determine the overall amplitude of the signal are complicated by the fact that the amplitude of the microphonic potential varies depending on the exact position of the electrode relative to the neuromast (see Fig. 4). To overcome this challenge, the amplitude of the evoked microphonic potential (in microvolts) can be calculated from the integral of a region of the microphonic waveform divided by the duration of that region (ie, microvolt·s/s). For example, integration of two microphonic traces recorded from two locations within a neuromast yield nearly identical magnitudes despite their differently shaped waveforms (see amplitudes reported in Fig. 4). This method can be used to quantify the amplitude of microphonic potentials across experiments or before and after drug applications during dose–response experiments.

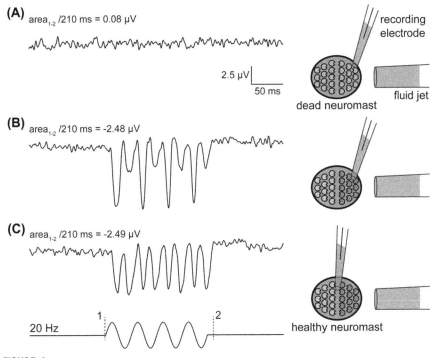

FIGURE 4

The waveform of the recorded microphonic potential varies with electrode placement. (A) In a control experiment, no microphonics are measured from a neuromast that had all its lateral-line hair cells killed by neomycin. (B) A recording from a healthy neuromast with the recording electrode placed toward one end of the neuromast. The peaks of the microphonic potential display two different amplitudes at the characteristic $2f$ frequency. The larger amplitude peaks result from activation of the closer, red hair cells, while the smaller peaks are from the more distal, yellow hair cells with opposite activation polarities. The drawings inaccurately simplify the segregation of the two populations of opposite polarity hair cells. (C) A recording with the electrode placed at the center of the neuromast, the $2f$ microphonic has equal amplitude peaks compared to those in B. The amplitudes noted for the microphonic potentials in A, B, and C were quantified from the absolute area of the waveform between stimulus onset (position 1 on the 20 Hz stimulus trace) and 10 ms after stimulus offset (position 2) divided by the duration of the position 1–2 interval. (See color plate)

4. IN VIVO HAIR CELL PHYSIOLOGY

In larval zebrafish, the lateral line consists of an anterior lateral line situated on the head and a posterior lateral line extending down the trunk and tail of the fish. Because the hair cells are more visible and accessible along the trunk, it is most straightforward to perform electrophysiological recordings on hair cells within primary neuromasts of the posterior lateral line (L1–L4) originating from the first primordium (primI; (Gompel et al., 2001; Pujol-Martí & López-Schier, 2013)). As mentioned above, these primary neuromasts show planar polarity along the anterior–posterior axis and contain hair cells that respond to stimuli of opposing polarity (López-Schier et al., 2004; Nicolson et al., 1998). The techniques described below have also been used to record from hair cells at different stages of maturation in the inner ear (Olt et al., 2014), the development and function of which is described in the chapter in this text by Baxendale and Whitfield.

4.1 IDENTIFICATION AND ACCESS

Sensory hair cells within each neuromast, are identified using Nomarski optics coupled to a 60× water immersion objective with additional magnification of 1.5× or 2× positioned (eg, via an optovar) before the 15× eyepieces. Each hair cell has one long kinocilium, which allows kinocilia count to be used as an estimate for the number of cells within each neuromast (Fig. 5A). Neuromasts are covered by skin cells (asterisks: Fig. 5A) that surround centrally located hair cells and support cells. Beneath the skin are the outer-most cells called mantle cells. Under mantle cells, hair cells are surrounded by support cells that span the basolateral membrane of hair cells. Hair-cell bodies are located in the most central part of the neuromast, while the majority of the supporting cell bodies lie beneath the hair cells (Fig. 5B).

Accessing lateral line hair cells for patch-clamp experiments is performed following a series of steps. Initially, a "cleaning" glass pipette with a diameter of ~3–4 μm (borosilicate glass: O.D. 1.5 mm; I.D. 1.17 mm, Harvard Apparatus) is positioned near the neuromast (Fig. 5C and D). This "cleaning" pipette is connected to a syringe via a silicone tube and is filled with the same extracellular solution bathing the fish. The application of positive and negative pressure through the syringe allows for the selective removal of one or two large skin cells at about 20–30 μm away from the neuromast (Fig. 5C and D), which are just outside the ring of smaller skin cells surrounding the hair cells. Next, the cleaning pipette is used to remove connective tissue surrounding the neuromast and a few supporting and mantle cells (Fig. 5E). This procedure allows the patch pipette to reach the basolateral membrane of hair cells (Fig. 5F). It is crucial that the "cleaning" method is performed with great care in order to avoid damaging the cupula. The cupula contains the hair bundles and kinocilia and its disruption can result in shortened kinocilia and damage to hair bundles.

4.2 ELECTROPHYSIOLOGY ELECTRODES AND PLACEMENT

Recording electrodes for whole-cell patch-clamp are pulled from soda glass capillary tubing (I.D. 0.94 mm, O.D 1.2 mm, Warner Instruments, the United States)

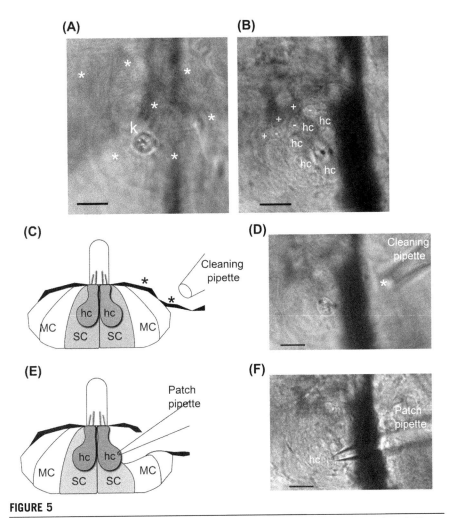

FIGURE 5

Morphological characteristics of the neuromast and hair-cell patching procedure in the larval lateral line. (A) Top view of a 3-day postfertilization neuromast showing the hair cell kinocilia (k) projecting from its center and surrounded by skin cells (asterisks). (B) View of the neuromast at the level of the hair cell (hc) basal pole. Hair cells are surrounded by supporting (−) and mantle (+) cells. (C, D) Diagram of the neuromast (C) showing the relative position of hair cells (hc), mantle cells (MC), supporting cells (SC), and skin cells (asterisks). Note that the cleaning pipette is approaching the skin cells (C and D) located at about 20–30 μm from the hair cells, which is evident by the presence of the kinocilia seen in D. (E, F) Diagram (E) and image (F) of the larval neuromast during the patching of a hair cell with a patch pipette. Scale bars in A, B, D, and F are 10 μm.

and have a resistance of 3–5 MΩ in normal extracellular solution. Soda glass is known as a "soft" glass for electrophysiology, which makes the sealing process onto the cell easier. The fast capacitative transient of the glass pipette is reduced by coating its shaft with surfboard wax (Mr. Zog's SexWax, Sexwax, Inc., Carpinteria, CA, the United States).

Once the basolateral membrane of hair cells is exposed (see above), the patch-pipette can be advanced toward hair cells perpendicularly to the length of the zebrafish, which allows access to cells of either planar polarity within individual neuromasts. Positive pressure applied through the patch-pipette prevents clogging of the tip. After generating a giga-ohm (GΩ) seal with a hair cell, the whole-cell configuration is achieved by applying gentle negative pressure, which ruptures the membrane patch at the tip of the patch pipette. Current and voltage responses are recorded using a patch-clamp amplifier (eg, Optopatch Amplifier, Cairn Research, UK), sampled at 5 kHz or 100 kHz, low pass filtered at 2.5 kHz or 10 kHz (8-pole Bessel) and collected using appropriate acquisition software (eg, PClamp, Molecular Devices, the United States). It is important to only use cells with a healthy appearance for electrophysiological recordings. Healthy hair cells are identified by intact hair bundles, cell membranes with a smooth surface, absence of vacuoles in the cytoplasm, and lack of Brownian motion of mitochondria.

4.3 ESTABLISHING WHOLE-CELL RECORDINGS

Morphologically hair cells are identified visually by the presence of the kinocilia and hair bundle at the cell's apical surface (see above). Electrophysiologically, hair cells can be recognized by the presence of voltage- and time-dependent current responses (Fig. 6). The neuromast of a day 5 larva contains hair cells at multiple developmental stages, ranging from immature with very small hair bundles to more mature with fully developed hair bundles (Kindt et al., 2012). In addition, lateral line hair cells are capable of regenerating within 72 h, even during adulthood (Pinto-Teixeira et al., 2015). Therefore, the age of the zebrafish is not indicative of the maturation state of a given hair cell. The majority of hair cells at larval stages have a current and voltage profile as that shown in Fig. 6A and B. This is characterized by the large-conductance Ca^{2+}-activated K^+ current ($I_{K,Ca}$), a delayed rectifier K^+ current ($I_{K,D}$), a small A-type K^+ current ($I_{A(s)}$) and in some cases an inward hyperpolarization-activated current (I_h) (Olt et al., 2014).

Supporting cells and mantle cells are visually distinguished from hair cells based on their position within the neuromast (see Fig. 5). Mantle cells only cover the very periphery of the tissue and supporting cells can then be identified based on the absence of hair bundles and the position of their cell bodies which are deeper, beneath the hair cells (see Fig. 5). Electrophysiologically, supporting cells can be recognized by both their near-linear current and voltage responses (Fig. 7).

Calcium currents and Ca^{2+}-induced exocytosis of presynaptic vesicles from hair cells of the larval zebrafish are recorded at 28.5°C using 2.8 mM extracellular Ca^{2+} in the extracellular solution (Fig. 8). Exocytosis can be estimated by changes in cell

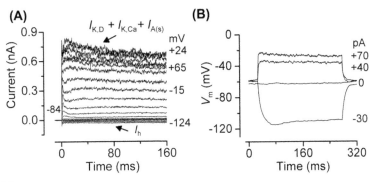

FIGURE 6

Whole-cell patch-clamp recordings from a single hair cell of a larval zebrafish neuromast. (A) Membrane currents recorded from a hair cell positioned in neuromast L4 of a 5-day postfertilization zebrafish. Currents were elicited by using depolarizing voltage steps in 10-mV nominal increments from −124 mV. The holding potential was −84 mV. (B) Voltage responses to hyperpolarizing and depolarizing current injections (values shown next to the traces) recorded from the same hair cell as in panel A.

membrane surface area, which is reflected in changes in membrane capacitance (ΔC_m) following depolarizing voltage steps near to the peak of the Ca^{2+} current. This is generally interpreted as a sign of neurotransmitter release from presynaptic cells (Johnson, Forge, Knipper, Münkner, & Marcotti, 2008; Johnson et al., 2013; Moser & Beutner, 2000; Olt et al., 2014). An example of Ca^{2+} current ΔC_m recorded from larval hair cells is shown in Fig. 8A and B, respectively.

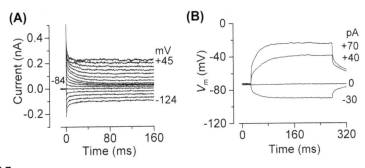

FIGURE 7

Whole-cell patch-clamp recordings from a supporting cell of a larval zebrafish neuromast. (A) Example of ohmic membrane currents recorded from a supporting cell from neuromast L3 of a 5-day postfertilization zebrafish. (B) Voltage responses to different current injections from the same supporting cell shown in (A). Voltage and current recordings were obtained as described for Fig. 6.

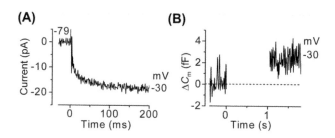

FIGURE 8
Ca^{2+} current and capacitance changes during patch-clamp of a zebrafish lateral line hair cell. (A) Ca^{2+} current recorded from hair cells of a 4-day postfertilization (dpf) larva in response to a 200-ms depolarizing voltage step from the holding potential of −79 mV to near −30 mV, which corresponds to the peak Ca^{2+} current. (B) Changes in membrane capacitance (ΔC_m) recorded from a 3-dpf hair cell obtained in response to 1-s voltage step from the holding potential of −79 mV to near −30 mV. The change in capacitance represents the fusion of synaptic vesicles at the presynaptic terminal.

4.4 ANALYSIS OF WHOLE-CELL RECORDING

Whole-cell patch-clamp recordings are stored on a computer for offline analysis using appropriate software (eg, Clampfit 10, Molecular Devices, USA and Origin, OriginLab Corp. Northampton, MA, the United States). Membrane potentials in voltage clamp must be corrected for the voltage drop across the uncompensated residual series resistance (R_s) and for a liquid junction potential (LJP), measured between the electrode and bath solutions (eg, −4 mV for the KCl based and −9 mV for Cs-glutamate-based intracellular solution). Current responses are normally recorded from a holding potential of −84 mV or −79 mV (including the LJP). Real-time ΔC_m is obtained using a patch-clamp amplifier with a 4 kHz sine wave (13 mV RMS) applied to hair cells from the holding potential that is interrupted for the duration of the voltage step. The capacitance signal from the amplifier is then amplified (50×), filtered at 250 Hz, sampled at 5 kHz and measured by averaging the C_m traces after the voltage step (around 200 ms) and subtracting from prepulse baseline (Johnson et al., 2013; Johnson, Thomas, & Kros, 2002).

5. AFFERENT NEURON ACTION CURRENTS

Hair bundle stimulation elicits a change in receptor potential that results in synaptic vesicle fusion at the ribbon synapse and the generation of action potentials (spikes) in the afferent neuron. Action potentials from an afferent neuron can be recorded using whole-cell patch-clamp (Liao, 2010; Liao & Haehnel, 2012) or with an extracellular "loose patch" in voltage clamp mode (v = 0) as action currents, which are equivalent to the first derivative of the neuron's action potential (dV/dt). Afferent activity can be used to examine hair cell transduction and encoding, vesicular release and reuptake within the hair cell, and spontaneous activity (Haehnel-Taguchi et al., 2014; Obholzer et al., 2008; Sheets, Trapani, Mo, Obholzer, & Nicolson, 2011; Trapani & Nicolson, 2011).

5.1 ACTION CURRENT ELECTRODES AND PLACEMENT

Glass microelectrodes are fabricated to a resistance of 8–12 MΩ in extracellular solution from filamented borosilicate glass (eg, BF150-86-10, Sutter Instrument Company, Novato, CA) using a Brown-Flaming style micropipette puller (eg, P-1000, Sutter Instrument Company, Novato, CA). The electrode is filled with 10 μL of extracellular solution and mounted to the headstage of a patch-clamp amplifier (eg, EPC9, HEKA Elektronik) that is attached to a micromanipulator (eg, MPC-385, Sutter Instrument).

During electrode placement and recordings, the tissue is visualized under a 40× water immersion lens. The cell bodies of afferent neurons that innervate hair cells in the posterior lateral line are clustered together within the posterior lateral line ganglion (PLLg). The PLLg is located posterior to the ear and has the appearance of a "bag of grapes." Visualization of the ganglion is challenging at first, but with practice one can observe the ganglion enclosed by an apparent membrane in a location just posterior to the inner ear underneath the larval epithelium. Visualization of the ganglion can also be achieved by using transgenic larvae that express fluorescent proteins in the afferent neurons (Faucherre, Pujol-Martí, Kawakami, & López-Schier, 2009; Obholzer et al., 2008). To access the PLLg for recording, the recording electrode must puncture through the skin and enter the PLLg (Fig. 9).

During the positioning of the recording electrode, ∼40 mmHg of positive pressure should be applied to the electrode using either a pneumatic transducer (DPM1B, Fluke Biomedical) or with a three-way stopcock with syringe to prevent the tip of the electrode from clogging. As soon as the electrode enters the PLLg, positive pressure is decreased to 20 mmHg to avoid overperfusing the PLLg with extracellular solution. The tip of the recording electrode is then positioned next to the cell body of an afferent neuron and approximately −20 mmHg of pressure is applied to obtain a loose-patch seal (30–50 MΩ) between the tip of the electrode and the membrane of the neuron (Fig. 9B).

5.2 ESTABLISHING AN ACTION CURRENT RECORDING

A quality loose-patch seal on a single neuron is achieved by observing spontaneous action potentials (spikes) of only a single amplitude and waveform. Observation of spikes of multiple amplitudes suggests that the electrode has formed a seal with membranes from more than one cell, and the different amplitude responses represent spiking from different cells. Because recordings from multiple spiking cells are difficult to analyze and may complicate experimental findings, isolated recordings from single units are strongly encouraged. After confirming the presence of a loose-patch

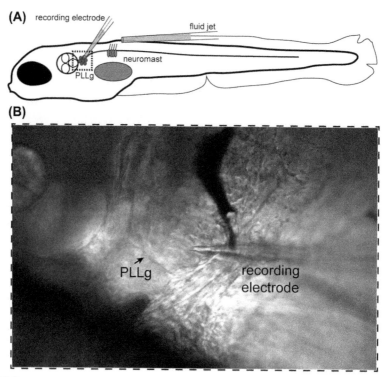

FIGURE 9

Appropriate placement and positioning of the recording electrode within the posterior lateral line ganglion (PLLg). (A) The PLLg consists of all the cell bodies (ganglion) of the afferent neurons of the PLL, lending it the appearance of a "bag of grapes." The recording electrode must pierce through the skin of the zebrafish and enter the PLLg under positive pressure. (B) Once the electrode is within the PLLg, it is positioned adjacent to a soma and negative pressure is applied to generate a loose seal with the membrane of cell. While obtaining seals is somewhat common, identifying the somas of afferent neurons that innervate primary neuromasts that are accessible with the fluid jet can be quite challenging. (See color plate)

seal, the innervated neuromast must be found. Beginning at the L1 neuromast and incrementally moving to more posterior neuromasts, each primary neuromast is stimulated using a fluid jet (see Section 2) until evoked spiking is observed. If experiments are performed on transgenic fish expressing optogenetic proteins in hair cells, then the stage of the microscope can be "scanned" along the lateral line of the fish, effectively moving the beam of excitation light from neuromast to neuromast. This optical scanning method is a rapid way to identify the innervated neuromast. Phase-locked spiking, the presence of action currents that occur in phase with the frequency of stimulation, indicates that the afferent neuron innervates the stimulated neuromast.

Obtaining a loose-patch seal and recording phase-locked action currents is complicated by the challenge of visualizing cells in the PLLg. Multiple attempts to locate an isolated soma can be made on a single larva as long as the preparation remains viable. To evaluate the viability of the larva, blood flow and condition of skin cells and hair cells should be monitored regularly. The resistance of the recording electrode can be obtained using a square-wave test seal pulse as would be used for patch-clamp experiments. If the electrode forms a high-resistance seal when negative pressure is applied, a brief burst of positive pressure should be applied with a syringe to clear the pipette tip before hunting for another suitable cell. Positive pressure can also occasionally be applied with a syringe to move cells and potentially improve the appearance of the ganglion before repeating the hunt and seal process. Since the ganglion contains neurons that innervate both the primary and secondary neuromasts for the entire side of the trunk, confirming the innervated neuromast is important prior to an experiment. In addition, some afferent neurons make contact with more than one neuromast, especially those that project most caudally to neuromasts toward the tail (Faucherre et al., 2009; Liao & Haehnel, 2012).

5.3 ANALYSIS OF AFFERENT FIBER SPIKING

In order to analyze spike trains elicited in response to mechanical or optical stimulation of hair cells, spikes within the afferent fiber recordings must be identified. If a recording is obtained with activity from a single neuron, then spikes can be identified using threshold detection. Spike detection and analysis can be performed using commercial software (eg, Clampfit from Molecular Devices; Fitmaster from HEKA Elektronik; and Spike2 from CED) or custom (often open-source) routines that are available for use in MATLAB (MathWorks, Natick, MA, the United States), Python, and Igor Pro (WaveMetrics Inc., Lake Oswego, OR, the United States). Following threshold detection, the identified spike's amplitude and arrival time during the recording can be obtained and tabulated.

Once information about spike timing has been extracted from the recordings, the data can be analyzed in a variety of ways depending on the experiment. To evaluate the degree to which spikes phase lock to a delivered stimulus, such as a sine- or square wave pulse, the vector strength of the spikes can be calculated. Vector strength (Goldberg & Brown, 1969; Joris, 2006) is determined by transforming each spike time to a unit vector whose direction is given by the phase of the spike time with respect to the onset of the stimulus. A mean vector of length r (also known as vector strength) is then calculated, with $r = 1$ indicating perfect phase locking and $r = 0$ indicating no relationship between the stimulus and response. Quantification of the timing and number of spikes elicited in response to a range of stimulus waveforms can be used to elucidate how properties of spike trains vary depending on input properties of the stimulus (Fig. 10).

In the absence of hair-cell stimulation, spontaneous spiking is seen in afferent neurons as a result of stochastic release of neurotransmitter from the innervated hair cells (for review see Heil and Peterson (2015)). The salient feature of

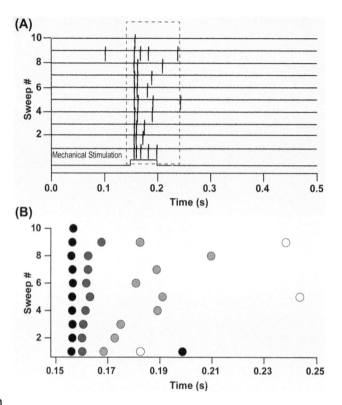

FIGURE 10

The timing of the first spike is highly reliable during mechanical stimulation. (A) Plot of 10 consecutive sweeps in response to a 50-ms mechanical stimulus (50-mmHg output pressure) recorded from a single afferent neuron innervating neuromast L1. (B) Individual action potentials within the red box in A were detected using event-detection software and are displayed as shaded symbols. Each consecutively evoked spike is coded as follows: first spike (black), second spike (dark grey), third spike (light grey), fourth spike (white), and fifth spike (black symbol on far right of sweep 1). Adaptation can be observed both within an individual sweep and across consecutive sweeps. (See color plate)

spontaneous spiking is the interspike interval (ISI), the time between consecutive spikes. ISI times derived from trains of spontaneous spikes recorded over durations sufficient to acquire ~2000 events can then be collectively plotted as event histograms. These data can then be fit by models or equations that describe the Poisson processes underlying the data (Peterson, Irvine, & Heil, 2014; Pfeiffer & Kiang, 1965; Trapani & Nicolson, 2011). Immature hair cells in the mammalian auditory system are also known to spontaneously spike in bursts (Johnson et al., 2013; Tritsch, Yi, Gale, Glowatzki, & Bergles, 2007), which leads to subsequent bursting patterns in afferent neurons (Tritsch et al., 2010). Whether this type of spiking occurs in the lateral line has not been determined, though occasionally cells with burst patterns of spiking are observed.

6. SUMMARY

In this chapter, we have described three methods for recording sensory transduction and transmission from hair cells of the posterior lateral line of larval zebrafish. The recently developed technique of single hair cell patch-clamp has made an important contribution toward an understanding of the biophysical properties of hair cells in vivo. Together with microphonic and action current recordings, these methods—along with other established cell physiology techniques in other animal models—represent a crucial asset in a toolbox to understand hair-cell processing and sensory signal encoding.

DISCUSSION

Zebrafish are an excellent vertebrate model to study hair-cell function and sensory transduction. A large number of auditory and vestibular mutants (Nicolson et al., 1998) and genome editing techniques including CRISPR-Cas9 (Hwang et al., 2013) are available for examining proteins of interest and continue to elucidate the molecular mechanisms underlying hearing and balance. Moreover, zebrafish are amenable for calcium imaging due to their optical transparency, which allows researchers to monitor calcium signaling inside several hair cells simultaneously (see chapter "Functional Calcium Imaging in Zebrafish Lateral-line Hair Cells" by Zhang, He, Wong, & Kindt, 2016). The electrophysiological methods described in this chapter add to the strength of the zebrafish for hearing research as they provide for study of hair-cell properties and allow for readout of the sensory encoding process across multiples stages of mechanosensation. Hair cell and afferent neuron electrophysiology has multiple applications. These methods can help determine the impact of an auditory or vestibular mutation and reveal the function of genes and proteins that are conserved from fish to humans (Fetcho & Liu, 1998; Whitfield, 2002). In addition, the techniques can be used to study the effects of ototoxic compounds or uncover how hair cells are able to encode mechanical stimuli into defined patterns of activity in afferent neurons. Moreover, the zebrafish is the only animal model to date where the activity of hair cells and their underlying circuitry can be studied in an undissected, in vivo preparation.

Hearing and balance are intricate sensations that require many functioning parts to be at the right place at the right time. Therefore, multiple interdisciplinary techniques are required to shed light on the precise mechanisms and molecules that are required for our auditory and vestibular systems. By combining the power of optophysiology across populations of cells together with electrophysiology, researchers may further understand how the lateral line functions as a somatotopic unit that provides the animal with a sense of distant touch. Furthermore, simultaneous dual recordings from single hair cells and afferent neurons, either at afferent boutons or from the afferent soma, could examine the contribution of hair cells within the neuromast to the activity of an afferent neuron.

The ability to perform electrophysiological and imaging experiments from hair cells of living zebrafish, together with the ability to combine the functional outcome of these techniques with the pliable genetic manipulation of this animal model, provides a unique experimental platform to further understanding of the molecular and cellular mechanisms underlying sensory transduction and signal encoding along the lateral line neuronal network. The conserved basic architecture of the ribbon synapses and mechanoelectrical transduction between hair cells of lower vertebrates and mammals (Maeda et al., 2014; Nicolson, 2005, 2015), indicates that zebrafish represents an additional model system to further our understanding of hair cell function and dysfunction.

ACKNOWLEDGMENTS

This work was supported by grants from the Wellcome Trust (102892) to WM and a Faculty Research Award from the H. Axel Schupf '57 Fund for Intellectual Life to JGT. JO was supported by a PhD studentship from the University of Sheffield to WM.

REFERENCES

Assad, J. A., Shepherd, G. M., & Corey, D. P. (1991). Tip-link integrity and mechanical transduction in vertebrate hair cells. *Neuron, 7*, 985–994.

Boyden, E. S., Zhang, F., Bamberg, E., Nagel, G., & Deisseroth, K. (2005). Millisecond-timescale, genetically targeted optical control of neural activity. *Nature Neuroscience, 8*, 1263–1268.

Buran, B. N., Strenzke, N., Neef, A., Gundelfinger, E. D., Moser, T., & Liberman, M. C. (2010). Onset coding is degraded in auditory nerve fibers from mutant mice lacking synaptic ribbons. *The Journal of Neuroscience, 30*, 7587–7597.

Corey, D. P., & Hudspeth, A. J. (1983). Analysis of the microphonic potential of the bullfrog's sacculus. *The Journal of Neuroscience: The Official Journal of the Society for Neuroscience, 3*, 942–961.

Faucherre, A., Pujol-Martí, J., Kawakami, K., & López-Schier, H. (2009). Afferent neurons of the zebrafish lateral line are strict selectors of hair-cell orientation. *PLoS One, 4*, e4477.

Fetcho, J. R., & Liu, K. S. (1998). Zebrafish as a model system for studying neuronal circuits and behavior. *Annals of the New York Academy of Sciences, 860*, 333–345.

Fettiplace, R., & Kim, K. X. (2014). The physiology of mechanoelectrical transduction channels in hearing. *Physiological Reviews, 94*, 951–986.

Ghysen, A., & Dambly-Chaudière, C. (2004). Development of the zebrafish lateral line. *Current Opinion in Neurobiology, 14*, 67–73.

Gleason, M. R., Nagiel, A., Jamet, S., Vologodskaia, M., López-Schier, H., & Hudspeth, A. J. (2009). The transmembrane inner ear (Tmie) protein is essential for normal hearing and balance in the zebrafish. *Proceedings of the National Academy of Sciences of the United States of America, 106*, 21347–21352.

Goldberg, J. M., & Brown, P. B. (1969). Response of binaural neurons of dog superior olivary complex to dichotic tonal stimuli: some physiological mechanisms of sound localization. *Journal of Neurophysiology, 32*, 613–636.

Gompel, N., Cubedo, N., Thisse, C., Thisse, B., Dambly-Chaudière, C., & Ghysen, A. (2001). Pattern formation in the lateral line of zebrafish. *Mechanisms of Development, 105*, 69–77.

Haehnel-Taguchi, M., Akanyeti, O., & Liao, J. C. (2014). Afferent and motoneuron activity in response to single neuromast stimulation in the posterior lateral line of larval zebrafish. *Journal of Neurophysiology, 112*, 1329–1339.

Harris, J. A., Cheng, A. G., Cunningham, L. L., MacDonald, G., Raible, D. W., & Rubel, E. W. (2003). Neomycin-induced hair cell death and rapid regeneration in the lateral line of zebrafish (*Danio rerio*). *Journal of the Association for Research in Otolaryngology: JARO, 4*, 219–234.

Heil, P., & Peterson, A. J. (2015). Basic response properties of auditory nerve fibers: a review. *Cell and Tissue Research, 361*, 129–158.

Howard, J., Roberts, W. M., & Hudspeth, A. J. (1988). Mechanoelectrical transduction by hair cells. *Annual Review of Biophysics and Biophysical Chemistry, 17*, 99–124.

Hwang, W. Y., Fu, Y., Reyon, D., Maeder, M. L., Tsai, S. Q., Sander, J. D. ... Joung, J. K. (2013). Efficient genome editing in zebrafish using a CRISPR-Cas system. *Nature Biotechnology, 31*, 227–229.

Johnson, S. L., Forge, A., Knipper, M., Münkner, S., & Marcotti, W. (2008). Tonotopic variation in the calcium dependence of neurotransmitter release and vesicle pool replenishment at mammalian auditory ribbon synapses. *The Journal of Neuroscience: The Official Journal of the Society for Neuroscience, 28*, 7670–7678.

Johnson, S. L., Kuhn, S., Franz, C., Ingham, N., Furness, D. N., Knipper, M. ... Marcotti, W. (2013). Presynaptic maturation in auditory hair cells requires a critical period of sensory-independent spiking activity. *Proceedings of the National Academy of Sciences of the United States of America, 110*, 8720–8725.

Johnson, S. L., Thomas, M. V., & Kros, C. J. (2002). Membrane capacitance measurement using patch clamp with integrated self-balancing lock-in amplifier. *Pflügers Archiv: European Journal of Physiology, 443*, 653–663.

Joris, P. X. (2006). A dogged pursuit of coincidence. *Journal of Neurophysiology, 96*, 969–972.

Kindt, K. S., Finch, G., & Nicolson, T. (2012). Kinocilia mediate mechanosensitivity in developing zebrafish hair cells. *Developmental Cell, 23*, 329–341.

Klapoetke, N. C., Murata, Y., Kim, S. S., Pulver, S. R., Birdsey-Benson, A., Cho, Y. K. ... Boyden, E. S. (2014). Independent optical excitation of distinct neural populations. *Nature Methods, 11*, 338–346.

Kroese, A. B., Das, A., & Hudspeth, A. J. (1989). Blockage of the transduction channels of hair cells in the bullfrog's sacculus by aminoglycoside antibiotics. *Hearing Research, 37*, 203–217.

Lenz, D. R., & Avraham, K. B. (2011). Hereditary hearing loss: from human mutation to mechanism. *Hearing Research, 281*, 3–10.

Liao, J. C. (2010). Organization and physiology of posterior lateral line afferent neurons in larval zebrafish. *Biological Letters, 6*, 402–405.

Liao, J. C., & Haehnel, M. (2012). Physiology of afferent neurons in larval zebrafish provides a functional framework for lateral line somatotopy. *Journal of Neurophysiology, 107*, 2615–2623.

López-Schier, H., Starr, C. J., Kappler, J. A., Kollmar, R., & Hudspeth, A. J. (2004). Directional cell migration establishes the axes of planar polarity in the posterior lateral-line organ of the zebrafish. *Developmental Cell, 7*, 401–412.

Maeda, R., Kindt, K. S., Mo, W., Morgan, C. P., Erickson, T., Zhao, H. … Nicolson, T. (2014). Tip-link protein protocadherin 15 interacts with transmembrane channel-like proteins TMC1 and TMC2. *Proceedings of the National Academy of Sciences of the United States of America, 111*, 12907–12912.

Marcotti, W. (2012). Functional assembly of mammalian cochlear hair cells. *Experimental Physiology, 97*, 438–451.

Marcotti, W., van Netten, S. M., & Kros, C. J. (2005). The aminoglycoside antibiotic dihydrostreptomycin rapidly enters mouse outer hair cells through the mechano-electrical transducer channels. *The Journal of Physiology, 567*, 505–521.

Monesson-Olson, B. D., Browning-Kamins, J., Aziz-Bose, R., Kreines, F., & Trapani, J. G. (2014). Optical stimulation of zebrafish hair cells expressing channelrhodopsin-2. *PLoS One, 9*, e96641.

Monesson-Olson, B. D., Troconis, E. L., & Trapani, J. G. (2014). Recording field potentials from zebrafish larvae during escape responses. *Journal of Undergraduate Neuroscience Education: JUNE: A Publication of FUN, Faculty for Undergraduate Neuroscience, 13*, A52–A58.

Moser, T., & Beutner, D. (2000). Kinetics of exocytosis and endocytosis at the cochlear inner hair cell afferent synapse of the mouse. *Proceedings of the National Academy of Sciences of the United States of America, 97*, 883–888.

Nicolson, T. (2005). The genetics of hearing and balance in zebrafish. *Annual Review of Genetics, 39*, 9–22.

Nicolson, T. (2015). Ribbon synapses in zebrafish hair cells. *Hearing Research*. http://dx.doi.org/10.1016/j.heares.2015.04.003.

Nicolson, T., Rüsch, A., Friedrich, R. W., Granato, M., Ruppersberg, J. P., & Nüsslein-Volhard, C. (1998). Genetic analysis of vertebrate sensory hair cell mechanosensation: the zebrafish circler mutants. *Neuron, 20*, 271–283.

Obholzer, N., Wolfson, S., Trapani, J. G., Mo, W., Nechiporuk, A., Busch-Nentwich, E. … Nicolson, T. (2008). Vesicular glutamate transporter 3 is required for synaptic transmission in zebrafish hair cells. *The Journal of Neuroscience: The Official Journal of the Society for Neuroscience, 28*, 2110–2118.

Olt, J., Johnson, S. L., & Marcotti, W. (2014). In vivo and in vitro biophysical properties of hair cells from the lateral line and inner ear of developing and adult zebrafish. *The Journal of Physiology, 592*, 2041–2058.

Pan, B., & Holt, J. R. (2015). The molecules that mediate sensory transduction in the mammalian inner ear. *Current Opinion in Neurobiology, 34*, 165–171.

Peterson, A. J., Irvine, D. R. F., & Heil, P. (2014). A model of synaptic vesicle-pool depletion and replenishment can account for the interspike interval distributions and nonrenewal properties of spontaneous spike trains of auditory-nerve fibers. *The Journal of Neuroscience: The Official Journal of the Society for Neuroscience, 34*, 15097–15109.

Pfeiffer, R. R., & Kiang, N. Y. (1965). Spike discharge patterns of spontaneous and continuously stimulated activity in the cochlear nucleus of anesthetized cats. *Biophysical Journal, 5*, 301–316.

Pickles, J. O., Comis, S. D., & Osborne, M. P. (1984). Cross-links between stereocilia in the guinea pig organ of Corti, and their possible relation to sensory transduction. *Hearing Research, 15*, 103−112.

Pinto-Teixeira, F., Viader-Llargués, O., Torres-Mejía, E., Turan, M., González-Gualda, E., Pola-Morell, L., & López-Schier, H. (2015). Inexhaustible hair-cell regeneration in young and aged zebrafish. *Biology Open, 4*, 903−909.

Pujol-Martí, J., & López-Schier, H. (2013). Developmental and architectural principles of the lateral-line neural map. *Frontiers in Neural Circuits, 7*, 47.

Ricci, A. J., Bai, J.-P., Song, L., Lv, C., Zenisek, D., & Santos-Sacchi, J. (2013). Patch-clamp recordings from lateral line neuromast hair cells of the living zebrafish. *The Journal of Neuroscience: The Official Journal of the Society for Neuroscience, 33*, 3131−3134.

Sheets, L., Kindt, K. S., & Nicolson, T. (2012). Presynaptic CaV1.3 channels regulate synaptic ribbon size and are required for synaptic maintenance in sensory hair cells. *Journal of Neuroscience, 32*, 17273−17286.

Sheets, L., Trapani, J. G., Mo, W., Obholzer, N., & Nicolson, T. (2011). Ribeye is required for presynaptic Ca(V)1.3a channel localization and afferent innervation of sensory hair cells. *Development, 138*, 1309−1319.

Sherman-Gold, R. (Ed.). (1993). *The axon guide for electrophysiology & biophysics: Laboratory techniques*. Foster City, CA: Axon Instruments.

Smedemark-Margulies, N., & Trapani, J. G. (2013). Tools, methods, and applications for optophysiology in neuroscience. *Frontiers in Molecular Neuroscience, 6*, 18.

Taberner, A. M., & Liberman, M. C. (2005). Response properties of single auditory nerve fibers in the mouse. *Journal of Neurophysiology, 93*, 557−569.

Trapani, J. G., & Nicolson, T. (2010). Physiological recordings from zebrafish lateral-line hair cells and afferent neurons. *Methods in Cell Biology, 100*, 219−231.

Trapani, J. G., & Nicolson, T. (2011). Mechanism of spontaneous activity in afferent neurons of the zebrafish lateral-line organ. *Journal of Neuroscience, 31*, 1614−1623.

Tritsch, N. X., Rodríguez-Contreras, A., Crins, T. T. H., Wang, H. C., Borst, J. G. G., & Bergles, D. E. (2010). Calcium action potentials in hair cells pattern auditory neuron activity before hearing onset. *Nature Neuroscience, 13*, 1050−1052.

Tritsch, N. X., Yi, E., Gale, J. E., Glowatzki, E., & Bergles, D. E. (2007). The origin of spontaneous activity in the developing auditory system. *Nature, 450*, 50−55.

Van Trump, W. J., & McHenry, M. J. (2008). The morphology and mechanical sensitivity of lateral line receptors in zebrafish larvae (*Danio rerio*). *The Journal of Experimental Biology, 211*, 2105−2115.

Westerfield, M. (2000). *The zebrafish book. A guide for the laboratory use of zebrafish (*Danio rerio*)* (4th ed.). Eugene: University of Oregon Press.

Whitfield, T. T. (2002). Zebrafish as a model for hearing and deafness. *Journal of Neurobiology, 53*, 157−171.

William Detrich, H., III, Zon, L. I., & Westerfield, M. (2009). *Essential zebrafish methods: Genetics and genomics*. Academic Press.

Zhu, P., Narita, Y., Bundschuh, S. T., Fajardo, O., Schärer, Y.-P. Z., Chattopadhyaya, B. … Friedrich, R. W. (2009). Optogenetic dissection of neuronal circuits in zebrafish using viral gene transfer and the tet system. *Frontiers in Neural Circuits, 3*, 21.

Zhang, Q. X., He, X. J., Wong, H. C., & Kindt, K. S. (2016). Functional calcium imaging in zebrafish lateral-line hair cells. In H. W. Detrich, III, M. Westerfield, & L. Zon (Eds.), *The Zebrafish: Cellular and Developmental Biology, Part A Cellular Biology* (Vol. 133, pp. 229−252).

Volumes in Series

Founding Series Editor
DAVID M. PRESCOTT

Volume 1 (1964)
Methods in Cell Physiology
Edited by David M. Prescott

Volume 2 (1966)
Methods in Cell Physiology
Edited by David M. Prescott

Volume 3 (1968)
Methods in Cell Physiology
Edited by David M. Prescott

Volume 4 (1970)
Methods in Cell Physiology
Edited by David M. Prescott

Volume 5 (1972)
Methods in Cell Physiology
Edited by David M. Prescott

Volume 6 (1973)
Methods in Cell Physiology
Edited by David M. Prescott

Volume 7 (1973)
Methods in Cell Biology
Edited by David M. Prescott

Volume 8 (1974)
Methods in Cell Biology
Edited by David M. Prescott

Volume 9 (1975)
Methods in Cell Biology
Edited by David M. Prescott

Volume 10 (1975)
Methods in Cell Biology
Edited by David M. Prescott

Volume 11 (1975)
Yeast Cells
Edited by David M. Prescott

Volume 12 (1975)
Yeast Cells
Edited by David M. Prescott

Volume 13 (1976)
Methods in Cell Biology
Edited by David M. Prescott

Volume 14 (1976)
Methods in Cell Biology
Edited by David M. Prescott

Volume 15 (1977)
Methods in Cell Biology
Edited by David M. Prescott

Volume 16 (1977)
Chromatin and Chromosomal Protein Research I
Edited by Gary Stein, Janet Stein, and Lewis J. Kleinsmith

Volume 17 (1978)
Chromatin and Chromosomal Protein Research II
Edited by Gary Stein, Janet Stein, and Lewis J. Kleinsmith

Volume 18 (1978)
Chromatin and Chromosomal Protein Research III
Edited by Gary Stein, Janet Stein, and Lewis J. Kleinsmith

Volume 19 (1978)
Chromatin and Chromosomal Protein Research IV
Edited by Gary Stein, Janet Stein, and Lewis J. Kleinsmith

Volume 20 (1978)
Methods in Cell Biology
Edited by David M. Prescott

Advisory Board Chairman
KEITH R. PORTER

Volume 21A (1980)
Normal Human Tissue and Cell Culture, Part A: Respiratory, Cardiovascular, and Integumentary Systems
Edited by Curtis C. Harris, Benjamin F. Trump, and Gary D. Stoner

Volume 21B (1980)
Normal Human Tissue and Cell Culture, Part B: Endocrine, Urogenital, and Gastrointestinal Systems
Edited by Curtis C. Harris, Benjamin F. Trump, and Gray D. Stoner

Volume 22 (1981)
Three-Dimensional Ultrastructure in Biology
Edited by James N. Turner

Volume 23 (1981)
Basic Mechanisms of Cellular Secretion
Edited by Arthur R. Hand and Constance Oliver

Volume 24 (1982)
The Cytoskeleton, Part A: Cytoskeletal Proteins, Isolation and Characterization
Edited by Leslie Wilson

Volume 25 (1982)
The Cytoskeleton, Part B: Biological Systems and In Vitro Models
Edited by Leslie Wilson

Volume 26 (1982)
Prenatal Diagnosis: Cell Biological Approaches
Edited by Samuel A. Latt and Gretchen J. Darlington

Series Editor
LESLIE WILSON

Volume 27 (1986)
Echinoderm Gametes and Embryos
Edited by Thomas E. Schroeder

Volume 28 (1987)
Dictyostelium discoideum: **Molecular Approaches to Cell Biology**
Edited by James A. Spudich

Volume 29 (1989)
Fluorescence Microscopy of Living Cells in Culture, Part A: Fluorescent Analogs, Labeling Cells, and Basic Microscopy
Edited by Yu-Li Wang and D. Lansing Taylor

Volume 30 (1989)
Fluorescence Microscopy of Living Cells in Culture, Part B: Quantitative Fluorescence Microscopy—Imaging and Spectroscopy
Edited by D. Lansing Taylor and Yu-Li Wang

Volume 31 (1989)
Vesicular Transport, Part A
Edited by Alan M. Tartakoff

Volume 32 (1989)
Vesicular Transport, Part B
Edited by Alan M. Tartakoff

Volume 33 (1990)
Flow Cytometry
Edited by Zbigniew Darzynkiewicz and Harry A. Crissman

Volume 34 (1991)
Vectorial Transport of Proteins into and across Membranes
Edited by Alan M. Tartakoff

Selected from Volumes 31, 32, and 34 (1991)
Laboratory Methods for Vesicular and Vectorial Transport
Edited by Alan M. Tartakoff

Volume 35 (1991)
Functional Organization of the Nucleus: A Laboratory Guide
Edited by Barbara A. Hamkalo and Sarah C. R. Elgin

Volume 36 (1991)
***Xenopus laevis*: Practical Uses in Cell and Molecular Biology**
Edited by Brian K. Kay and H. Benjamin Peng

Series Editors
LESLIE WILSON AND PAUL MATSUDAIRA

Volume 37 (1993)
Antibodies in Cell Biology
Edited by David J. Asai

Volume 38 (1993)
Cell Biological Applications of Confocal Microscopy
Edited by Brian Matsumoto

Volume 39 (1993)
Motility Assays for Motor Proteins
Edited by Jonathan M. Scholey

Volume 40 (1994)
A Practical Guide to the Study of Calcium in Living Cells
Edited by Richard Nuccitelli

Volume 41 (1994)
Flow Cytometry, Second Edition, Part A
Edited by Zbigniew Darzynkiewicz, J. Paul Robinson, and Harry A. Crissman

Volume 42 (1994)
Flow Cytometry, Second Edition, Part B
Edited by Zbigniew Darzynkiewicz, J. Paul Robinson, and Harry A. Crissman

Volume 43 (1994)
Protein Expression in Animal Cells
Edited by Michael G. Roth

Volume 44 (1994)
***Drosophila melanogaster*: Practical Uses in Cell and Molecular Biology**
Edited by Lawrence S. B. Goldstein and Eric A. Fyrberg

Volume 45 (1994)
Microbes as Tools for Cell Biology
Edited by David G. Russell

Volume 46 (1995)
Cell Death
Edited by Lawrence M. Schwartz and Barbara A. Osborne

Volume 47 (1995)
Cilia and Flagella
Edited by William Dentler and George Witman

Volume 48 (1995)
Caenorhabditis elegans: Modern Biological Analysis of an Organism
Edited by Henry F. Epstein and Diane C. Shakes

Volume 49 (1995)
Methods in Plant Cell Biology, Part A
Edited by David W. Galbraith, Hans J. Bohnert, and Don P. Bourque

Volume 50 (1995)
Methods in Plant Cell Biology, Part B
Edited by David W. Galbraith, Don P. Bourque, and Hans J. Bohnert

Volume 51 (1996)
Methods in Avian Embryology
Edited by Marianne Bronner-Fraser

Volume 52 (1997)
Methods in Muscle Biology
Edited by Charles P. Emerson, Jr. and H. Lee Sweeney

Volume 53 (1997)
Nuclear Structure and Function
Edited by Miguel Berrios

Volume 54 (1997)
Cumulative Index

Volume 55 (1997)
Laser Tweezers in Cell Biology
Edited by Michael P. Sheetz

Volume 56 (1998)
Video Microscopy
Edited by Greenfield Sluder and David E. Wolf

Volume 57 (1998)
Animal Cell Culture Methods
Edited by Jennie P. Mather and David Barnes

Volume 58 (1998)
Green Fluorescent Protein
Edited by Kevin F. Sullivan and Steve A. Kay

Volume 59 (1998)
The Zebrafish: Biology
Edited by H. William Detrich III, Monte Westerfield, and Leonard I. Zon

Volume 60 (1998)
The Zebrafish: Genetics and Genomics
Edited by H. William Detrich III, Monte Westerfield, and Leonard I. Zon

Volume 61 (1998)
Mitosis and Meiosis
Edited by Conly L. Rieder

Volume 62 (1999)
Tetrahymena thermophila
Edited by David J. Asai and James D. Forney

Volume 63 (2000)
Cytometry, Third Edition, Part A
Edited by Zbigniew Darzynkiewicz, J. Paul Robinson, and Harry Crissman

Volume 64 (2000)
Cytometry, Third Edition, Part B
Edited by Zbigniew Darzynkiewicz, J. Paul Robinson, and Harry Crissman

Volume 65 (2001)
Mitochondria
Edited by Liza A. Pon and Eric A. Schon

Volume 66 (2001)
Apoptosis
Edited by Lawrence M. Schwartz and Jonathan D. Ashwell

Volume 67 (2001)
Centrosomes and Spindle Pole Bodies
Edited by Robert E. Palazzo and Trisha N. Davis

Volume 68 (2002)
Atomic Force Microscopy in Cell Biology
Edited by Bhanu P. Jena and J. K. Heinrich Hörber

Volume 69 (2002)
Methods in Cell–Matrix Adhesion
Edited by Josephine C. Adams

Volume 70 (2002)
Cell Biological Applications of Confocal Microscopy
Edited by Brian Matsumoto

Volume 71 (2003)
Neurons: Methods and Applications for Cell Biologist
Edited by Peter J. Hollenbeck and James R. Bamburg

Volume 72 (2003)
Digital Microscopy: A Second Edition of Video Microscopy
Edited by Greenfield Sluder and David E. Wolf

Volume 73 (2003)
Cumulative Index

Volume 74 (2004)
Development of Sea Urchins, Ascidians, and Other Invertebrate Deuterostomes: Experimental Approaches
Edited by Charles A. Ettensohn, Gary M. Wessel, and Gregory A. Wray

Volume 75 (2004)
Cytometry, 4th Edition: New Developments
Edited by Zbigniew Darzynkiewicz, Mario Roederer, and Hans Tanke

Volume 76 (2004)
The Zebrafish: Cellular and Developmental Biology
Edited by H. William Detrich, III, Monte Westerfield, and Leonard I. Zon

Volume 77 (2004)
The Zebrafish: Genetics, Genomics, and Informatics
Edited by William H. Detrich, III, Monte Westerfield, and Leonard I. Zon

Volume 78 (2004)
Intermediate Filament Cytoskeleton
Edited by M. Bishr Omary and Pierre A. Coulombe

Volume 79 (2007)
Cellular Electron Microscopy
Edited by J. Richard McIntosh

Volume 80 (2007)
Mitochondria, 2nd Edition
Edited by Liza A. Pon and Eric A. Schon

Volume 81 (2007)
Digital Microscopy, 3rd Edition
Edited by Greenfield Sluder and David E. Wolf

Volume 82 (2007)
Laser Manipulation of Cells and Tissues
Edited by Michael W. Berns and Karl Otto Greulich

Volume 83 (2007)
Cell Mechanics
Edited by Yu-Li Wang and Dennis E. Discher

Volume 84 (2007)
Biophysical Tools for Biologists, Volume One: In Vitro Techniques
Edited by John J. Correia and H. William Detrich, III

Volume 85 (2008)
Fluorescent Proteins
Edited by Kevin F. Sullivan

Volume 86 (2008)
Stem Cell Culture
Edited by Dr. Jennie P. Mather

Volume 87 (2008)
Avian Embryology, 2nd Edition
Edited by Dr. Marianne Bronner-Fraser

Volume 88 (2008)
Introduction to Electron Microscopy for Biologists
Edited by Prof. Terence D. Allen

Volume 89 (2008)
Biophysical Tools for Biologists, Volume Two: In Vivo Techniques
Edited by Dr. John J. Correia and Dr. H. William Detrich, III

Volume 90 (2008)
Methods in Nano Cell Biology
Edited by Bhanu P. Jena

Volume 91 (2009)
Cilia: Structure and Motility
Edited by Stephen M. King and Gregory J. Pazour

Volume 92 (2009)
Cilia: Motors and Regulation
Edited by Stephen M. King and Gregory J. Pazour

Volume 93 (2009)
Cilia: Model Organisms and Intraflagellar Transport
Edited by Stephen M. King and Gregory J. Pazour

Volume 94 (2009)
Primary Cilia
Edited by Roger D. Sloboda

Volume 95 (2010)
Microtubules, in vitro
Edited by Leslie Wilson and John J. Correia

Volume 96 (2010)
Electron Microscopy of Model Systems
Edited by Thomas Müeller-Reichert

Volume 97 (2010)
Microtubules: In Vivo
Edited by Lynne Cassimeris and Phong Tran

Volume 98 (2010)
Nuclear Mechanics & Genome Regulation
Edited by G.V. Shivashankar

Volume 99 (2010)
Calcium in Living Cells
Edited by Michael Whitaker

Volume 100 (2010)
The Zebrafish: Cellular and Developmental Biology, Part A
Edited by: H. William Detrich III, Monte Westerfield and Leonard I. Zon

Volume 101 (2011)
The Zebrafish: Cellular and Developmental Biology, Part B
Edited by: H. William Detrich III, Monte Westerfield and Leonard I. Zon

Volume 102 (2011)
Recent Advances in Cytometry, Part A: Instrumentation, Methods
Edited by Zbigniew Darzynkiewicz, Elena Holden, Alberto Orfao, William Telford and Donald Wlodkowic

Volume 103 (2011)
Recent Advances in Cytometry, Part B: Advances in Applications
Edited by Zbigniew Darzynkiewicz, Elena Holden, Alberto Orfao, Alberto Orfao and Donald Wlodkowic

Volume 104 (2011)
The Zebrafish: Genetics, Genomics and Informatics
3rd Edition
Edited by H. William Detrich III, Monte Westerfield, and Leonard I. Zon

Volume 105 (2011)
The Zebrafish: Disease Models and Chemical Screens 3rd Edition
Edited by H. William Detrich III, Monte Westerfield, and Leonard I. Zon

Volume 106 (2011)
Caenorhabditis elegans: Molecular Genetics and Development 2nd Edition
Edited by Joel H. Rothman and Andrew Singson

Volume 107 (2011)
Caenorhabditis elegans: Cell Biology and Physiology 2nd Edition
Edited by Joel H. Rothman and Andrew Singson

Volume 108 (2012)
Lipids
Edited by Gilbert Di Paolo and Markus R Wenk

Volume 109 (2012)
Tetrahymena thermophila
Edited by Kathleen Collins

Volume 110 (2012)
Methods in Cell Biology
Edited by Anand R. Asthagiri and Adam P. Arkin

Volume 111 (2012)
Methods in Cell Biology
Edited by Thomas Müler Reichart and Paul Verkade

Volume 112 (2012)
Laboratory Methods in Cell Biology
Edited by P. Michael Conn

Volume 113 (2013)
Laboratory Methods in Cell Biology
Edited by P. Michael Conn

Volume 114 (2013)
Digital Microscopy, 4th Edition
Edited by Greenfield Sluder and David E. Wolf

Volume 115 (2013)
Microtubules, in Vitro, 2nd Edition
Edited by John J. Correia and Leslie Wilson

Volume 116 (2013)
Lipid Droplets
Edited by H. Robert Yang and Peng Li

Volume 117 (2013)
Receptor-Receptor Interactions
Edited by P. Michael Conn

Volume 118 (2013)
Methods for Analysis of Golgi Complex Function
Edited by Franck Perez and David J. Stephens

Volume 119 (2014)
Micropatterning in Cell Biology Part A
Edited by Matthieu Piel and Manuel Théry

Volume 120 (2014)
Micropatterning in Cell Biology Part B
Edited by Matthieu Piel and Manuel Théry

Volume 121 (2014)
Micropatterning in Cell Biology Part C
Edited by Matthieu Piel and Manuel Théry

Volume 122 (2014)
Nuclear Pore Complexes and Nucleocytoplasmic Transport - Methods
Edited by Valérie Doye

Volume 123 (2014)
Quantitative Imaging in Cell Biology
Edited by Jennifer C. Waters and Torsten Wittmann

Volume 124 (2014)
Correlative Light and Electron Microscopy II
Edited by Thomas Müller-Reichert and Paul Verkade

Volume 125 (2015)
Biophysical Methods in Cell Biology
Edited by Ewa K. Paluch

Volume 126 (2015)
Lysosomes and Lysosomal Diseases
Edited by Frances Platt and Nick Platt

Volume 127 (2015)
Methods in Cilia & Flagella
Edited by Renata Basto and Wallace F. Marshall

Volume 128 (2015)
Building a Cell from its Component Parts
Edited by Jennifer Ross and Wallace F. Marshall

Volume 129 (2015)
Centrosome & Centriole
Edited by Renata Basto and Karen Oegema

Volume 130 (2015)
Sorting and Recycling Endosomes
Edited by Wei Guo

Volume 131 (2016)
The Neuronal Cytoskeleton, Motor Proteins, and Organelle Trafficking in the Axon
Edited by K. Kevin Pfister

Volume 132 (2016)
G Protein-Coupled Receptors: Signaling, Trafficking and Regulation
Edited by Arun K. Shukla

Index

'*Note*: Page numbers followed by "f" indicate figures and "t" indicate tables.'

A

Abbe's diffraction limit, 110
Acyltransferase (ACAT), 170
Adult donor cells, 29–32, *see also* Embryonic donor cells
   irradiation, 31
   protocols for isolating hematopoietic cells, 30–31
   transplantation, 32
     cells into irradiated adult recipients, 31
     WKM, 31
Adult hematopoiesis, 18–23
Afferent fiber spiking analysis, 273–275
Afferent neuron action currents, 270–275
   afferent fiber spiking analysis, 273–275
   electrodes and placement, 271
   recording, 271–273, 272f
AGM, *see* Aorta, gonads, and mesonephros (AGM)
Aldehyde dehydrogenases (Aldhs), 142–143
aldh1a2 expression, 143
Aldhs, *see* Aldehyde dehydrogenases (Aldhs)
Alkaline phosphatase staining (AP staining), 71
   for 3 DPF embryos, 82
     materials, 82
     notes, 83
     protocol, 83
ALL, *see* Anterior lateral line (ALL)
Amiloride, 237
4-Aminopyridine (4-AP), 258
Anterior lateral line (ALL), 262
Anterior macula, 187
Anterior–posterior patterning (A–P patterning), 142
Anti-enhanced green fluorescent protein (Anti-EGFP), 83
Antibodies, 194
   antibody staining, 211
Antigen retrieval, 199
Aorta, gonads, and mesonephros (AGM), 15–18
4-AP, *see* 4-Aminopyridine (4-AP)
A–P patterning, *see* Anterior–posterior patterning (A–P patterning)
AP staining, *see* Alkaline phosphatase staining (AP staining)
Arched continuous stripes (ArCoS), 117
Arl13b-GFP fusion, 200–202
Arl13b-mKate2 fusions, 200–202
"9 + 0"Arrangement of microtubules, 180, 190
"9 + 2"Arrangement of microtubules, 180, 184–185, 187–188
Auditory system, 187, 254
   hair cells, 212–213
Axial resolution in wide field detection, 110
Axopatch 200B, 261–262

B

basic fibroblast growth factor (bFGF), 3, 6f
BCIP, *see* 5-Bromo-4-Chloro-3-indolyl-phosphate (BCIP)
Beam splitter, 231–232
bFGF, *see* basic fibroblast growth factor (bFGF)
Bicoid in *Drosophila*, 141–142
Big data, 120
Blastomere cell culture, 3–5, *see also* Neural crest cell culture
   materials and reagents, 4
   muscle gene expression changes, 6f
   plating zebrafish blastomeres, 4–5
   representative results, 5
Blastomeres, 3
Blocking Solution (BS), 194
Blood vessels, 70–71
BODIPY, *see* 4,4-Difluoro-4-bora-3a, 4a-diaza-S-indacene (BODIPY)
BODIPY fatty acid analogs
   BODIPY C12/C16, 174
   BODIPY C2, 173–174
   BODIPY C5, 174
   BODIPY-cholesterol, 169–171
   lipid metabolism using, 171
     BODIPY C12/C16, 174
     BODIPY C2, 173–174
     BODIPY C5, 174
     excitation/emission maxima, 171–173
5-Bromo-4-Chloro-3-indolyl-phosphate (BCIP), 82
BS, *see* Blocking Solution (BS)
Bungarotoxin, larval immobilization with, 256–257

C

Calcium indicator
   calcium signal validation, 236–237, 236f

Calcium indicator (*Continued*)
　comparison, 232–236
　parameters for three zebrafish hair-cell calcium imaging systems, 235t
　selection, 231–232
　widefield images of neuromasts, 234f
Calcium signal validation, 236–237, 236f
Calmodulin–calcium binding domain, 231–232
Camera-based systems, 239–240
Cardiac asymmetry, 184
Cardiac jogging process, 184
Caudal fin amputation, 62–63
*cd*41, *see itga2b*
CFP, *see* Cyan fluorescent protein (CFP)
CFU, *see* Colony forming unit (CFU)
Channelrhodopsin-2 (ChR2), 255, 260
Chemical screening of zebrafish blastomeres, 3
ChR2, *see* Channelrhodopsin-2 (ChR2)
Cilia, 180
　analytical tools for morphology and motility, 190
　　ciliary transport analysis using inducible transgenes, 204–208
　　detection of ciliary proteins using immunohistochemistry, 190–199
　　live imaging of cilia and basal bodies, 200–202
　　live imaging of cilia movement, 202–203
　　live imaging of cilia using light sheet microscopy, 203–204
　in embryo, 181
　markers and ciliated cells in zebrafish, 191t–193t
　mutants in zebrafish, 208
　　evaluation of heart position in live embryos, 209–210
　　evaluation of kidney function, 210
　　neuromast hair cells staining in live specimen, 213–214
　　olfaction tests, 214–215
　　olfactory neurons labeling by DiI incorporation, 214
　　sensory cell morphology analysis, 211–213
　in zebrafish organs, 181, 182f–183f
　　KV, 183–184
　　pronephros, 184–185
　　sensory organs, 185–190
　　spinal canal, 190
Ciliary transport analysis using inducible transgenes, 204–208
Classical SPIM, 111
Cleaning method, 266
Clonal methylcellulose-based assays, 42–46, *see also* Stromal cell culture assays
　CFU enumeration, 45
　methylcellulose, 44
　　clonal assays, 44–45
　　picking and analyzing colonies from, 45–46
　　stock preparation, 44
CLSM, *see* Confocal laser scanning microscopy (CLSM)
Collagen, 56
　fibers, 56
　reorganization, 57
Collagen, 56
　fibers, 56
　reorganization, 57
Colony forming unit (CFU), 45
　enumeration, 45
Confined primed conversion, 128–129, 130f–131f
　unraveling single neuron morphology with, 129–135, 133f
　　individual labeling of single zebrafish neurons, 134f
　　neural morphology analysis, 132–135
　　photoconversion of single cells in vivo, 129–131, 132f
　　spatially confined primed conversion, 133f
Confocal laser scanning microscopy (CLSM), 126, 129
Confocal microangiography, 83–84, 86, 90
"Connecting cilium", 185
Conversion beam, 128–129
Crabps, 150–151
Crestin:GFP + cells, 9
Crestin:GFP reporter fish, 6–7
CRISPR/Cas9
　CRISPR/Cas9-mediated targeted gene inactivation, 117
　nuclease system, 181
Cryosectioning, 207
Crypt cells, 189
Cyan fluorescent protein (CFP), 231–232
Cyp26a1 expression, 143–146
CYP26S, 143–146
Cytoplasmic Dendra2, 129–131

D

Dab2, *see* Disabled-2 (Dab2)
Danio rerio, *see* Zebrafish (*Danio rerio*)
DASPEI, *see* 2-[4-(Dimethylamino) styryl]-Methylpyridinium Iodide (DASPEI)
Data
　acquisition, 117–119
　analysis process, 120

Index

collection, 207
handling, 117–119
processing, 120
days post fertilization (dpf), 25, 184, 254–255
Definitive hematopoiesis, 13–18
Dendra2 expressing cells, 129–131
Differential dye efflux, 34–35
Differential interference contrast optics, 233–234
4,4-Difluoro-4-bora-3a, 4a-diaza-S-indacene (BODIPY), 171
Digitally scanned light-sheet microscope (DSLM), 106–107
Dihydrostreptomycin, 237, 262–264
DiI incorporation, olfactory neurons labeling by, 214
2-[4-(Dimethylamino) styryl]-Methylpyridinium Iodide (DASPEI), 213
N-((6-(2,4-Dinitro-phenyl)amino)hexanoyl)-1-palmitoyl-2-BODIPY-FL-pentanoyl-*sn*-glycerol-3-phosphoethanolamine (PED6), 168
Disabled-2 (*Dab*2), 71–74
Divergence, 115
Dorsal lines, 188–189
Dorsolateral lines, 188–189
DPBS, *see* Dulbecco's phosphate-buffered saline (DPBS)
dpf, *see* days post fertilization (dpf)
3DPF embryos, alkaline phosphatase staining for, 82
   materials, 82
   notes, 83
   protocol, 83
Drosophila melanogaster embryos, 116
DSLM, *see* Digitally scanned light-sheet microscope (DSLM)
Dulbecco's phosphate-buffered saline (DPBS), 26
Dye injection method, 74, *see also* Resin injection method
   materials, 80
   protocol
      embryos and early larvae, 80–81
      juvenile and adult zebrafish, 81–82
Dye-filling assay, 214

E

Electron microscopy, 211
Embryonic cell culture in zebrafish, 2
   blastomere cell culture, 3–5
   neural crest cell culture, 6–9
Embryonic donor cells, 24–29, *see also* Adult donor cells
   protocol for isolating hematopoietic cells from embryos, 26–28
   transplanting cells into
      blastula recipients, 28
      48 hpf embryos, 29
   transplanting purified cells into embryonic recipients, 28
Embryonic stem cell (ESC), 2
EMP, *see* Erythromyeloid progenitor (EMP)
ENU, *see* Ethylnitrosourea (ENU)
*ephb*4 gene, 71–74
Epo, *see* Erythropoietin (Epo)
Erythromyeloid progenitor (EMP), 12
Erythropoietin (Epo), 42–44, 43f
ESC, *see* Embryonic stem cell (ESC)
Ethylnitrosourea (ENU), 167
etv2/etsrp gene, 71–74
Extracellular, loose-patch recordings, 255
Extracellular morphogens, 141–142

F

2f response, *see* Twice the frequency response (2f response)
FA, *see* Free fatty acid (FA)
FACS, *see* Fluorescence-activated cell sorting (FACS)
FCS, *see* Fetal calf serum (FCS)
Femtosecond pulsed lasers, 128
FEP, *see* Fluorinated ethylene propylene (FEP)
Fetal calf serum (FCS), 26
Fetal liver (FL), 15–18
Fgf, *see* Fibroblast growth factor (Fgf)
Fibrillar collagen, 56
Fibroblast growth factor (Fgf), 140
Field of view (FOV), 126
FIJI, 65
Fish embryos, 116
FITC-labeled lectins, 34
Fixation buffer, 82
Fixed samples preparation, 63–64
FL, *see* Fetal liver (FL)
fli1a genes, 71–74, 90–91
*flk*1, *see kdrl*
*flt*4 genes, 71–74
Fluid jet system, 259
Fluorescence resonance energy transfer (FRET), 147, 231–232
Fluorescence-activated cell sorting (FACS), 19–23
Fluorescent lipids
   forward genetic screening with BODIPY-cholesterol, 169–171

Fluorescent lipids (*Continued*)
  NBD-cholesterol, 169–171
  PED6, 168
 lipid metabolism, 166
 visualizing digestive organ uptake and transport, 169f
 visualizing lipid metabolism using BODIPY fatty acid analogs, 171–174
 whole animal studies of lipid metabolism, 166–168
Fluorescent protein (FP), 126
Fluorinated ethylene propylene (FEP), 116–117
FM-dye labeling, 230–231
Forward scatter (FSC), 19–23
FOV, *see* Field of view (FOV)
FP, *see* Fluorescent protein (FP)
Free fatty acid (FA), 167–168
"French Flag" model, 140
FRET, *see* Fluorescence resonance energy transfer (FRET)
FRET-based calcium indicators, 231–232
FSC, *see* Forward scatter (FSC)

G

GARP, *see* Golgi-associated retrograde protein (GARP)
gata2a, 24–25
Gaussian laser beams, 115
Gcsf, *see* Granulocyte colony stimulating factor (Gcsf)
GECI, *see* Genetically encoded calcium indicator (GECI)
Genetic labeling methods, 126
Genetically encoded calcium indicator (GECI), 230–231
Genetically Encoded reporter Proteins for RA (GEPRA), 147
GFP, *see* Green fluorescent protein (GFP)
Giga-ohm (GOhm), 268
Glass microelectrodes, 271
Glass microinjection needles, 87
Glass needles, 76–77, 76f
GOhm, *see* Giga-ohm (GOhm)
Golgi-associated retrograde protein (GARP), 168
Granulocyte colony stimulating factor (Gcsf), 42–44, 43f
Green Dendra2, 135
Green fluorescent protein (GFP), 90–91
Green-to-red photoconversion, 128–129
Grid/Collection Stitching, 65

H

Hair bundle stimulation, 270–271
Hair cell
 ciliogenesis, 189
 patch-clamp recordings, 258
HCT, *see* Hematopoietic cell transplantation (HCT)
Heart position evaluation in live embryos, 209–210
Heat shocking, 207
Hematopoietic cell transplantation (HCT), 24
 adult donor cells, 29–32
 embryonic donor cells, 24–29
 methods in zebrafish, 24f
Hematopoietic progenitors, in vitro culture and differentiation of, 35–46
 clonal methylcellulose-based assays, 42–46
 stromal cell culture assays, 36–42
Hematopoietic stem and progenitor cell (HSPC), 24
Hematopoietic stem cell (HSC), 12
 enrichment, 32–35
 potential methods of stem cell enrichment, 33f
High-speed pressure clamp (HSPC-1), 259
Holding pipettes, 87
Hours post-fertilization (hpf), 143, 183–184
hoxb1a, 156–157
hpf, *see* Hours post-fertilization (hpf)
HSC, *see* Hematopoietic stem cell (HSC)
HSPC, *see* Hematopoietic stem and progenitor cell (HSPC)
HSPC-1, *see* High-speed pressure clamp (HSPC-1)

I

IAUC, *see* Institutional Animal Care and Use Committee (IAUC)
ICM, *see* Intermediate cell mass (ICM)
IFT, *see* Intraflagellar transport (IFT)
Iguana-GFP fusion, 200–202
Image acquisition, 64–65
Image processing, 242–243
 and analysis, 65–66
 components, 61
 image registration using ImageJ, 243–244
 live imaging of neuromast RGECO calcium signals, 243f
 signal detection and representation in ImageJ, 244
 spatial detection and visualization using MATLAB, 244–248
Image quality, 110
ImageJ

image registration using, 243−244
plotting temporal curve of selected ROI using, 245f
signal detection and representation in, 244
Imaging blood and lymphatic vessels
  imaging vascular gene expression, 71−74
  nonvital blood vessel and lymphatic vessel imaging, 74−83
  vital imaging of blood and lymphatic vessels, 83−98
Imaging vascular gene expression, 71−74
  marker genes used in zebrafish vasculature research, 73t
Immature hair cells, 273−275
Immunohistochemistry, ciliary proteins detection using, 190−194
  staining of cryosections, 197−199
  staining of dissected adult tissues, 196−197
  staining of whole embryos, 194−196
  staining of whole larvae, 196
In vitro
  culture and differentiation of hematopoietic progenitors, 35−46
    clonal methylcellulose-based assays, 42−46
    stromal cell culture assays, 36−42
  differentiation experiments, 13−15
  studies, 166−167
In vivo afferent neuron recordings, 254
In vivo hair cell physiology, 265−270
  hair-cell electrophysiology electrodes and placement, 266−268
  hair-cell identification and access, 266
  morphological characteristics, 267f
  whole-cell recordings
    analysis, 270
    Ca^{2+} current and capacitance changes, 270f
    establishment, 268−269, 269f
Inducible transgenes, ciliary transport analysis using, 204−208
Institutional Animal Care and Use Committee (IAUC), 255−256
Intact larval zebrafish, 255
Intermediate cell mass (ICM), 13
Interspike interval (ISI), 273−275
Intraflagellar transport (IFT), 187
ISI, see Interspike interval (ISI)
itga2b, 18

K

Kaede, 127−128
kdrl genes, 24−25, 71−74
Kidney function evaluation, 210
Kinocilia, 187
Kinocilium, 188
*krox*20, 156−157
Kupffer's vesicle (KV), 181, 183−184

L

Laboratory for Optical and Computational Instrumentation (LOCI), 59−61
lacZ, 146−147
Larval immobilization with bungarotoxin, 256−257
Larval mounting, 257−258
Larval tissue preparation, 256
Larval zebrafish, 56−57, 256, 257f
Lateral lines, 188−189
  electrophysiology methods, 255−258
    ethics statement, 255−256
    larval immobilization with bungarotoxin, 256−257
    larval tissue preparation, 256
    physiological solutions, 258
    recording chamber and larval mounting, 257−258
  lateral-line hair cells, 213, 232
LCFA, see Long-chain fatty acid (LCFA)
LD, see Lipid droplets (LD)
LED source, 260−261
Left−right symmetry (LR symmetry), 181
Light attenuation of immersion medium, 114−115
Light-sheet fluorescence microscopy (LSFM), 106
Light-sheet microscopy, 106, see also Second harmonic generation microscopy (SHG microscopy)
  challenges and perspectives, 119−120
  data acquisition and handling, 117−119
  history, 106−107
  imaging zebrafish vasculature using, 97−98
  live imaging of cilia using, 203−204
  microscope for sample or sample for microscope, 114−117
    mounting of sample, 116−117
    refractive index and light attenuation of immersion medium, 114−115
    spatial resolution, 115
    specimen size *vs.* field of view and movement range of stages, 114
    temporal resolution, 115
  SPIM, 107−114, 118f
Lipid droplets (LD), 171−174
Lipid metabolism
  using BODIPY fatty acid analogs, 171−174
  whole animal studies of, 166−168

Liquid junction potential (LJP), 270
Live imaging
  cilia and basal bodies, 200—202
  cilia movement, 202—203
  of cilia using light sheet microscopy, 203—204
LJP, *see* Liquid junction potential (LJP)
*lmo*2, 24—25
Local neural circuits, 126
LOCI, *see* Laboratory for Optical and Computational Instrumentation (LOCI)
Long-chain fatty acid (LCFA), 174
Long-term mounting for time-lapse imaging, 91—93
  materials, 93
  mounting animals in imaging chambers, 95
  preparation of imaging chambers, 93—95
Looping, 184
LR symmetry, *see* Left—right symmetry (LR symmetry)
LSFM, *see* Light-sheet fluorescence microscopy (LSFM)
Lymphangiography, 80—81
Lymphatic vessels, 70—71
Lymphatics, 70—71
Lyve-1 gene, 71—74

M

Macrophage-specific protein expression, 62
Macrophages, 57
Mantle cells, 266, 268
MATLAB, spatial detection and visualization using, 244—248
Matrix metalloprotease (MMP9), 57
MCFA, *see* Medium-chain fatty acid (MCFA)
Mean vector of length, 273
Mechanical stimulation, 259, 260f
Mechanoelectrical transducer channels (MET channels), 255
Mechanosensory hair cells, 187—189
Mechanotransduction channel blockers, 237
Medium-chain fatty acid (MCFA), 174
MET channels, *see* Mechanoelectrical transducer channels (MET channels)
Methylcellulose, 44
  clonal assays, 44—45
  picking and analyzing colonies from, 45—46
  stock preparation, 44
Microangiography, 86
  materials, 87
  protocol
    experimental procedure, 88—90
    preparation of apparatus, 87—88

Microdye injection, 74—75
  dye injection method, 80—82
  resin injection method, 75—80
Microinjection, 207
Microphonic potentials recording, 255, 261—264, 263f
  analysis, 264
  microphonics equipment and setup, 261—262
  positioning microphonic recording electrode, 262
Micropipette, 259, 261
Microresin injection, 74—75
  dye injection method, 80—82
  resin injection method, 75—80
Microvillous olfactory neurons, 189
MMP9, *see* Matrix metalloprotease (MMP9)
Mono-and diacylglycerols, 167—168
Monociliated cells, 184—185
Morphogen(s), 140, *see also* Retinoic acid (RA)
  challenges for morphogen gradient studies, 140—142
  dynamics and regulation, 141f
  gradients, 143
Motile cilia, 183—184
Mounting of sample, 116—117, 118f
MS-222, 256—257
Multiciliated cells, 184—185
Multiphoton microscope, 60—61
Multiphoton time-lapse imaging, 84, 96—97
Multiview recordings, 119
Murine system, 15
Mutagenesis methods, 167
*myf*5:GFP, 5
*mylz*2:mCherry, 5
Myogenic differentiation of pluripotent cells, 3

N

22-NBD-cholesterol, *see* 22-[N-(7-Nitronbenz-2-oxa-1,3-diazol-4-yl) amino]-23,24-bisnor-5-cholen-3-ol (22-NBD-cholesterol)
NBD-cholesterol, 169—171
NBT, *see* 4-Nitro Blue Tetrazolium (NBT)
nEGFF, *see* nuclear-targeted EGFP (nEGFF)
Nephric system, 185
Neural connectivity, 126
Neural crest cell culture, 6—9, *see also* Blastomere cell culture
  materials and reagents, 7
  plating zebrafish neural crest cells, 7—9
  representative results, 9
Neural crest medium, 7, 7t
Neural morphology analysis, 132—135
Neural plasticity studies, 135

Neuromast hair cells
  staining in live specimen, 213–214
  stimulation, 258–261
    mechanical stimulation, 259, 260f
    optical stimulation, 260–261
Neuromasts, 188, 230, 234f, 254–255, 266
4-Nitro Blue Tetrazolium (NBT), 82
22-[N-(7-Nitronbenz-2-oxa-1,3-diazol-4-yl)
    amino]-23,24-bisnor-5-cholen-3-ol
    (22-NBD-cholesterol), 169–170
Noise in RA signaling, 156–157
Noise-induced switching, 151, 154f–155f, 157
Non-FRET-based indicators, 231–232
Nonsensory motile multiciliated cells, 189–190
Nonvital blood vessel and lymphatic vessel
    imaging, 74, see also Vital imaging of
    blood and lymphatic vessels
  alkaline phosphatase staining for 3 DPF embryos,
    82–83
  microdye injection, 75–82
  microresin injection, 75–82
nuclear-targeted EGFP (nEGFF), 84
Numerical aperture (NA), 110

O

Olfaction tests, 214–215
Olfactory neurons labeling by DiI incorporation,
    214
Olfactory sensory neurons, 189–190, 213
Open-SPIM wiki, 98
Opsin, 204–205
Optical scanning method, 271–272
Optical stimulation, 260–261
Optogenetic stimulation, 258–259
ORCA-Flash4.0 V2 Digital CMOS camera, 119
"Outer segment", 185

P

Paraffin bed, 75, 76f
Paraformaldehyde (PFA), 63, 194
PBS, see Phosphate-buffered saline (PBS)
Peanut agglutinin (PNA), 34
PED6, see N-((6-(2,4-Dinitro-phenyl)amino)hex-
    anoyl)-1-palmitoyl-2-BODIPY-FL-penta-
    noyl-sn-glycerol-3-phosphoethanolamine
    (PED6)
PFA, see Paraformaldehyde (PFA)
Phalloidin, 212
Phase-locked spiking, 271–272
Phenylthiourea (PTU), 207
Phosphate-buffered saline (PBS), 7, 26
Phospholipase A_2 (PLA$_2$), 168

Photoactivatable FP, 127
2P-Photoactivation, 129–131
Photoconversion of single cells in vivo, 129–131,
    132f
Photoconvertible fluorescent proteins, 127–128
Photomultiplier tube (PMT), 60, 239
Photoreceptor(s), 185–187
  cells, 211–212
Phototoxicity, 109–110
Physiological recordings from zebrafish lateral
    line
  afferent neuron action currents, 270–275
  auditory system, 254
  lateral line electrophysiology methods,
    255–258
  microphonic potentials, 255
    recording, 261–264
  neuromast hair cells stimulation, 258–261
  neuromasts, 254–255
  stereocilia, 255
  in vivo hair cell physiology, 265–270
Physiological solutions, 258
Physiological temperatures, 258
Piezoelectric devices, 258–259
PLA$_2$, see Phospholipase A$_2$ (PLA$_2$)
Plasticity, 151–156
PLL, see Posterior lateral line (PLL)
PLLg, see Posterior lateral line ganglion (PLLg)
2PLSM, see Two-photon laser scanning
    microscopy (2PLSM)
"Plug and play" instrument, 120
PMT, see Photomultiplier tube (PMT)
PNA, see Peanut agglutinin (PNA)
Point-or line-scanning confocal systems, 239
35-mm Polystyrene bottom petri dishes
    preparation, 62
Posterior lateral line (PLL), 262
Posterior lateral line ganglion (PLLg), 271
Posterior macula, 187
Potato lectin (PTL), 34
Primary neuromasts, 265–266
Priming beam, 128–129
Primitive hematopoiesis, 13
Pronephros, 184–185
Prox-1 gene, 71–74
PSF of detection objective (PSF$_{det}$), 115
PSF$_{det}$, see PSF of detection objective (PSF$_{det}$)
PSF$_{sys}$, see System's point spread function
    (PSF$_{sys}$)
PTL, see Potato lectin (PTL)
PTU, see Phenylthiourea (PTU)
*pu.*1 expression, see *spi*1 expression

Q

Quantitative RT-PCR (qRT-PCR), 40–41
Quantum dot (QD), 83–84

R

RA, see Retinoic acid (RA)
RA-response element (RARE), 146–147
RARE, see RA-response element (RARE)
3xRARE:eYFP analysis, 146–147
Raw data, 120
Rayleigh range, 115
RBI, see Rostral blood island (RBI)
RD, see Rhodamine dextran (RD)
Realization, 110–111
Recording chamber, 257–258
Refractive index, 114–115
Region of interest (ROI), 64, 210, 243–244, 245f–246f
Resin injection method, 74, see also Dye injection method
   adult zebrafish, 79f
   materials, 75
   protocol
      experimental procedure, 77–80
      preparation of apparatus, 75–77
      preparation of resin for injection, 78f
Resistance (Rs), 270
Resolution, 110
retinal homeobox factor 2 (rx2), 117
Retinoic acid (RA), 142
   binding proteins, 150
   boundaries and morphogens, 157–158
   as graded morphogen, 142–143
      in hindbrain patterning, 145f
   gradient
      CYP26S as key regulators, 143–146
      noise-induced switching, 154f–155f
      sharpening boundaries of gene expression, 151–156
      visualization, 146–147, 148f–149f
   signal robustness, 150–151
      negative feedback through Cyp26a1 and Crabp2a, 153f
      noise, 156–157
   transgenic RA reporters in zebrafish, 149t
Reverse transcription polymerase chain reaction (RT-PCR), 40
RGECO calcium signal
   live imaging of neuromast, 243f
   spatiotemporal patterns of calcium signal using, 247f
   temporal representation, 246f
Rhodamine dextran (RD), 210
Rinse buffer, 82
Robustness, 150
ROI, see Region of interest (ROI)
Room temperature (RT), 83, 194
Rostral blood island (RBI), 13
RT, see Room temperature (RT)
RT-PCR, see Reverse transcription polymerase chain reaction (RT-PCR)
rx2, see retinal homeobox factor 2 (rx2)

S

Saccular macula, see Posterior macula
SCFA, see Short-chain fatty acid (SCFA)
scl genes, 71–74
Second harmonic generation microscopy (SHG microscopy), 56, see also Light-sheet microscopy
   collagen reorganization, 57
   endogenous nature, 57–58
   imaging of fibers, 59f
   larval zebrafish, 56–57, 58f
   materials, 59–61
      buffers, other reagents, and tools for sample preparation, 61
      image processing components, 61
      microscope supplies and components, 59–61
      multiphoton excitation of Dendra-expressing macrophages, 60f
      zebrafish embryos, 59
   methods, 62–66
      caudal fin amputation, 62–63
      fixed samples preparation, 63–64
      image acquisition, 64–65
      image processing and analysis, 65–66
      macrophage-specific protein expression, 62
      35-mm polystyrene bottom petri dishes preparation, 62
   protocols, 58
Selective plane illumination microscopy (SPIM), 106–114, 126
   components, 112–114
   phototoxicity, 109–110
   point scanning vs. light-sheet illumination and detection, 108f–109f
   popular arrangements from large collection of SPIM setups, 113f
   realization, 110–111
   resolution and image quality, 110
   speed, 107–109
   static vs. digitally scanned light-sheet, 111f
Sensory cell morphology analysis, 211

auditory system hair cells, 212–213
lateral line hair cells, 213
olfactory sensory neurons, 213
photoreceptor cells, 211–212
Sensory hair cells, 266
Sensory organs, 185
mechanosensory hair cells, 187–189
olfactory sensory neurons, 189–190
photoreceptors, 185–187
SFC, *see* Swept-field confocal (SFC)
SHG microscopy, *see* Second harmonic generation microscopy (SHG microscopy)
Short-chain fatty acid (SCFA), 173–174
Short-term mounting for time-lapse imaging, 96, *see also* Long-term mounting for time-lapse imaging
Side population (SP), 34–35
Side scatter (SSC), 19–23
Signal-to-noise ratio (SNR), 240–241
Single cell patch-clamp, 255
Single plane illumination microscopy (SPIM), *see* Light sheet microscopy
Smad2:Venus reporter, 157–158
SNR, *see* Signal-to-noise ratio (SNR)
Soda glass, 266–268
SP, *see* Side population (SP)
Sparse labeling of cells, 117
Spatial resolution, 115
Speed, 107–109
*spi*1 expression, 13
Spike timing, 273, 274f
SPIM, *see* Selective plane illumination microscopy (SPIM)
Spinal canal, 190
Spontaneous spiking, 273–275
SSC, *see* Side scatter (SSC)
SSTR3-GFP fusions, 200–202
StackReg plugin, 243–244
Staining buffer, 82
Static light-sheet, 111
Stereocilia, 188, 255, 262
Stimulus frequency, 264
Stitches, 188–189
Stromal cell culture assays, 36–42, *see also* Clonal methylcellulose-based assays
ZEST cells
generation of, 38
maintenance and culture of, 38–39
protocols for in vitro proliferation and differentiation assays on, 40–42
ZKS cells
generation of, 37
maintenance and culture of, 37–38
protocols for in vitro proliferation and differentiation assays on, 40–42
Styryl dyes, 230–231
Superresolution microscopy, imaging zebrafish vasculature using, 98
Supporting cells, 268
Swept-field confocal (SFC), 237, 239, 242
System's point spread function (PSF_{sys}), 115

T

TAG, *see* Triacylglycerol (TAG)
*tal*1 gene, 13
TALEN system, 181
Teleostean kidney, 19
Temporal resolution, 115
Tether cells, 187
Tg(myo6b:ChR2-EYFP), 260
TGFb, *see* Transforming growth factor beta (TGFb)
Thrombopoietin (Tpo), 42–44, 43f
Time-lapse imaging
long-term mounting, 91–93
materials, 93
mounting animals in imaging chambers, 95
preparation of imaging chambers, 93–95
mounting zebrafish embryos and larvae, 92f–93f
multiphoton time-lapse imaging, 96–97
short-term mounting, 96
Tip link, 255
Tpo, *see* Thrombopoietin (Tpo)
Trachyphyllia geoffroyi (*T. geoffroyi*), 127–128
Transforming growth factor beta (TGFb), 142
Transgenic lines, live imaging of cilia and basal bodies using, 200–202
Transgenic zebrafish
imaging blood and lymphatic vessels in, 90
imaging zebrafish vasculature using light sheet microscopy, 97–98
imaging zebrafish vasculature using superresolution microscopy, 98
long-term mounting for time-lapse imaging, 91–95
multiphoton time-lapse imaging, 96–97
short-term mounting for time-lapse imaging, 96
lines, 16t, 23
Transistor–transistor logic (TTL), 239, 260–261
Triacylglycerol (TAG), 167–168
Trunk kidney, 19
TTL, *see* Transistor–transistor logic (TTL)
TurboReg, 243–244

Twice the frequency response (2*f* response), 262, 264
Two-photon laser scanning microscopy (2PLSM), 126

U

Ultra violet illumination (UV illumination), 127–128
Unraveling single neuron morphology with confined primed conversion, 129–135, 133f
   individual labeling of single zebrafish neurons, 134f
   neural morphology analysis, 132–135
   photoconversion of single cells in vivo, 129–131, 132f
   spatially confined primed conversion, 133f
Utricular macula, *see* Anterior macula
UV illumination, *see* Ultra violet illumination (UV illumination)

V

Vascular endothelial growth factor (vegf), 71–74
Vascular system, 70–71
Vector strength, *see* Mean vector of length
vegf, *see* Vascular endothelial growth factor (vegf)
Ventral lines, 188–189
Vertebrate sensory neurons, 185
Vessel-specific gene 1 (*vsg*1), 71–74
Vestibular system, 254
Vital imaging of blood and lymphatic vessels, 83–84, *see also* Nonvital blood vessel and lymphatic vessel imaging
   double transgenic lines, 86
   microangiography, 86–90
   in transgenic zebrafish, 90–98
   zebrafish *fli*1a:*EGFP*, 84
   zebrafish transgenic lines, 85t–86t
Voltage signals, 261–262
Vps51, 168
*vsg*1, *see* Vessel-specific gene 1 (*vsg*1)

W

Whole animal studies of lipid metabolism, 166–168
Whole kidney marrow (WKM), 19
Whole-cell patch-clamp recordings, 270
Whole-cell recordings
   analysis, 270
   establishment, 268–269, 269f
Widefield imaging system, 239
Wimbledon medaka fish, 117
WiscScan image acquisition software, 64–65
WKM, *see* Whole kidney marrow (WKM)
Wolpert's model, 140
Wound healing, 57–58

X

X-Phosphate, *see* 5-Bromo-4-Chloro-3-indolyl-phosphate (BCIP)

Y

Yellow fluorescent protein (YFP), 231–232
Yolk sac (YS), 13

Z

z-projections, 65
Z-stacks, 65–66
ZebraBox system, 214–215
Zebrafish (*Danio rerio*), 56–58, 71, 181, 254
   cell cultures, 2
   cilia in, 181, 182f–183f
      ear, 186f
      KV, 183–184
      pronephros, 184–185
      sensory organs, 185–190
      spinal canal, 190
   cilia mutants in, 208
      evaluation of heart position in live embryos, 209–210
      evaluation of kidney function, 210
      neuromast hair cells staining in live specimen, 213–214
      olfaction tests, 214–215
      olfactory neurons labeling by DiI incorporation, 214
      sensory cell morphology analysis, 211–213
   embryo, 59, 107–109
   enterocytes, 167–168
   hematopoiesis, 12
      adult hematopoiesis, 18–23, 20f
      blood lineage, 22f
      definitive, 13–18, 21f
      functional in vitro differentiation studies, 17f
      HCT, 24–32
      HSCs enrichment, 32–35
      model of hematopoietic ontogeny, 14f–15f
      primitive, 13
      transgenic zebrafish lines, 16t
      in vitro culture and differentiation of hematopoietic progenitors, 35–46
   innate immune system, 56–57
   larvae, 230
   model system, 230

olfactory system, 214–215
3xRARE:eYFP analysis, 146–147
renal system, 184
transgenic RA reporters in, 149t
trigeminal sensory ganglion, 129–131
vasculature, imaging
  using light sheet microscopy, 97–98
  using superresolution microscopy, 98
Zebrafish embryonic stem cell (zESC), 4, 4t
Zebrafish embryonic stromal trunk cells (ZEST cells), 36
  generation, 38
  maintenance and culture, 38–39
  protocols for in vitro proliferation and differentiation assays, 40–42
Zebrafish kidney stromal cells (ZKS cells), 36
  generation, 37
  maintenance and culture of, 37–38
  protocols for in vitro proliferation and differentiation assays on, 40–42
Zebrafish lateral-line hair cells, 230
  calcium indicator selection and comparison, 231–237
  image processing, 242–248
  imaging systems and optimal parameters, 237, 239–242
    comparison of hair-cell evoked calcium responses, 241f
    general microscope and equipment requirements, 237–239
    imaging setup for calcium imaging and hair-cell stimulation, 238f
    synchronizing stimulus with image acquisition, 242
  methods, 231
  styryl dyes, 230–231
zESC, *see* Zebrafish embryonic stem cell (zESC)
ZEST cells, *see* Zebrafish embryonic stromal trunk cells (ZEST cells)

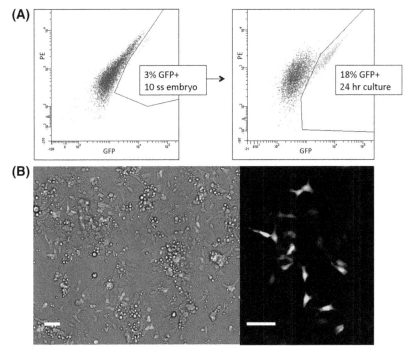

C.A. CIARLO AND L.I. ZON, FIGURE 3

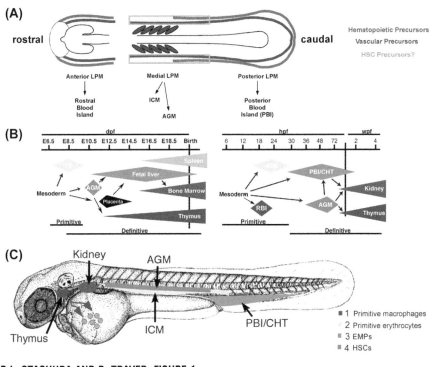

D.L. STACHURA AND D. TRAVER, FIGURE 1

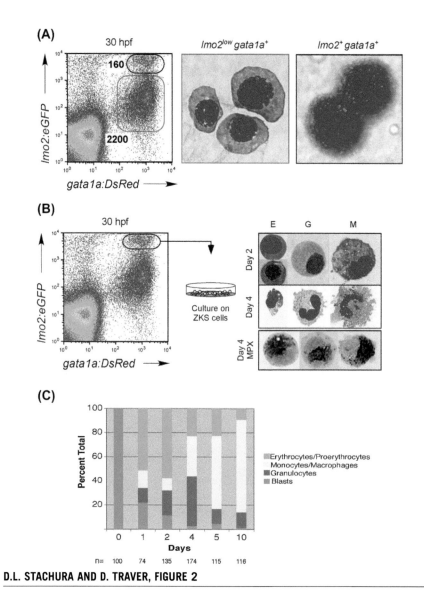

D.L. STACHURA AND D. TRAVER, FIGURE 2

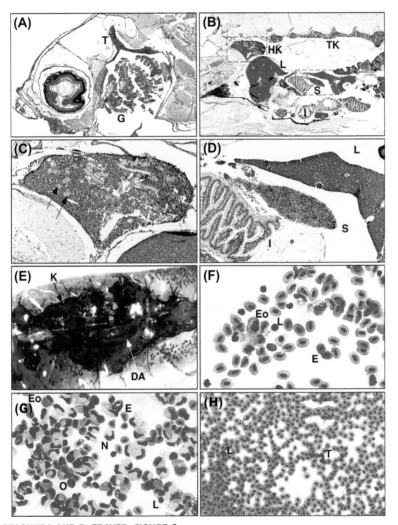

D.L. STACHURA AND D. TRAVER, FIGURE 3

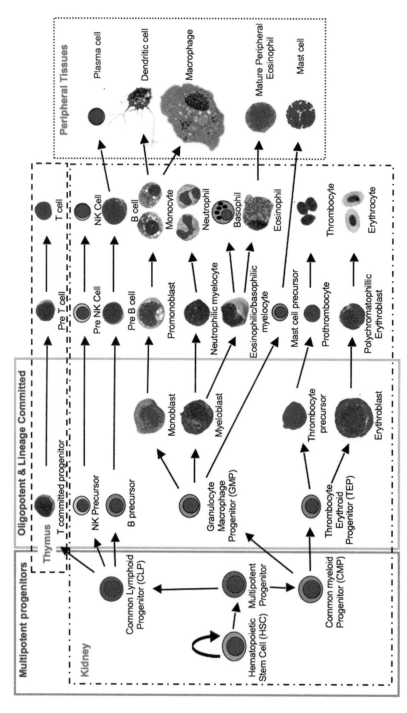

D.L. STACHURA AND D. TRAVER, FIGURE 4

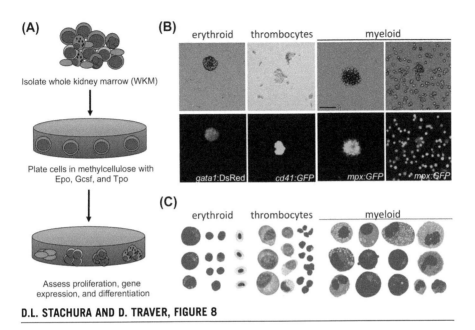

D.L. STACHURA AND D. TRAVER, FIGURE 8

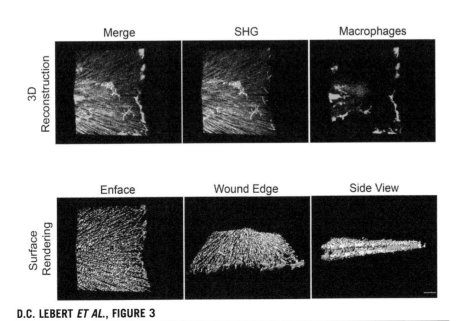

D.C. LEBERT ET AL., FIGURE 3

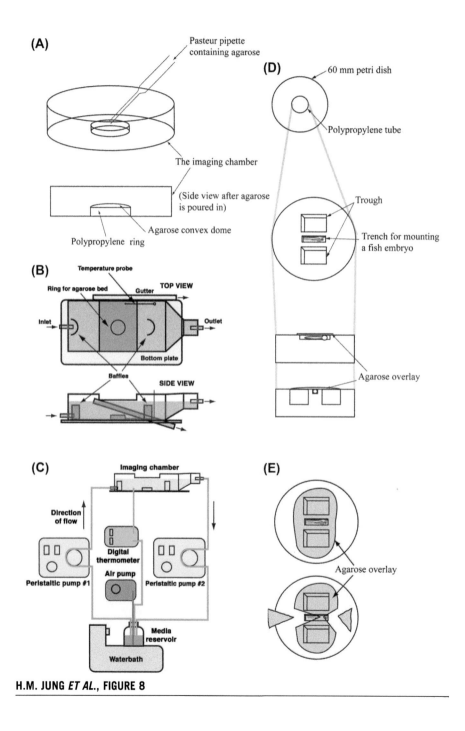

H.M. JUNG ET AL., FIGURE 8

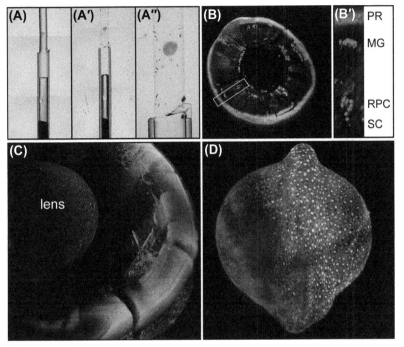

D. KROMM ET AL., FIGURE 4

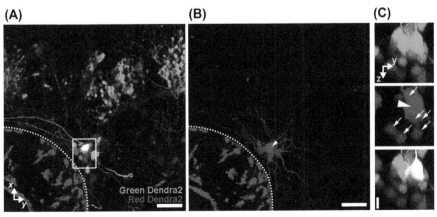

M.A. MOHR AND P. PANTAZIS, FIGURE 3

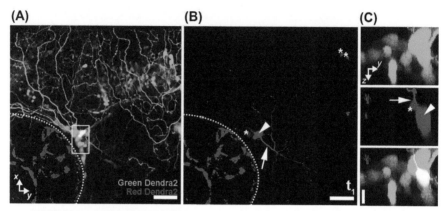

M.A. MOHR AND P. PANTAZIS, FIGURE 4

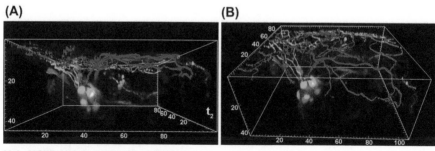

M.A. MOHR AND P. PANTAZIS, FIGURE 5

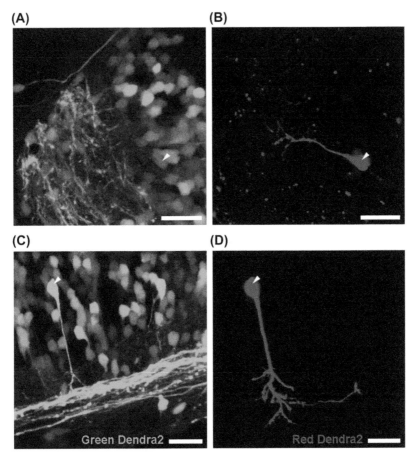

M.A. MOHR AND P. PANTAZIS, FIGURE 6

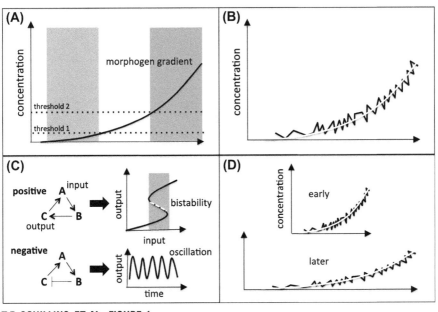

T.F. SCHILLING ET AL., FIGURE 1

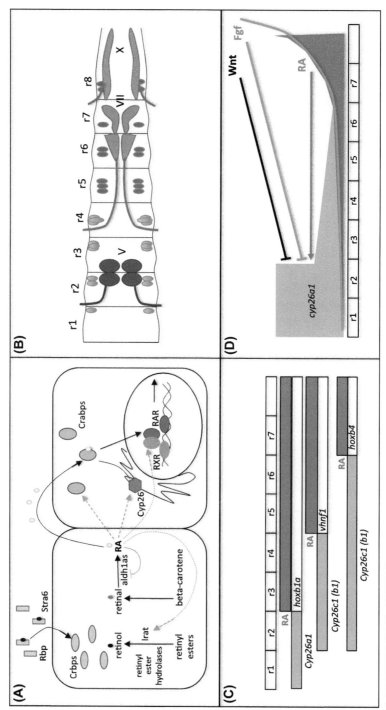

T.F. SCHILLING *ET AL.*, FIGURE 2

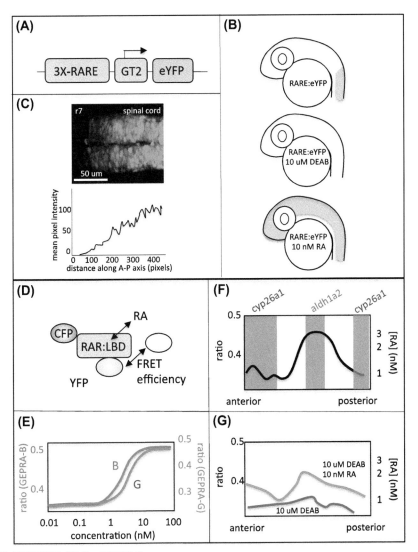

T.F. SCHILLING ET AL., FIGURE 3

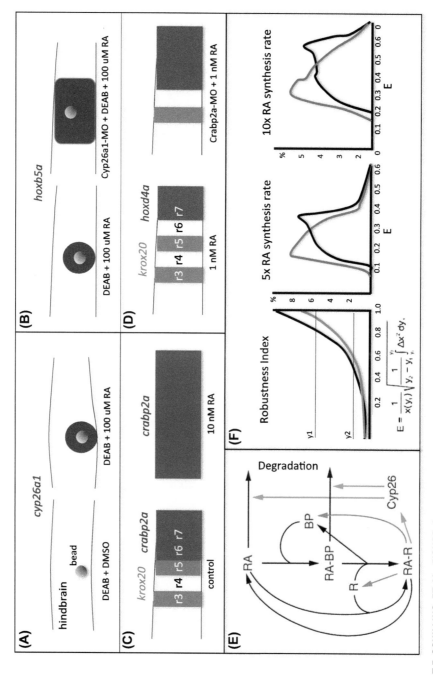

T.F. SCHILLING *ET AL.*, FIGURE 4

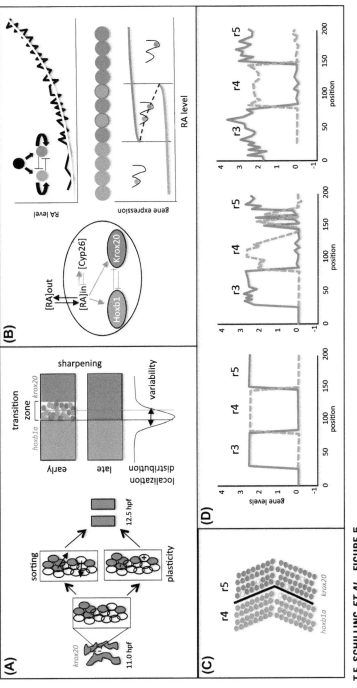

T.F. SCHILLING ET AL., FIGURE 5

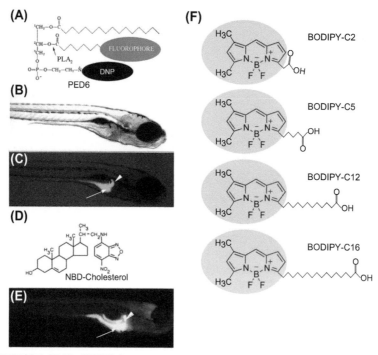

J.L. ANDERSON ET AL., FIGURE 1

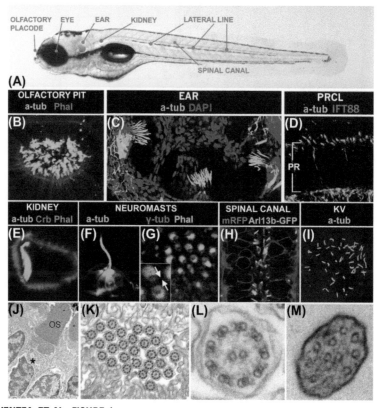

E. LEVENTEA ET AL., FIGURE 1

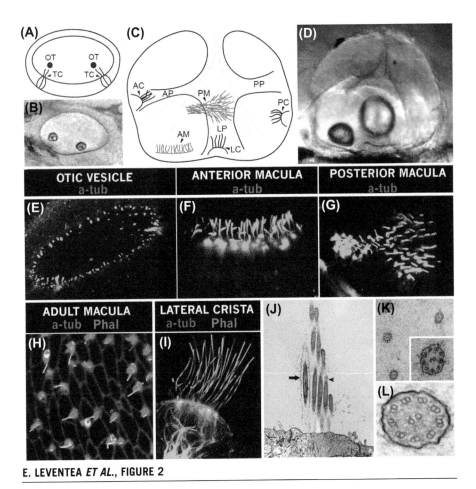

E. LEVENTEA ET AL., FIGURE 2

E. LEVENTEA ET AL., FIGURE 3

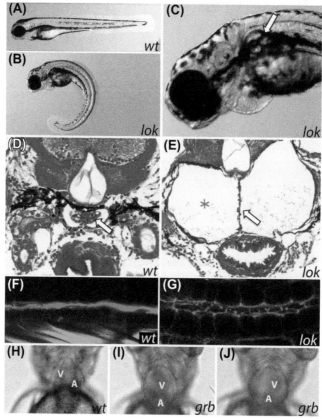

E. LEVENTEA ET AL., FIGURE 4

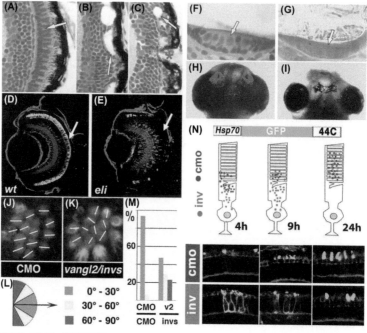

E. LEVENTEA ET AL., FIGURE 5

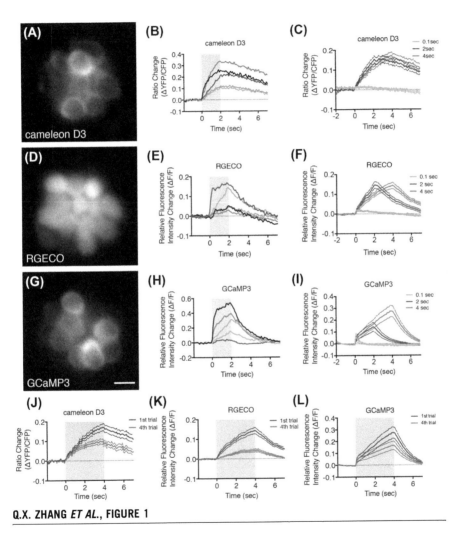

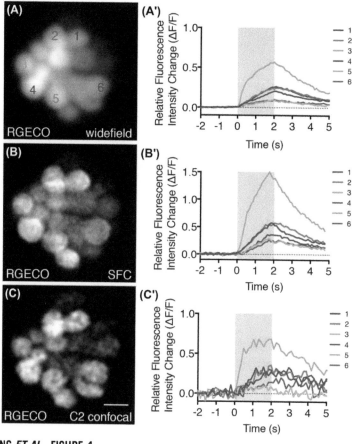

Q.X. ZHANG ET AL., FIGURE 4

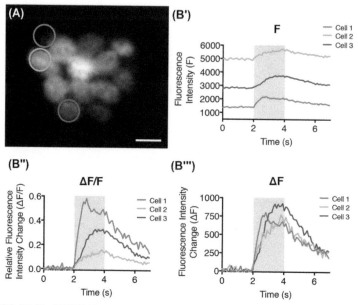

Q.X. ZHANG ET AL., FIGURE 7

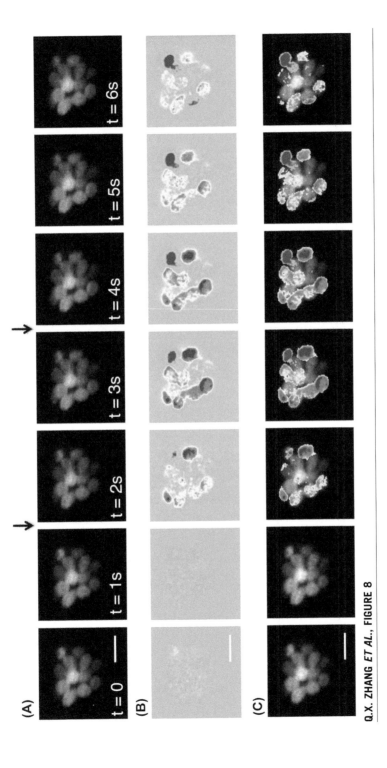

Q.X. ZHANG ET AL., FIGURE 8

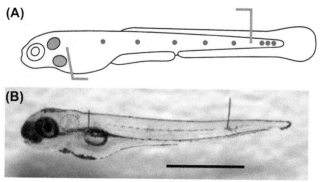

J. OLT ET AL., FIGURE 1

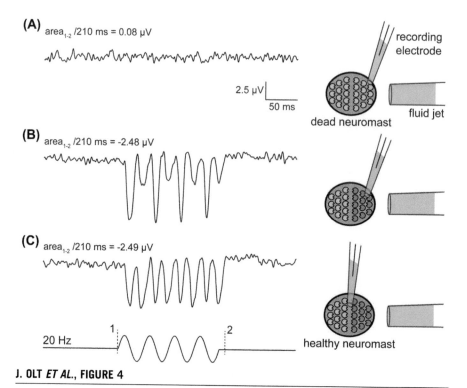

J. OLT ET AL., FIGURE 4

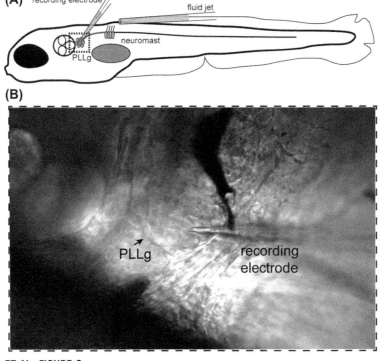

J. OLT ET AL., FIGURE 9

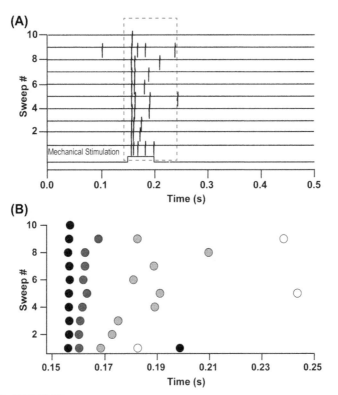

J. OLT ET AL., FIGURE 10

CPI Antony Rowe
Chippenham, UK
2016-06-14 14:42